Human Architecture: Journal of the Sociology of Self-Knowledge

Edited Collection Series

OKCIR: The Omar Khayyam Center for Integrative Research in Utopia, Mysticism, and Science (Utopystics)

OKCIR (est. 2002) is an independent research, pedagogical, and publishing initiative dedicated to exploring, in a simultaneously world-historical and self-reflective framework, the human search for a just global society.

Since the world's utopian, mystical, and scientific movements have been the primary sources of inspiration, knowledge, and/or practice in this field, OKCIR aims to critically reexamine the limits and contributions of these world-historical traditions—seeking to clearly understand why they have failed to bring about the good society, and what each can integratively contribute toward realizing that end.

The center aims to develop new conceptual (methodological, theoretical, historical), practical, curricular, inspirational and disseminative structures of knowledge whereby the individual can radically understand and determine how world-history and her/his selves constitute one another.

OKCIR promotes creative exercises in liberating sociology and alternative pluriversities of knowledge production and publication in the global cyberspace. As a virtual research center, its publications are available in part freely online in its open-stacks digital library, in part via subscription to its own or other academic database member-stacks, and others for purchase online via the Okcir Store and other online distributors. Selected publications are also available in print for online purchase by libraries, institutions, and interested print readers.

OKCIR pursues innovative editorial and virtual and print publishing practices reflecting its substantive goals, and is the publisher of *Human Architecture: Journal of the Sociology of Self-Knowledge* (ISSN: 1540-5699, est. 2002) which explores issues pertaining to the center's interests. For more information about OKCIR and its journal's Edited Collection as well as Monograph and Translation series, visit the center's homepage.

www.okcir.com

Editor's Perspective

Human architecture and the sociology of self-knowledge relate to one another as practice to research—as whole to part.

Human architecture is about tearing down walls of human alienation, and building integrative human realities in favor of a just global society. The sociology of self-knowledge explores how everyday personal self-knowledges of multiple selves and singular world-historical social structures constitute one another. And the present journal seeks to chronicle the creatively evolving spiral of their dialectical journey toward untapped human potentialities.

Human Architecture maintains that all human failures at self and broader social change in favor of the good life are rooted in the problem of habituation, i.e., the human propensity to become subconsciously attached to sensations, ideas, feelings, things, relations, and processes. Decisive among these habituations are the dualisms of theory/practice, self/society, and matter/mind—by-products of dualistic oppositions of materialist and idealist world outlooks lasting for millennia. These dualisms are responsible for the world-historical fragmentation of the essentially creative human search for the good life into mutually alienated and thereby failing paradigms of philosophy, religion, and science—giving rise to equally fragmented and mutually alienated Western utopian, Eastern mystical, and global scientific movements.

The splitting of the inherently artistic and creative human spirit into its ideological components more or less corresponds to the world-historical transitions of ancient civilizations to classical political, medieval cultural, and modern economic empires—for which the dialectics of nomadic vs. settled modes of life paved the way in the course of an increasingly synchronous global development. The postmodern condition today is the general crisis of all fragmented paradigmatic structures, modern and/or traditional. It follows, then, that the good life will not be the gift of a wise few, of supernatural forces beyond, or of an objectively preordained natural or historical progress. Human de-alienation can only be an artistic endeavor by each and all—only within a creative humanist framework can the habituated dualisms and fragmentations of philosophy, religion, and science be overcome while preserving their true meanings and contributions.

It will be demonstrated that all dualisms can be effectively transcended through their conscious and intentional re-articulation as diverse manifestations of part-whole dialectics. The habituated common sense definition of society as multiple ethno-national and civilizational systems of relations among "individuals"—based on ahistorical presumptions of human "individuality"—will be rejected in favor of its definition as a singular world-historical ensemble of multiple intra-, inter-, and extrapersonal self relations. It will be argued that human life can be harmonious only when it is a world-system of self-determining individualities. Contributions of Western utopianism, Eastern mysticism, and science to an otherwise singular movement in humanist utopystics will be critically explored within an integrative framework. Human architecture will be introduced as the spatiotemporal art of design and construction of part-whole dialecticities in everyday life—of building alternative world-historical realities amid the personal here and now.

Human Architecture provides a forum for the exploration of personal self-knowledges within a re-imagined sociological framework. It seeks to creatively institutionalize new conceptual and curricular structures of knowledge whereby critical study of one's multiple selves within a singular and increasingly world-historical framework is given educational and pedagogical legitimacy. The journal is a public forum for those who seek to radically understand and, if need be, change their world-historically constructed selves. It is a utopystic research and educational landscape for fostering de-alienated and self-determining human realities.

Human Architecture will transcend the habituated dualisms of young and old, undergraduate and graduate, student and teacher, in and outside classroom, on- and off-campus, academic and non-academic, knowledge and feeling, mind and body, private and public, society and nature, reality and imagination, and philosophy, religion, science, and the arts—East and West. It will disempower the social stratifications of class, status, and power arising from economy, culture, and politics in favor of recognizing the all-encompassing stretch of human alienation—fostering new sociological imaginations more conducive to a shared human liberation project.

The foe claims in error that a philosopher I am.
God knows I am not what he says *I am.*
But, having endured this sorrow's nest, I ask:
Why should I *not know at least what I am?*

—Omar Khayyam

HUMAN ARCHITECTURE

Journal of the Sociology of Self-Knowledge

Editor:
Mohammad H. (Behrooz) Tamdgidi
Assistant Professor of Sociology
UMass Boston

Human Architecture: Journal of the Sociology of Self-Knowledge (ISSN # 1540-5699) is published by OKCIR: the Omar Khayyam Center for Integrative Research in Utopia, Mysticism, and Science (www.okcir.com, info@okcir.com) and printed by the Okcir Press, an imprint of Ahead Publishing House (APH), located at 6 Hobbs Road, Medford, MA, 02155, U.S.A., tel/fax: 781.874.1448. *Human Architecture* is indexed in CSA Illumina's *Sociological Abstracts*® and included in EBSCO's SocINDEX with Full-Text® and ProQuest's "Social Science Journals" full-text database.

Submissions: *Human Architecture* publishes both submitted and invited manuscripts as well as the working papers of OKCIR: the Omar Khayyam Center for Integrative Research in Utopia, Mysticism, and Science—an independent research and educational project. Contributors extend permission to *Human Architecture* for the publication of their work in the journal. They retain copyrights to their work and may publish them elsewhere. If the submitted manuscript has been published elsewhere before, written permission from both the author(s) and publication(s) where it earlier appeared should accompany submission to *Human Architecture*.

Editorial decisions: *Human Architecture* adheres to the peer reviewing principle for advancing scholarship—seeking innovative ways to meet the need in favor of liberatory scholarly practices most conducive to the aim and purpose of the journal. Selection of papers from submitted or invited manuscripts are based on their substantive relevance and the coherence and innovativeness of their argument in consideration of the mission of the journal. Views expressed in the journal by contributors are those of their authors and may not necessarily coincide with one another, or with the views of the editor, members of the Editorial Advisory Board, or the institutions with which any of the above are affiliated. Authors are responsible for the accuracy and integrity of factual, bibliographic, and referential materials used in their own articles, and for obtaining permissions for using copyrighted material in their manuscripts. Methodological, theoretical, historical, empirical, practical, as well as literary and artistic contributions relevant to the mission of the journal are all encouraged. The primary language used is English, but material in other languages may be included if relevant to the purpose of the journal.

What to submit: All manuscripts should be submitted in electronic format. They should preferably be double-spaced in Times 12 typeface., with 1 inch margins all around. Footnotes, endnotes, or reference lists may be single-spaced. In general, authors may follow a consistent bibliographic and citation style of their choice throughout the manuscript. Using the ASA (American Sociological Association) style is preferred by the editor.

Where to submit: The Editor, *Human Architecture*, Okcir Press, 6 Hobbs Road, Medford, MA, 02155, U.S.A., tel/fax: 781.874.1448, e-mail: mohammad.tamdgidi@umb.edu

Subscriptions: *Human Architecture* is a quarterly publication, published in either single-issue or double-issue formats, all issues for each volume becoming usually available concurrently at the end of every summer. Individual and institutional rates are $15 and $30 for single-issues and $30 and $60 for double-issues respectively. *Individual and institutional subscription rates* per year beginning from the most recently published issue (when subscription order is received) are $60 and $120 respectively. Back issues or additional copies of the journal are available upon request at the same rates as indicated above. Rates include domestic shipping and sales tax, where applicable. For international or bulk orders please inquire for special rates & shipping charges. Make checks payable in U.S. dollars to Ahead Publishing House, and send payments to Ahead Publishing House, 6 Hobbs Road, Medford, MA, 02155, U.S.A. Contributors each receive one free copy of the issue in which their articles appear. Rates are subjected to change without notice.

Advertisements: Current rates and specifications may be obtained by contacting the Okcir Press, 6 Hobbs Road, Medford, MA, 02155, U.S.A., tel/fax: 781.874.1448, e-mail: mohammad.tamdgidi@umb.edu

Inquiries: Address all correspondence and requests to *Human Architecture*, Okcir Press, 6 Hobbs Road, Medford, MA, 02155, U.S.A., tel/fax: 781.874.1448, e-mail: mohammad.tamdgidi@umb.edu

Changes of address: Six weeks' advance notice must be given when notifying change of address. Please include both the old and the new addresses in your request. **Postmaster**: Send address changes to Ahead Publishing House, 6 Hobbs Road, Medford, MA, 02155, U.S.A.

Human Architecture:
Journal of the Sociology of Self-Knowledge
Volume VI, Issue 2, Spring 2008
ISSN: 1540-5699

Editorial Advisory Board

A Peer-Review*ing* Journal

Contributions to *Human Architecture: Journal of the Sociology of Self-Knowledge* pass through a rigorous selective process with respect to their fit, relevance, coherence of argument, and innovativeness in consideration of the scope, nature, and intended purpose of the journal. The journal adheres to the peer-reviewing principle for advancing scholarship, but aims to design and build new scholarly avenues to meet this requirement—seeking mechanisms that foster openness of inquiry and evaluation; mechanisms that invite constructive judgments subject to free, open, and mutually interactive, not blinded and one-sided, peer reviewing practices; mechanisms that can be employed as widely and dynamically as possible among specialist and interested scholars in the field who value the need for the proliferation of new, critical, and innovative personal and global insights and transformations.

To meet the highest standards of scholarship, liberatory editorial practices need to transition from static peer review*ed* to dynamic peer review*ing* models that de-couple publication from defective pre-publication peer review requirements, and engage in alternative peer review practices that remain open to all those wishing to review a manuscript at any time in the post-publication phase—encouraging expanded and deepening exchanges among scholars, authors and readers alike. They need to invite critical thinking about prevailing and dominant paradigms and inflame creative spirits to forge new scholarly horizons and intellectual landscapes. And they need to embrace the subaltern voices in the academia and beyond, voices of those who have been deprived of cultivating their sociological imaginations through formal scholarly publishing avenues.

Human Architecture warmly invites contributors and readers to peer review the articles herein and to openly share their critical and constructive insights with one another in the future chronicles of this journal.

Contents

HUMAN ARCHITECTURE

Journal of the Sociology of Self-Knowledge

Volume VI Issue 2 Spring 2008

Contents

HUMAN ARCHITECTURE

Journal of the Sociology of Self-Knowledge

Volume VI Issue 2 Spring 2008

Editor's Note

Toward Sociological Re-Imaginations of Science and Peer Reviewing

Mohammad H. Tamdgidi

University of Massachusetts Boston

mohammad.tamdgidi@umb.edu

Abstract: To what extent does the traditional and prevailing models and practices of "science" and especially of academic peer-reviewing enhance or hinder the advancement of the sociological imagination? More specifically, if the sociological imagination centrally involves the cultivation of an ever deepening link between our knowledges of our personal troubles in everyday life to our knowledges of the ever widening public issues arising from the structures of contemporary capitalism in a world-history context, how does a science still biased in favor of classical notions of impersonal "objectivity," and a peer-review mechanism structurally bent on "blind" peer reviewing and subtraction of the "personal" from the texture of reviewed manuscripts, help us advance the sociological imagination? The juxtaposition—in this issue of *Human Architecture*—of a symposium of papers on the public issues related to the malfunction of science and peer reviewing on the one hand, and a series of innovative, reflective self-explorations by students and/or faculty of their experiences of learning and teaching the sociological imagination on the other hand, can hopefully and "empirically" illustrate the difference our sociological re-imaginations of science and peer reviewing can make for students and faculty alike in understanding and transforming their everyday lives, on- and off-campus.

To use the parlance of peer reviewing, it may initially appear as "unfitting" to include a symposium of important articles by two international scholars and specialists on the sociology of science and peer reviewing in this issue of *Human Architecture* titled "Sociological Imaginations From the Classroom"—a journal issue which also includes innovative papers by various international faculty and UMass Boston students and faculty focusing on teaching and learning of the sociological imagination. How can these two themes relate to one another and "fit" into the same thematic issue of a journal?

The sociological imagination according to C. Wright Mills[1], readers may recall, is a quality of mind that enables its holder to relate his or her own and others' personal troubles to the ever larger public issues facing society and humanity as a whole. It is the ability to relate reflections on the here-and-now dynamics of one's everyday life and personal troubles to the larger social is-

[1] See C. Wright Mills, *The Sociological Imagination* (New York: Oxford University Press, 1959).

Mohammad H. Tamdgidi, the founding editor of *Human Architecture*, is Assistant Professor of Sociology, teaching social theory at UMass Boston; most recently he is the author of *Advancing Utopistics: The Three Component Parts and Errors of Marxism* (Paradigm Publishers, 2007).

sues of the prevalent society, of the times, and in the context of ever wider world-historical landscapes.

As editor of this journal and this issue, however, I find the two to be quite intimately linked and fitting with one another. In fact, I would argue that exploring and understanding the nature of this intimate link between the symposium and other substantive papers published in this issue is a much needed effort in cultivating our sociological imaginations of science and of peer reviewing.

The question that may direct us to see more clearly the link between the two kinds of inquiry may best be expressed as follows: To what extent does the traditional and prevailing models and practices of "science" and especially of academic peer-reviewing enhance or hinder the advancement of the sociological imagination? More specifically, if the sociological imagination centrally involves the cultivation of an ever deepening link between our knowledges of our personal troubles in everyday life to our knowledges of the ever widening public issues arising from the structures of contemporary capitalism in a world-history context, how does a science still biased in favor of classical notions of impersonal "objectivity," and a peer-review mechanism structurally bent on "blind" peer reviewing and subtraction of the "personal" from the texture of reviewed manuscripts, help us advance the sociological imagination?

The way we view and practice science as a public issue and phenomenon, and how we come to understand our personal troubles in their intimately private and inter/intrapersonal dimensions are closely interlinked. In prevailing traditional practices of peer reviewing where "blind" reviews[2] are treated as somehow more "scientific" than personally transparent ones, the very point of the sociological imagination involving an exploration of intimate events and issues of the scholars' own personal lives are sacrificed in the name of scienticity and presumed scientific functionalism. The two symposium papers by Klaus Fischer and Lutz Bornmann—kindly and especially translated from German and Swiss for publication in this issue of *Human Architecture*—shed significant light on the extent to which the taken-for-granted structures of science and peer reviewing have been and continue and need to be problematized in favor of more liberatory scientific and peer reviewing practices that, I would argue, can be also much more conducive and favorable to the advancement of the sociological imagination.

Human Architecture has had a continuing commitment to publishing undergraduate and graduate student papers as scholarships of learning, in addition to papers from faculty and more senior scholars, both local and international.[3] As noted in the founding statement of the journal, "*Human Architecture* provides a forum for the exploration of personal self-knowledges within a re-imagined sociological framework. It seeks to creatively institutionalize new conceptual and curricular structures of knowledge whereby critical study of one's selves within an increasingly world-historical framework is given scholarly and pedagogical legitimacy. The journal is a public forum for those who seek to radically understand and, if need be, change their world-historically constituted selves. It is a re-

[2] I have previously argued that in fact, strictly speaking and from a sociology of knowledge and science perspective, no peer reviewing can be "blind." See my previous editor's note, "Peer Reviewing the Peer Review Process" (*Human Architecture*, vol. IV, issues 1&2, pp. vii-xv).

[3] See the table of contents of previous issues of *Human Architecture* at the end of this volume, full-texts of which are freely available online at http://www.okcir.com. Student and other papers published in *Human Architecture* are regularly used as required or recommended readings in course instruction. Contents of *Human Architecture* are also compiled in the Sociological Abstracts and SocINDEX with Full-Text (and, forthcoming, in ProQuest's "Social Sciences Journals").

search and educational microcosm for fostering de-alienated and globally concerned, self-determining human realities."

Such a substantive commitment on the part of the journal to helping advance the sociological imagination requires new and innovative structures of scientific investigation and peer reviewing that do not habitually take for granted the traditionally narrow definitions of "objectivity" and "blindness," but seek to expand the standards of "objectivity" and of peer reviewing to a level that makes the scholars' and authors' globally informed personal self-reflexivity a transparent requirement of the scientific and peer reviewing process.[4] Such functional alternatives to the prevailing regimes of peer reviewing may initially seem odd and "unfitting" in the broader context of what has come to be known as science and peer reviewed scholarship. But it is the very questioning of such presumptions that can open new horizons for the re-imagination not only of our personal lives, but also of what our public scientific and scholarly pursuits can achieve.

In this issue and in his sociology of science reflections on general notions and dys/functional practices of science and peer reviewing, Klaus Fischer[5] provides us with new reasons to continue publishing and appreciating the voices of students and faculty that chronicle the pages of *Human Architecture*. Klaus Fischer aptly writes,

> The refusal of all attempts to support radically new ways of thought and points of view, let alone to develop them, preserves the disciplinary matrix (cf. Kuhn) and the informal social structure attached to it. At the same time it also impedes the career of those scientists, usually young, who are connected to what is new (which always involves an informal social structure) and leads to an unjust distribution of recognition for achievements. Normal science rewards the guardians of the ruling consensus and punishes those who threaten to violate it. As a rule those succeed best who appear competent while adhering to current conventions of thought. Only in the best of all scientific worlds, in which many different paradigms compete with one another, would the negative effects of these innovation-impeding reward structures be sufficiently mollified. (p. 3)

Fischer continues:

> ... This enmity toward innovation manifests itself at many levels. Many examples show that phe-

[4] In fact, turning the tables around, "objective" science can and should be conducted with a self-reflective awareness of how the scholar is doing research. See the interesting discussion of "Interactive Ethnography" by Frank Nutch and Dick Butcher in "Reflections on Doing Interactive Ethnography" (*Discourse of Sociological Practice*, v. 2, no. 1, Fall, 10-11), and Frank Nutch's "In the Field: Reflections on Selected Aspects of Sociological Field-Work" (*Discourse of Sociological Practice*, v. 5, n. 1, Spring, 15-19). As I have noted elsewhere reflecting on Nutch's approach, interactive ethnography involves "collective and reflective participation with a research partner in a common filed work in such a way that the representational goal of understanding the field 'object' is deemed secondary and a means to the end of self-reflective study of what goes on in the cross- and inner dialogues of researchers as they prepare for, conduct, and evaluate their work" (Tamdgidi, "Rethinking Sociology: Self, Knowledge, Practice, and Dialectics in Transitions to Quantum Social Science," *Discourse of Sociological Practice*, v. 6, no. 1, Spring 2004, pp. 61-81). In the latter, appreciating Nutch's perspective, I continue to problematize the requirement of doing interactive ethnography "with a research partner" and argue that, given the multiplicity of selves constituting the inner lives of persons, including scholars, an interactive *auto*ethnography is also possible and required.

[5] Fischer, Klaus. "Science and Its Malfunctions." *Human Architecture: Journal of the Sociology of Self-Knowledge*, v. VI, no. 2, pp. 1-22).

> nomena which institutionalized science at first and in some cases for a long time had vehemently rejected originally, because they were not compatible with the disciplinary matrix of current research and hence had appeared unbelievable, were later able to prove themselves to be genuine ... (p. 3)
>
> ... Complementary to those "unbelievable" phenomena are those which science had viewed as certain for years, decades, even centuries, until finally, sometimes in a very short time, they literally dissolved as theoretical frameworks changed... (p. 4)

I think the best compliment one can provide to the above observation and to the student and faculty voices contained in this volume is that they are "unbelievably" courageous and pioneering in advancing their sociological imaginations, taking in consideration, of course, their differing backgrounds and levels of preparation ranging from freshmen undergraduate students to senior international faculty. Likewise, it is important to note what Lutz Bornmann[6] also observes in this issue regarding the prevailing structures of peer-reviewing in the world today and how more rigorous theoretical reflection and intervention in the taken-for-granted scientific and peer reviewing habitus can contribute to widening and deepening our knowledges of our selves, of our world, and their interrelations. Bornmann writes,

> ... [T]he theories in sociology of science (especially the more recent theories) have remained unconsidered in empirical analysis of the scientific peer review process. The suppositions of the North American school are often viewed as outdated, and a more extensive theoretical analysis of peer review from the perspective of social constructivism and social systems theory has so far failed to appear. Therefore, the goal of the present contribution is to present a theoretical framework for peer review that also considers more recent theory developments and that can serve as a basis for future empirical peer review research.

The present issue of *Human Architecture* chronicles thoughtful papers by students aimed at cultivating their sociological imaginations. The authors of these papers (see the two papers by Jacquelyn Knoblock, the essays by Henry Mubiru, David Couras, Dima Khurin, Kathleen O'Brien, Nicole Jones, Nicole [pen name], Eric Reed, Joel Bartlett, Stacey Melchin, Laura Zuzevich, Michelle Tanney, Lora Aurise, and Brian Ahl) make serious efforts at developing their theoretically informed sociological imagination of gender, race, ethnicity, learning, adolescence and work in dialogue with scholarly sources, relevant theoretical frameworks and concepts, and various other texts such as films and works of art.

The volume also includes papers by faculty (see Satoshi Ikeda, Karen Gagne, Leila Farsakh) who self-reflectively explore their own life and pedagogical strategies for the cultivation of sociological imaginations regardless of the disciplinary field in which they do research and teach. Two joint student-faculty papers and essays (see Khau & Pithouse, and Mason, Powers, & Schaefer) also imaginatively and innovatively explore their own or what seem at first to be "strangers'" lives in order to develop a more empathetic and pedagogically healing sociological imaginations for

---

[6] Bornmann, Lutz. "Scientific Peer Review: An Analysis of the Peer Review Process from the Perspective of Sociology of Science Theories." *Human Architecture: Journal of the Sociology of Self-Knowledge*, v. VI, no. 2, pp. 22-38).

their authors and subjects. These papers include innovative classroom exercises that help cultivate the sociological imagination in their students, and can in turn serve as aids to the narrative explorations and presentations of the pedagogical approaches invented and used by faculty in the course of their teaching career. These papers amply demonstrate the extent to which scholarships of teaching and learning can complement and enrich one another while advancing the sociological imagination in comparative and international frameworks.

I would like to take this opportunity to especially thank the dedicated care, attention, and energies that Anna Beckwith, lecturer of sociology at UMass Boston, offered to this issue of *Human Architecture* as its guest co-editor. As noted in each of the contributing student papers, nine of the student papers published in this volume were written in the courses Beckwith taught at UMass Boston during the 2007-8 academic year. These included courses on the Sociology of Work, Youth and Adolescence, and Race and Ethnic Relations. Six other papers by UMass Boston students were written in various sections of the course, Elements of Sociological Theory, taught by Tamdgidi at UMass Boston.

The juxtaposition—in this issue of *Human Architecture*—of a symposium of papers on the public issues related to the malfunction of science and peer reviewing on the one hand, and a series of innovative, reflective self-explorations by students and/or faculty of their experiences of learning and teaching the sociological imagination on the other hand, can hopefully and "empirically" illustrate the difference our sociological re-imaginations of science and peer reviewing can make for students and faculty alike for understanding and transforming their everyday lives, on- and off-campus.

Science and Its Malfunctions

Klaus Fischer

Universität Trier, Germany

fischer@uni-trier.de

Abstract: This article explores the nature, causes, and genesis of malfunctions of science. Topics explored are: deception; subordination to dominant spirit and paradigms of the times; impacts of dogmatism and social interests; ingroup-outgroup behavior and other effects of the social structure of science; inappropriate peer review and malfunctions of prevailing academic review system; misappreciation of innovative research; impact of networking and loyalties on scientific progress; impact of economic interests and liaisons; pathological interpretations of scienticity and validity of scientific results; defects of scientometric measures as indicators of achievement, as a manifestation of the impact of medial code on science; trustworthiness or lack thereof of Science Citation Index as measure of research and scientific achievement. The impact of social, economic, political, cultural, linguistic, medial, religious, and other codes on scientific process and behavior of scientists are explored.

1. The Purpose and Form of Science

(1) Before speaking of the malfunctions of science, its functions need to be identified. This alone could become a very lengthy task. For what follows it appears that a very general characterization is sufficient, without exhaustive clarification of detail. According to this minimal definition, science can be functionally characterized as that part of society, which aims at the acquisition of "certified" information, i.e., information tested with a positive result according to the currently accepted methodological rules, such as widens our knowledge of certain aspects of reality. Although the daily tasks of a business consultant, a lawyer, an artist or a priest are partly of the same kind, the extension of knowledge of a certain part of the world is the scientist's primary goal and *not only* an instrument for

Professor Dr. Klaus Fischer holds a faculty chair in the department of philosophy at the University of Trier, Germany. His research interests are in the fields of history and philosophy of science, sociology of science, sociology of knowledge, and philosophy of nature and in selected areas of modern philosophy such as Neurophilosophy and Neoconstructivism. More specific priorities include thematic foci on specific individuals and schools in science and scientific theory such as Popper, Kuhn, Feyerabend, and the Vienna Circle. In the area of history of science, he is interested in Galileo and Einstein. Other interests include history of Empiricism, history of university, innovation processes, possibilities of science, and history of science in the 20th century (nuclear physics, molecular genetics, cosmology, chaos theory, and artificial intelligence). The present article is an abridged English translation of an article by the author, entitled "Fehlfunktionen der Wissenschaft," recently published in German in the journal *Erwägen Wissen Ethik* (Vol. 18, Issue 1, 2007, pp. 3-16).

reaching another goal.

(2) In the empirical history of science, this goal takes on various forms. Depending on the basic attitudes of an epoch, it can metamorphose into the intention of reading God's thoughts, finding the signature of the world, uncovering the beauty of nature or demystifying it, spreading enlightenment, unlocking the cosmic secret, investigating incredible phenomena, satisfying curiosity, or fulfilling personal ambition.

(3) Even researchers for whom science is primarily a means to something else, for example to be famous or rich, to attain social recognition or authority, to gain power over others or over nature, to increase utility for themselves or society, to find peace of mind, to reach certitude of salvation, to pursue religious goals, to further political ideologies, or to save the world, can also make significant scientific accomplishments. The specific mixture of motives, though, leads to conflicts of purpose and to behavioral consequences in which the acquisition of verified information does not serve as highest maxim, but recedes before other ideals.

2. A Phenomenology of Scientific Malfunction

2.1. *Deception*

(4) One of the most marked signs of scientific malfunction would be the increase in the amount and varieties of scientific misbehavior. Recent surveys from the natural sciences in the United States point out that questionable research behavior is more prevalent than usually assumed. A third of all scientists make anonymous admission of questionable behavior and almost forty percent of young researchers report being the victim or the witness of unethical behavior on the part of others (Martinson 2005; Physik-Journal 2005; Kirby).

(5) Whereas science views the uncovering of great instances of deception as success in scientific control which strengthens the system, habituation to many small daily instances of human and methodical failure leads to a gradual erosion of scientific standards. The struggle against bad scientific ethics is not decided by public investigations of "cases" such as that of Friedhelm Herrmann and Marion Brach, Jan Hendrik Schön or Hwang Woo-Suk. It is instead carried out in the gray zone between what is methodically or ethically ideal and what is manifest deception (Fischer 2004a).

2.2. *The Spirit of the Times*

(6) The effects of the spirit of the times or of political correctness are a subtle source of scientific malfunction. Their negative effects are partly due to the misperceptions it causes, partly due to a false evaluation of scientific theories, problems, arguments, and data, partly due to a degradation or even defamation of those who think otherwise, possibly in good faith. If this game is played in a consistent and industrious manner it can do much more harm to research than direct deception.

(7) From today's point of view those psychologists, therapists and analysts were subject to the influence of the spirit of the times who (especially in the U.S.) thought they had uncovered millions of cases of misused children (Ofshe & Watters 1996), multiple personality (Hacking 1996), abduction by beings from outer space. Elaine Showalter (1997) has invented the term "hystoria" for this kind of intellectual confusion, which can take on an epidemic scale, basing the word on the term "hysteria," the name for a sickness which a hundred years ago the physicians of the time thought upper class women tended to have because of a disposition of their gender.

(8) We do not wish to become involved in current debates, but rather to exemplify the workings of the spirit of the times. Its effect consists in self-attribution of knowledge not really had, as well as in considering some theories, hypotheses, models and interpretations to be far more certain, or far more flawed, than they are. In some cases they take on such a degree of certainty in the minds of those subject to the spirit of the times that those who are of differing opinions are even considered indecent, dishonest, disbelievers or heretics. Sometimes it takes on the form of self-attribution of ignorance (Willgerodt 2004), when available data and arguments are not taken into account, but are considered inadmissible, unreliable, or even as ideological. Areas which are susceptible today are studies of risk, climatic change, intelligence, especially regarding gender and race, atomic energy, mind and brain, and free will.

2.3 Dogmatism and Social Interests

(9) Organized interests and social stratification in scientific systems (hierarchies of status among persons, laboratories, nations) interfere in various ways with the development of research. The *formal* structures of science usually contribute to stability. They tend to open up to new developments only slowly and under pressure from external competition. One particular aspect of social structuring is that which is related to the binding effect of a paradigm. Other than in the case of the institutionalized structures of science, this is rather of an *informal* nature. Like organized interests and hierarchies of status in laboratories, institutes and other organizational units, strong commitments to paradigms tend to hinder innovation, because they impede the free play of arguments, ideas and perspectives, the emergence of alternatives, and the recognition of unusual phenomena. Although bound to the rhetoric of epistemic progress, the critical attitude and the unbiased examination of novelty, research governed by a paradigm rewards only *internal* criticism, small progress within the given framework of thought, the formulation and examination of hypotheses which do not disturb this framework, but rather protect, improve, expand, embellish it. The refusal of all attempts to support radically new ways of thought and points of view, let alone to develop them, preserves the disciplinary matrix (cf. Kuhn) and the informal social structure attached to it. At the same time it also impedes the career of those scientists, usually young, who are connected to what is new (which always involves an informal social structure) and leads to an unjust distribution of recognition for achievements. Normal science rewards the guardians of the ruling consensus and punishes those who threaten to violate it. As a rule those succeed best who appear competent while adhering to current conventions of thought. Only in the best of all scientific worlds, in which many different paradigms compete with one another, would the negative effects of these innovation-impeding reward structures be sufficiently mollified.

(10) This enmity toward innovation manifests itself at many levels. Many examples show that phenomena which institutionalized science at first and in some cases for a long time had vehemently rejected originally, because they were not compatible with the disciplinary matrix of current research and hence had appeared unbelievable, were later able to prove themselves to be genuine. Examples are: the Greek Antikythera mechanism, meteors, solar wind, ball lightning, continental drift, reverse transcriptase, infectious proteins, life at temperatures above boiling temperature and life under extreme pressure, large impact craters on the earth, non-classical inheritance (epigenetic memory), oscillating chemical reactions (Belousov-Zhabotinsky Reaction), giant snowballs hitting the earth from outer space, or strange ionospheric

phenomena accompanying storms (sprites, "elves," blue jets).

(11) Complementary to those "unbelievable" phenomena are those which science had viewed as certain for years, decades, even centuries, until finally, sometimes in a very short time, they literally dissolved as theoretical frameworks changed: Schiaparelli's "Martian canals," van Maanen's "unwinding" galaxies, phlogiston, the symptoms of hysteria, neurasthenia and vegetative dystonia, the planet "vulcan," "polywater." An even more impressive witness of the connection between systems of theory (paradigms), dogmatic attitude (Gernert 1999), and social interests are given by those once dominating theories, paradigms, and world views whose active opponents suffered from lack of recognition of their research achievements or the granting of research opportunities by the system of institutionalized science. Although these theories are now themselves rotting in the junkyards of history, the damage which their faithful adherents caused by the oppression of concurrent ideas and of their representatives can not be made up for.

(12) There are obviously mechanisms which protect conforming ("good") information and fend off dissonant ("dangerous") novelties. Well-documented instances in the history of the sciences show two things: 1) the almost limitless facility with which scientists believe whatever fits their world view and possibly gives it more support (Broad & Wade 1984, 136ff.; for the humanities cf. the so-called Sokal-affair: Sokal & Bricmont 1999); 2) the complementary constellation where a practically impenetrable wall of scepticism is erected whenever anything (e.g., "out-of-place artefacts" in archaeology) is reported which might undermine the currently accepted scientific world view, its standard models, and its conventional ways of seeing things. Especially the latter is a strong deterrent to the development of science. It sometimes has tragic effects on individual scientists who refuse to fit into the mainstream and hence it represents a serious malfunction of science for several reasons.

2.4. Ingroup-Outgroup Behavior and Other Effects of the Social Structure of Science

(13) It is well known that many ideas and theories are not attributed to their first discoverers, because they were originally rejected, then forgotten, and later had to be thought out again by others. In these cases it is not the pioneers, but rather the second discovers or imitators who are rewarded and cited for their achievement. An example is Chandrasekhar and his theory of the evolution of stars (Wali 1991, Chap. 6). The young Indian astrophysicist did not publish his theory and turned to a new research field, after he encountered strong resistance from established scientists. More than twenty years later the theory was developed again by others. This case stands for many. It contradicts the spirit of an enterprise, in the internal logic of which only the acquisition of certified information (in the sense explained above) is decisive, not the age, status, nationality, gender, cultural background or attractiveness of those who originate information and technical achievements.

(14) There is another important aspect of such processes, which is of no small significance in connection with the politically propagated goal of rewarding scientists according to their achievements. According to the criteria of assessment of achievement, which does not evaluate retrospectively, but measures contemporary perceptions, the first discoverers wasted their time on "worthless" ideas, instead of occupying themselves with things which were of in-

terest to their mostly higher-ranking and more well-known colleagues. With respect to the opportunity costs of the "wasted time," they and their home institution would have been punished by reduced funding for their farsightedness and their creativity, which bore fruit not immediately, but only after twenty years. Despite its rhetoric of innovation, research politics, intoxicated by the promises of a growing evaluation business, at the beginning of the third millennium furthers such hidden innovation-inhibiting effects by relying too much on the consensus of the scientific community—which as a rule is the consensus of those who are in key organizational positions and hence tied to the political interests of their field.

(15) This innovation-impeding malfunction of science is the expression of the multidimensionality of scientific activity, which is not only a cognitive process, but also a social, political, economic, medial, cultural, etc., one (Fischer 2005 and 2006). This multidimensionality has the effect that scientific activities and products are also subject to social, medial, and political evaluation (to only name these three dimensions), in which the prestige, the publicity, the network of loyalties, the fame of a researcher, institute or university, and its past successes and failures, play a decisive role (Collins 1975 and 1981). It is a matter of trust, of focusing attention, of winning allies, of warding off the danger of potential opponents. Differences of position in the social and political systems of science effect different perceptions of problems and hence different reactions to conflicts and contradictions on the part of those involved (Bloor 1978; Caneva 1981). Which attempts at resolving a problem prevail at a given time, which alternatives one is tolerant, skeptical or hostile toward, will depend largely, although not exclusively, on the distribution of power within the scientific social system.

(16) This coupling of social, political and cognitive structure in science modifies the process of knowledge acquisition in a way which is basically foreign to it. Such interference of systems becomes pathological when the research code is dominated by the code of economics, politics, society, culture, media, religion, etc. (see below).

(17) One of the *social* bases for the development and success of new ideas consists in the subgroup of those lower-ranked researchers who do not expect to raise their status by allying with the mainstream, and by conforming to its rules and thinking. They are eager to latch onto and attempt to support that which is new and, from a social and cultural point of view, is seen as dissidence, even independently of its inherent qualities (see also Douglas & Wildavsky 1982). This is to be understood in a neutral sense. Novelty means merely that the ideas which are prevalent in this social milieu dissent from those which are presently in the middle of the attention of a field and are as a rule associated with higher-ranking researchers. It does not mean that these ideas are more of an aid to the goal of science than those which are current within the institutionalized centers of their field. In such marginal social areas of the same institutions not only do ideas flourish which further research and science but also strange creatures which are innovative and unusual, but do not carry any other advantages.

(18) How is it possible that certain phases or areas of research can be dominated by cohesive social networks and the code of "society" which are based on solidarity but not on methodologically certified knowledge? The reason is that "networking" appears to bring many advantages to the individual researcher. His inclusion in scientific networks satisfies elementary needs for contact with colleagues, opportunities to cooperate, gaining access to restricted in-

formation, and receiving support in the fight for resources. It is a matter of give and take in which the community offers protection and services to the individual in return for his solidarity with the community.

(19) This barter appears more advantageous to the individual than it is, a part of its costs being hidden. Among these costs (or latent functions) of group-building are that the social pre-structuring of contacts effectively reduces the number of information channels which the member of a group generally makes use of. Because of our limited capacity to receive and process information the intrusion of dissonant information into the internal group will be effectively impeded thereby. This is a process essential for maintaining the disciplinary matrix and its social structure, depending in its effectiveness on how coherent and dense the network is.

(20) Since a part of the distorted or hidden information would be able to modify the cognitive matrix of a paradigmatic network much faster than it actually does, the filtering and altering of such information, although inherent to sciences bound to such networks, must be seen as a serious malfunction of science. It lies in the nature of socially organized science. Its negative consequences would be negligible only in the best of all scientific worlds, in which a great number of different paradigms, both on the side of supply and of demand (institutions, journals, funders, evaluators, etc.) would compete with each other and make the borders between the corresponding networks more permeable.

(21) The higher density of internal as opposed to external communication has another function besides the limitation of the quantity of "uncontrolled" and hence possibly dissonant information. It shows the group member how others think and exercises, by subtle reinforcements and sanctions, an orienting influence on his own thought, in particular on the disciplinary matrix which he accepts and on his scientific world view.

(22) A researcher in this position can scarcely be described as a lonely and critical seeker of truth, but rather as a narrow-minded, prejudiced, ideologically inclined member of a "herd in the sphere of thought" (Nietzsche), who believes in a world view which in essential respects has been provided by others and who can only rarely permit himself the luxury of doubt, among other things, because he feels such doubt only rarely, given the way information is selected, and because he knows that open criticism might cost him the solidarity of the network. As long as it can successfully guide the process of research (such as progress is seen by the network) the disciplinary matrix provides for the cohesion of the network just as social consensus provides for the perpetuation of the matrix. This filter-and-reinforce function of the social network in which the individual researcher is embedded is central to an understanding of the dynamics of science. It is also central to an understanding of the causes of intellectual intolerance in any science characterized by firmly knit social networks. Networks are the social bearers of cognitive structures, with the latter providing for a level of information-processing which is optimal under the given circumstances (Fischer 1987, 1988), and in turn solidifying the networks.

2.5. *Inappropriate Peer-Review*

(23) Further symptoms of malfunctioning in science can be found in the system, almost universal today, of peer review (cf. Fröhlich 2002; Fischer 2004b). Among the field experiments, which have been devoted to determining the reasons for the problems of this evaluation system, that of Peters and Ceci from 1982 is the most clear

and the most disturbing (Peters & Ceci 1982). The sobering result is that when secondary indications of quality are absent, the psychological referees seem to become blind to the inherent cognitive qualities of a piece of work, and the experimentally determined rate of reacceptance of articles which had already been accepted 18 to 32 months before, by the same journals, sinks to that quota which might be statistically expected. John Ziman's reaction shows the consternation and the helplessness of many commentators regarding this result (Ziman 1982, p. 245).

(24) One possible attitude toward such empirical studies is a flight to sarcasm. An example thereof is the "author's formula" which one of the investigators of the peer-review system, J. Scott Armstrong, also coeditor of the *Journal of Forecasting*, has presented on the basis of his experiences and the available empirical evidence. Authors who want to improve their chances and accelerate the acceptance of their manuscript should "(1) *not* pick an important problem, (2) *not* challenge existing beliefs, (3) *not* obtain surprising results, (4) *not* use simple methods, (5) *not* provide full disclosure, and (6) *not* write clearly" (Armstrong 1992, p. 197).

(25) Is this pessimistic view justified? What could a crucial experiment, which might answer this question, look like? How it treats its elite is decisive for the development of a science. According to a more narrow definition which is based on the goal of research, those belong to the elite of science who contributed decisively, in some way or another, to the acquisition of new knowledge. The real difficulty lies less in the determination of achievements far in the past than in those of the present or the recent past. Is today's system of evaluating scientific achievement capable of dealing with this task? According to the experiences of a considerable number of high-level researchers we must say: very often not!

(26) From the biographies and autobiographies of scholars, researchers and inventors we know of many examples in which the peer-review system has led to sorry misestimations of achievements of the highest-level. In a substantial number of cases referees of scientific journals have refused central contributions of later nobel prize winners or others who have become world-renowned, because they did not recognize their value. What trust does a system deserve which cannot even recognize quality of the highest kind in a reliable manner?

(27) In other cases work was in fact published in spite of difficulties, but not given recognition for a long time. Decades can pass until high-level work is recognized as such and correspondingly honored in the form of citations. For the political program of implementing a market-oriented instrument of guidance into science which would reward scientists according to their achievements, this amounts to a disaster. It is just the most innovative contributions which are most often misrecognized, and as a rule the young talents are those who bear the consequences and turn away from research in disappointment.

(28) To summarize, the following conclusions can be drawn from the available empirical investigations. One can say of the system of referral and evaluation that:

- It does not have clear standards of scientific quality or it does not apply them consistently. In the past, dozens of pioneers of future science and technology have been rejected, defamed, ridiculed, which sufficiently proves its inability to objectively judge the achievement of independent and creative researchers with unconventional ideas (many examples in Fischer 2002; Horrobin 1990; Sommer 2001; Hook 2002; Thiel 1944).

Empirical studies seem to indicate that the main problem might consist less in the criteria themselves than in their practical application (cf. Neidhardt 1988, Tab. A12, p. 148). From a historical perspective even many criteria are also questionable.

- It rewards conformity or existing alliances with certain theoretical and methodical orientations, while insisting that this conformity should not be complete, but should show individual nuances—a proof of originality. The main function of this tactical subordination consists in demonstrating the connectibility of the scientists' research and to convince referees that the author or applicant is not an intellectual adventurer, but understands himself as a part of the community effort of science. From a sociological point of view, it is a matter of social control in the system of science, which takes on the form of self-control in anticipation of the consequences of violating a norm.

- It favors certain paradigms, themes and types of argumentation. A comparison of articles submitted to the *Zeitschrift für Soziologie* between 1972 and 1980 with those published by this journal shows that the chance of publication depends very strongly on the theoretical orientation of an article. The average rate of diffusion of articles from the circle of functionalism, for example, was eleven times higher than that of articles from that of critical theory and five times higher than that of articles from the area of symbolic interactionism (cf. Sahner 1982, p. 88, Table 4). Sahner conjectures that journals with a system of referees are more impenetrable for innovations than those where editors are granted the decision about manuscripts. Altogether this publication system disadvantages above all those theoretical orientations that are only insufficiently represented by the preferences of editors and referees. Individual preferences are thus translated into quality judgements.

- It respects, or even submits to, the social status, the political power and the fame of applicants, institutions, journals and countries. This phenomenon has been called the "Matthew Effect" by Robert K. Merton. It can be found at all levels of the scientific system. Although it does not always have a negative effect on research, it must be viewed as a serious malfunction of the scientific system, because it rests on effects which, being mainly social, political and medial, are foreign to the system. In extreme cases it can lead to misuse of office and of position (Martin 1992; Sommer 2001).

- It avoids risk and disadvantages projects which are innovative, explorative, speculative, interdisciplinary as against those which, because they can be located in a clear methodical and theoretical framework, will bring about small, but certain scientific results.

2.6 The Misappreciation of Innovative Research

(29) Empirical results concerning the function of the peer review system reveal that it encounters particular difficulties with novelty. This becomes especially visible when novelty consists not only in the application or improvement of a known set of instruments, but when novelty departs from the framework and does in fact venture into terra incognita (Fischer 2004b). Although moving into new territory is one of the deepest sources of motivation for research, the system of evaluation tends in general to impede novelty, as soon as it has gone be-

yond a certain format. This is not its intention, but it is a consequence of the effort to minimize the risk of misjudgment. Consequently the referees for the evaluation of applications for resources (outside funding, publication venues, etc.) will prefer small, but certain gains to the uncertain chance of a jackpot. This means that well-known names and institutions, as well as projects with a certain conventional dress already have a bonus at the beginning of the race. A sign of the conservatism of the system of evaluation is that the average referee prefers articles with positive and conventional results to those with controversial ones (Sahner 1979; Mahoney 1976), putting manuscripts at a disadvantage in which a hypothesis which is currently of high regard is falsified (Armstrong 1997, p. 71; Martin 1997, Chap. 5).

(30) Among the researchers, well-known today, whose innovative ideas once experienced vehement rejection or disinterest on the part of the "scientific community" are Alfred Wegener, Alan Turing, Konrad Zuse, Konstantin Ziolkowsky, Hermann Oberth, Peyton Rous, Mitchell Feigenbaum, Frank Rosenblatt, Barbara McClintock, Stanley Prusiner, Andrei Linde, Günter Blobel, Noam Chomsky, Karl Popper (further examples in Fischer 2002; Horrobin 1990; Sommer 2001, Nissani 1995). It affects first of all younger, unknown, non-established scientists who have distanced themselves too far from actual consensus, such as Polly Matzinger (Hanisch 1997), Lynn Margulis (Margulis 1999), Candace B. Pert (Pert 1999) or Bonnie Bassler (Breuer 2003), also such as have violated the borders of their discipline. Occasionally it also hits established scientists, even nobel prize recipients when they depart too far from the consensus of the majority (e.g., Brian Josephson), or if they too openly admit that they themselves do not know what will come out of their investigations.

(31) This openness regarding results is the natural characteristic of innovative research, but it is unsettling to those always eager to have everything "under control" and be accountable to others, for example for granting outside funding. This structure of research funding does not adequately correspond to the manner in which successful science functions, which must always count on what is unforeseen and be able to react to it flexibly. It always has chaotic (non-calculable) properties, because it cannot predict its own future, in particular the many intervening accidents, the dynamics of interactions arising in certain situations, the new factors emerging in the course of the game, the surprising non-linearities evolving from chaos; as Günter Blobel says: If you can predict what you're going to do for five years, its probably going to be bad" (Goodman 1995).

(32) That does not mean that research and science are not plannable per se. Although it may be that a great part of science conducted today—and perhaps an even greater part of technology—rests on planning, a strategically central aspect of science and technology seems to be inaccessible to control: the scouting of new paths in unknown territory. Fleming could not plan to discover penicillin, Galilei to see the moons of Jupiter, Columbus to find America, Kamerlingh-Onnes to discover superconduction, Rutherford to hit upon a supermassive nucleus in the atom, Penzias and Wilson to notice cosmic background radiation, Bednorz and Müller to hit upon high-temperature superconduction, Perlmutter and Schmidt to discover the accelerated expansion of the universe.

(33) It can be argued that many discoveries do indeed have plannable elements and are based on theoretical expectations, which have in part been confirmed. There are discoveries and developments the conditions for which are so well-defined that they

could have been called *intentional* if there had been the necessary scientific far-sightedness. In these may be counted flying machines, automobiles, electronmicroscopes, scanning tunnelling and atomic force microscopy, transistors, integrated circuits, computers, atom bombs, fusion reactors, genetically altered creatures, nanomachines, as well as electromagnetic waves and the structure of DNA. On close inspection, each case exhibits, all planning notwithstanding, aspects of the unforeseeable. As an example one may take the development of flying machines or the endless history of fusion reactors. Cases like these show that chance, "serendipity," subjective drives and hopes, but also adjunct innovations in neighbouring fields (Sommer 2001; Roberts 1989), played a greater role than the planners of science would like to concede. The mania which can be observed today for wanting to have all aspects of research under control and to restrict its freedom in order to avoid, because of considerations of cost, "unnecessary repetitive research"—although the latter is an important tribunal for testing and improvement—is responsible for many of the most unsettling malfunctions of science: the growth of monocultures, the withering of variety and hence of the presupposition for the selection of the optimal variant, and the absence of fundamental innovations.

2.7 Misbehavior toward Young Scientists

(34) In well-documented cases researches have misused positional and institutional authority to credit themselves with achievements which were really attained by coworkers of lower status. Ombudspersons for good scientific practice can confirm that sensibility for misbehavior toward young scientists is still very underdeveloped, at least in the German system. Many supervisors, department heads or other superiors still seem to regard it as their right to use, pass on, or include in publications results by doctoral and postdoctoral students and by subordinates, even without their consent (Tröger 2005). In some cases this usurpation of property rights goes so far that results of persons from the groups mentioned have been passed on by superiors or supervisors even against the will of their originators or without indication of origin. Power of position, status, threat of repression are used, in order to keep those affected from attempting to maintain their rights by appealing to university committees, ombudspersons, or if necessary to courts.

(35) Another type of misbehavior, which is especially harmful to the careers of young scientists, is the use of formal authority and positional power in fighting against other paradigms, research approaches or scientific positions. Since new research approaches are often supported by young scientists, a political structure which allows or tolerates such things leads to lack of innovation or even to animosity toward such in the countries, disciplines, organizations or departments involved.

3. CAUSES AND GENESIS OF MALFUNCTIONS

(36) Studies of science have not yet made human deficiencies as causes of specific malfunctions of science a subject of systematic study. This reservedness could be unjustified. It may be that the disinterested search for information which is certified in the sense explained above—such as has no other motive than the investigation of the structure of the world—is expecting too much of human beings from a moral point of view. That may be an ideal which surpasses the measures of an entity guided by considerations of utility which stand in the service of interests, emotions, and desires:

in the interest of social recognition, bettering of economic position, control of the personal environment, warding off of perceived dangers, and resistance to dissonant information. To an entity addicted to fame, sex, money, power, the search for security and sensory stimulation, striving towards truth about the fabric of the world may seem an unprofitable tenet. Biologically based insufficiencies might also play a role in the failure of man to come to grasps with the classical ideal of the scientist's role. Perhaps the causal texture of the world is too complex for our brains—at least at the present stage of evolution. The British empiricists, as well as Descartes, have seen it this way, although they were of the opinion that our faculty of knowledge is still enough for practical life and to find our way to the markets. Research in its originally intended form as the search for that which holds the world together in its innermost core might turn out to be vain from an evolutionary point of view. The success of this enterprise played no role in the selection of our genes so far, our genetic equipment not being optimized with regard to this goal.

3.1 The New Network Ideology: Loyalty or Scientific Progress?

(37) Other factors appear, however, to be more significant for the explanation of present-day malfunctions of the institutionalized systems of science and research. If we consider results from recent sociology of science, we find indications of the sources of misjudgment, distortion, discrimination, and favoritism, which we have already referred to in the paragraph on the phenomenology of malfunctions. The most important factor seems to be the particular embedding of a researcher with given paradigmatic and social commitments in the institutional framework of science. This location creates certain orientations of interest. In one case it can be connected to higher formal authority and to the power of distributing research resources and employment, granting or denying access to publication opportunities, directing organizations, initiating decisions in the system of peer review and establishing interpretations which are binding. In other cases this may all be lacking. The position of a researcher in scientific networks, in editorial boards, scientific associations, allocation commissions, laboratories, or government boards sets up positive or negative mechanisms which can lead to unfair evaluations. Unfair evaluations can also hit those who carry undesirable personal characteristics, the wrong nationality, ethnic group membership, political ideology, gender, or an "unfavorable prehistory" in any desired sense.

(38) One of the conditions for these causes of malfunctions to work in an unmitigated way is the very restricted competition between institutes, laboratories, financers, journals, publishers. In addition to the processes of concentration which already exist, politicians encourage monocultures by openly pursuing the goal of avoiding "double research" in order to save money. This goal is to be reached only in the short term, for without comparative research there is no possibility of examination, by socially impartial and critical judges, of the strengths and weaknesses of the what has been attained. In a system in which oligopolistic or monopolistic structures prevail there cannot develop those variety of ideas, theories, perspectives and arguments which would be necessary for a fruitful culture of scientific discussion and hence for a selection of the best variety of solution to a problem. Even worse, it is only in monopolistic and oligopolistic markets that many of the deformations mentioned above (deception, dogmatism, obsequiousness to the spirit of the times, inappropriate judgement of colleagues, rejection of novelty, op-

pression and exploitation of young scientists) can exert themselves in full force. In markets over which the demanders of research have no control, the suppliers can, if buyers act unfairly, look elsewhere and seek out another institute or laboratory, another boss, another sponsor, another journal, another publisher, etc. In the present-day research landscape these suppliers must, however, realize that variety is only apparent: in many branches of research there are only very few laboratories, sponsors, and journals which really come into question, and they are all bound up in one and the same network of referees and deciders, who believe themselves to be "guardians of standards," but from a sociological point of view are guarding the purity of their world view. Anyone who has spoiled his relationship with a part of the network will not get any hearing elsewhere in it. He is "excommunicated" and can at best pass a marginal existence on the rim of the system as a notorious outsider, eccentric, or fool.

(39) The political tendency, which has recently become manifest both at national levels and in the European Union, of supporting primarily research projects carried by "networks" leads, whether intended or not, to the replacement of knowledge as a motive by social loyalties, the immediate goal of which is to form and foster social ties in the service of economic advantage. Such loyalties can, especially when the networks involved have common "skeletons in the closet" after fighting together, be cemented as "social capital" and take on mafia-like traits. To put it lightly, one could call this *publicly organized colonization of science by society.* Its highest goal is no longer the attainment of knowledge which has been certified in the present sense, but rather solidaristic action toward the greatest possible economic advantage. It should not be a surprise that "networking" has become one of the most important conditions for advancing a career and obtaining outside funding. If the networks are tightly knitted enough and if they know how to work secretly, "metering of achievement" provides no correction to science based on loyalty and orientation toward profit. The potential controllers are either themselves part of the network or they hesitate to criticize it out of rational fear of the network's power to influence opinion. The existence of a competing network improves the situation only at first glance. The judgement of the evaluators is also determined by non-scientific motives in this case. What they do in a particular case depends on the constellation of powers. If there is hope of dealing a clear and permanent defeat to a competitive network, they will not shy from issuing intrinsically, although not formally, defamatory reports. If the competition is equally strong, they will prefer to proceed according to the principle "live and let live," since it must be feared that retaliation in kind will be given. The principle that criticism should be as objective, impartial, comprehensive and open as possible has lost its institutional backing in a science which is permeated by networks demanding cohesive loyalty. It turns into an atavistic reminiscence of a "heroic era" in which science was pursued more "for its own sake." Only a great variety of competitors with open boundaries and weak internal controls could save it.

3.2 "Dangerous Liaison": The Economic Connection

(40) There is also another way in which contemporary scientific policy unintentionally contributes to the furtherance of malfunctions such as proneness to deceit, dogmatism, inappropriate peer evaluation, subservience to the spirit of the times, and rejection of novelty. Under budget strains it calls upon science to seek funding from private sponsors and from industry. In the same breath it undertakes to curtail the

freedom of science to "simply research at will" and insists on social "relevance." Which research is relevant from a political viewpoint is indeed difficult to determine. According to one particular political world view or another, this or that appears more important, the main risk appearing to lie here or there. Since there is no clear criterion for non-scientific relevance, this reorientation makes science into a plaything of conflicting non-scientific agents. The internal motivations of science which under more favorable conditions see to it that research always presses forward ("plus ultra"—Francis Bacon) atrophy and makes space for pure utilitarian research which does not increase knowledge but only applies and exploits it until the reservoir of fundamental knowledge is drained empty.

(41) From the point of view of system theory, the politically intended penetration of research by social, economic and political goals and standards represents a pathological interpenetration of the subsystems of society (cf. Fischer 2005 and 2006). This process is pathological because science should be called upon or even forced to make values and codes foreign to its system into its primary ones. For example, only such research as "pays off" is socially "relevant" or politically correct is to be financed by society. It becomes clear how questionable this demand is if one takes a step back from daily affairs and looks at other times and contexts. In the former Eastern Block, in national socialist Germany, in Christian countries before the Enlightenment (and even up to modern times), in Islamic countries today, we find analogous forms of pathological interpenetration, where in these cases the goals, standards and values of politics or religion were what penetrated science and restricted its freedom.

(42) One malfunction of science which is becoming more and more important derives from economization of research which politics is striving to realize. Why is this intrusion dangerous to science? Let us consider the implications for science of the code specific to the economic system. Just as in politics, truth is only a secondary code in business and is always subordinate to the primary goal, getting the best return within the horizon of given planning. What strategies does business use to get there? The following means have proven themselves: not to mention or to play down own research results when they are at odds with expectations; to make research by independent institutes appear doubtful through obfuscation in the media and by court action; to defame and marginalize critical researchers (Martin 1997); to use non-disclosure clauses which prevent researchers in one's own laboratories or in subsidiary institutions from publishing negative results which could harm the marketing of products! That the collateral harm of these strategies might concern the health and life of many buyers or users of the products pumped into the market is of interest to their producers or distributors only to the extent that liability claims or image problems follow which could reduce the profit margins hoped for.

(43) What helps the strategy of the enterprises is the fact that the scientific results used are often ambivalent and open to interpretation—hence share zones of interpenetration with the realm of culture—and that science itself is methodically complicated and full of snares. By the choice of control substances with strategic advantages, by adjusting the dosage of test and control substances, by selection of experimental subjects from suitable subgroups, by the choice of suitable statistical procedures and finally by an art of interpretation which is rhetorically armored, many substances can be proven to be effective and having few side effects, as long as, in the phrase of sarcastic critics, they have effects "not much worse than a drink of water dis-

tilled three times" (Paulus 2004; Montori 2004). The surprising fact that the scientific results of such test series depend "decisively on the kind of sponsor" (Lutterotti 2003) has a plausible explanation therein. Astonishing correlations like these make cooperation between business and science a balancing act for both sides. For business, they bear the danger of letting capital sink in the sand—for science, of suspending its inherent rules. This balancing act can be especially delicate when researchers themselves become businessmen. In this case they must, at least in the area where they act as businessmen, let themselves be led by its code. This is not a question of dishonest intentions or unethical behavior. They *must* adopt this code, for otherwise they will not remain entrepreneurs for long. Yet as soon as they enter the laboratory, the rules of research apply again. How long can a scientist do such acrobatics before starting to make "compromises"?

(44) Providing services and accepting tasks for other systems is not at all per se damaging for science. Knowledge is good for many things. It may be the development of de-escalation strategies for police, bunker-busting "mini-nukes," the conception of sales strategies for detergents, bio-tomatoes, big macs and tampons. It may be the organization of election campaigns for parties of every persuasion or conception of a laser show for a municipal cultural event. The same is true of law, business, politics or culture. They also provide instruments and skills which can be useful or essential to other subsystems. One of the most important functions of systemic interpenetration consists in the possibility of letting specialists take over necessary, but difficult tasks of the macrosystem.

(45) From the internal perspective of a system, for example of business, interpenetration means that what counts are not only capital and effective production, but also reliable information, power structures, legal rules, cultural resources, religious values and social connections. They are necessary components of the system. But in a well-functioning system they are only means (secondary codes), which are always subordinate to the primary goal of business, making profit, cultivating good returns, or increasing capital. It is not a question of motivation or of more or less "good will" in those involved. An entrepreneur is just as little free to strive for profit or not as a politician is free to strive for means of influence (hence power) or not, or as a researcher is free to strive for valid knowledge or not. The entrepreneur who has other primary goals soon leaves the game because of lack of capital. The politician who has other primary goals than the preservation or augmentation of power will soon lose it and exit the political game. The scientist and researcher who has other *primary* goals than the development of valid knowledge leaves the field of research and enters into the functional logic of another system.

3.3 Pathological Interpenetrations

(46) For science, interpenetrations with other subsystems means that research also has secondary codes even under "normal conditions," hence that it is marked by power structures, social networks, competitive fights for prestige and attention, and struggle for rights of interpretation. The economic aspect of science is for example visible in the fact that certain elements of competition have exceptional importance *within the bounds of the system of science.* Scientists as well as laboratories, disciplines, sciences and universities compete for property rights, publication success, funding, qualified personnel, etc. The political dimension of science is found in the hierarchies which structure its institutions and communication systems. The social element in the system of science becomes visi-

ble when loyalties and group membership become important to the process of science. The medial dimension becomes evident when scientist strive to place their publications in "high-impact" journals, when they try to win as many citations as possible, or when they quote some names or publications much more frequently than others, because certain works of certain authors have aroused especial attention in the past. The juristic dimension of research is founded in the fact that the activity of research is bound by the constitutional law of the macro-system as well as by its more specific laws, insofar as they are of significance for it. The cultural dimension of science consists in its results being not merely interpretation-free information devoid of purpose, but rather yielding means for the most various attempts at giving a meaningful perspective on all areas of reality and life. That science has a religious dimension today is not recognizable at first sight. An analysis of the most recent controversies in the field of cosmology, artificial intelligence, neurobiology, psychology (free will) or genetics makes it clear, however, that it is not exclusively a matter of clarifying matters of fact, but also of problems of transcendent meaning.

Systems can yield achievements of great value for other systems only as long as they follow their own steerage and not foreign rules. When the latter occurs, they degenerate. Law, business, or science suffer a process of decay when they stand under the guardianship of politics and ideology; religion does so when it is understood mainly as a field of ethics or of social concern; politics does, if it only obeys economic imperatives; and science does, if it is above all seen as a cultural event or as subordinate to and an appendix of institutionalized religion, etc. The invasion of science by foreign standards and goals effects a disturbance of its internal mechanism of rules, which in an extreme case can lead to collapse of the system. That which presents itself today as an "offer of aid" to replace lack of state support is in effect the attempt at a "hostile acquisition" of science by economics, politics, society, media, and religion.

(47) The deformation of the symbolic code of science by economic, political, cultural, mass medial and social factors can therefore lead to serious functional disturbances of the system—in the extreme case to justification of its colonization by other systems. Summarizing, we may recall the following observable phenomena, which deeply contradict the functional norms of science:

- strategies of secrecy in genetic and military research, with a view to securing profit and power (dominance of the codes of politics, of the military and of economics);
- concealment of important information in experimental reports in order to prevent verification of the results by others and in order to secure one's own advantage (dominance of the code of the economy);
- unfair treatment of a scientist lower in the hierarchy by someone of higher rank (dominance of the code of politics);
- ritual admiration of the "decision-makers" of the system of research by those of lower rank, who hope to receive a benefit from this gesture of submission (dominance of the code of politics);
- "in-group out-group" behavior: exclusion or rejection of outsiders by means of undue criticism, mild "solidarical" critic of group members (dominance of the code of society);
- decisions of the "peer review system" on the basis of the reputation or recognition value of a person, institution or a paradigm —and not on the basis of the intrinsic quality of the project which is to be evaluated, of that of the essay, the

report or the book (dominance of the code of politics, of society or of the media);

- formation of communities of citation (dominance of the code of society);
- building of networks mainly in order to exploit sources of funding (dominance of the codes of society and of economics);
- confusion of productivity (emission of "papers" per time unit) with contribution to the process of "discovery of truth," or scientific achievement in science (dominance of the code of the mass media);
- confusion of the number of citations (of an author or of a publication) with the contribution of this person or publication to the process of "discovering truth" in science (dominance of the code of the mass media);
- cultural reevaluation, to be observed in many historical and contemporary examples, which sees science as *one* system of interpretation of reality among many possible other ones (dominance of the code of culture);
- confusion of the consensus of those involved, which is a result of decision-making, with an indication of "truth" or scientific quality (dominance of the code of politics or of society).

3.4 Scientometric Measures as Indicators of Achievement—a Medial Code for Science?

(48) Each of these forms of pathological interpenetration would deserve extensive treatment. We close by limiting attention to one point which in future will be more and more important in many countries: the evaluation of achievement by scientometric standards. Measures of publication and citation are already used for this purpose in some sciences (for example medicine in Germany) now. We want to ask whether this use is legitimate (Frank 2005, Chap. 3), or whether it documents an overflow of medial standards and hence a case of pathological interpenetration. Nowhere else does the sentence "To be is to be perceived" prevail as it does in the media. Should science adopt the medial code as a means of governance?

3.4.1 How Trustworthy is the Science Citation Index?

(49) If one wants to use citation statistics as a basis for evaluations, one must at least suppose that the database is essentially reliable and stable and that citations can be viewed as measures of achievement. Are these suppositions justified? Even the most comprehensive file of citations, the Science Citation Index, is subject to distortions both random and systematic. An internal audit showed that up to 10% of the citations were erroneous and not attributable. According to an independent empirical investigation of biochemistry in the Netherlands again roughly 10% of the expected data in the SCI were missing (or could not be found because of programming errors). These 10% were, moreover, not distributed evenly. For some research groups half of the data were missing. In the case of another it was just the most cited article which was missing, an article which was named as frequently as all the other articles of the research group together (Moed et al. 1985, p. 139f). It is evident that the mistake is to be noticed only if an undistorted sample can be used for comparison. This is a rare happening, since it presupposes that the work of the Institute for Scientific Information (ISI) has been replicated. The ISI is a private monopoly which dictates its prices, is not subject to quality control by competitors and evaluators, and must not let anyone know what is going on. The use of its data is at one's own risk.

(50) An incalculable distortion of the database already results from the actual process of citation as articles are produced. An empirical study of papers on the history of genetics has shown that on the average only about 30% of those sources which one would ideally need to cover the information presented were actually named. The variance lay between 0% and 64% (MacRoberts & MacRoberts 1986, p. 166). Secondly, one can empirically observe a tendency to cite by example. In the humanities and social sciences one has recourse to "classics" as a rule, even though these are out-of-date methodically and theoretically in comparison to more recent, non-cited articles which are in fact drawn from (Shadish et al. 1995, p. 488). In the natural sciences the most recent research articles in a sequence on which one is dependent are generally those which are cited. These citing habits contrast with the obliteration phenomenon: that certain authors and their work are not named or directly cited any more, since everyone knows anyway whom one is referring to. If MacRoberts's result can be generalized, then it is the following: the data basis for measuring achievement by enumeration of citations contains only a small part of what would be necessary for a complete evaluation of sources. In the case of a random sample that would perhaps be enough. But is the sample random?

(51) The kinds of offences which can be hidden in the references of a scientific treatise are familiar to most scientists from their own work: important publications and publications clearly used are often not cited. Some are simply forgotten, others are supposed to be well-known. Some people are negligent, others do not indicate some sources or do so only in a secretive manner. Yet others cite authors whose books they have never read. Certain texts are cited because "everyone knows them," but an insider knows that such sources can often only be known by hearsay.

(52) As long as such offenses remain in the domain of randomly distributed personal idiosyncrasies, they do not distort the results of the scientometricians in a systematic way, but rather only increase the "noise level" (which does indeed make it difficult to read a signal). However, what the sociology of science knows about social processes in the scientific system, about the exclusion of outsiders, about the non-perception of research in certain languages, countries, institutions, and about the spread of books and journals, does not support the inference that there is a random distribution. Citing is a social action. There are citations of politeness and ingratiation, which are due to social obligations, career hopes, etiquette, submission rituals, contact search or scientific self-localization. In the times of SCI, social networks are also recognizable by a specific politics of citation. Citation cartels are formed, the members of which regularly cite one another, in order to raise the citation rates therein with respect to the outside. In a time in which the SCI is used to evaluate rank this is not unreasonable behavior. But the opposite effect is also observable. Many sources are consciously ignored because one does not know or like the authors or because their home institutions have a lower prestige, are considered to be competitors, or lie in the wrong country.

(53) Citation numbers can in principle correlate with many variables. The dependencies which can dominate at any time depend on contingent circumstances, for example:

- the language an article is written in;
- the prestige (impact factor) of the journal in which an article appears;
- the circulation (actual availability) of the journal;
- the citation rules of the journals;

- the extent and structure of the research area;
- its dynamics of development ("hot" or "cold" field);
- the habits ("politics") of citation now current;
- the "followability," or "connectibility," of the articles concerned;
- the prestige of author and coauthors;
- the previous presence of the authors in the public arena of research;
- the scientific genealogy of the authors;
- the worldview (ideological position) of the authors;
- the disciplinary matrix which the authors are tied to;
- the reputation of the institution in which the authors work;
- the position of an author in the social and political system of science.

(54) Citation rates are also subject to short-term variations. The "estimated lifetime citation rates" of articles in an area of specialization can vary from year to year so that a direct comparison of citations of articles which have appeared in different years is no longer possible even in the same area of speciality (let alone in the same discipline). Different tendencies can be ascertained as well when a comparison is made among different disciplines regarding the probability of an article being cited (Moed et al. 1985, p. 155). Work from different areas of specialization, disciplines, times, etc.—the list of possible factors is open—are hence incomparable regarding the analysis of citations. Standardizations are also hardly possible because of the multitude of factors and their variability, because the number of cases in the statistical categories successively formed in search for exactness keeps decreasing, until only very few authors are left. Since many authors work in different specialities, it is practically impossible to compare them.

3.4.2. *Citations as Indicators of Research Achievement?*

(55) According to Robert K. Merton's classical approach, the citation of an author can be compared to the payment of a debt. Whoever cites uses the intellectual property (ideas, results, etc.) of another and must, according to the rules of scientific ethics, indicate the source or the inventor (Merton 1979). When applied to a discipline as a unit, the totality of citations of a researcher is an index of the use of the latter's published work.

(56) This view is based on certain assumptions. It assumes that one is aware of an intellectual debt. It assumes that it is recognized as necessary, judicious or opportune to repay the debt. Citations cannot, unless they are literal quotations, be made the subject of litigation. There are so many possibilities for combining ideas and coming to conclusions that a direct genealogy is hard to prove. This points not only to the central function of perception in this process, but also at the significance of the "ecology" in the science concerned. In order to be cited, someone must appear on the "radar screen" of the citer and be deemed a worthy object of attention. Even in ideal cases, citations reflect not the actual, but rather the perceived value of research for those who use its information for their own activity. In real science it seems more plausible to assume that citations mainly measure the utility which the members of a scientific community expect from the *act of citation* itself. That a publication is not useful to others at present or that a researcher does not see any use in citing it and does not believe it necessary to pay off an intellectual debt does not say the least regarding its inherent quality.

(57) There are many reasons for not perceiving or citing a publication. One of these

is language. Certain languages are systematically disadvantaged, others totally ignored. According to an empirical study by Montada et al. (1995), the publications of German psychologists in English-language journals are cited, *by German psychologists*, much more often than similar publications of the same authors in German-language journals. If there is such a thing as falsification in science, then *this* is a direct falsification of the assertion that rates of citation could be a standard of scientific quality. Identical contributions, which differ only by language and by journal of publication are by consequence cited about ten times more often. Does the quality of an article really increase by 1000% when it is translated into English and published in the "Journal of Abnormal and Social Psychology" rather than in the "Zeitschrift für Sozialpsychologie"?

(58) That an article is not cited frequently does not mean that it did not really have any influence and that it has not been used. But even if it has been cited, this does not mean that this is recognizable in a scientometric evaluation. It is possible that many of the articles which do cite it are not contained in the database of the SCI. Especially in the humanities a large part of primary literature still consists of monographs. The restriction of all three parts of the index to "high-impact, refereed journals" (mostly English-language) entails a further narrowing of the base of citations. As a justification of this restriction it is often pleaded that the other journals are of no importance, since the really significant papers appear in the aforementioned group of journals. This argument is wrong. On the one hand, it is quite unclear which conditions a journal must fulfil in order to be considered "refereed." Two mutually independent lists of 784 peer-reviewed journals from clinical medicine had only 54% in common (Eldrege 1997), that is to say that almost half of the journals were only in one list or the other. On the other hand, evaluations of the ISI show that of the most cited works of a discipline or of a time period many have been published in journals with a relatively low impact factor (e.g., Garfield 1988, p. 58), whereas even in high-impact journals most articles reach only low levels of citation rate. Since citations follow the Lotka distribution, the impact-values of journals can go upwards quickly because of a few highly cited articles, but also fall again, if such articles are less frequent in the following years. The great majority of articles in these journals thus profit without cause, in an evaluation based on impact values, from the presence of a relatively few highly cited works just as they suffer without cause from the absence of such works.

It is furthermore to be noted that the impact factors of journals can be different in the special areas which they serve. Medical journals whose articles on AIDS research have only a moderate level of impact may for example considerably surpass or lag behind these values in other specialities such as immunology and Alzheimer dementia research. Such differences are not taken into account by the calculation of the impact factor, but are rather smoothed out.

The impact factor can be manipulated to a large degree. A journal with a comparatively high number of editorial articles (such as obituaries, short notices, marginal glosses, reviews, commentaries, letters, corrections) can leave these out in the calculation of the number of published and citable articles of the last two years, although contributions in these categories are indeed cited. Since the impact factor is defined as the number of citations of a journal in the year in question divided by the number of "genuine articles" in the two previous years, this kind of polishing artificially drives up the impact.

(59) If high citation rates were indicators of scientific accomplishment, there should be a high correlation between the citations of certain articles in the time period following their publication and the judgements of experts years later concerning their significance for the development of their field. Yet the correlation is much poorer than expected! From fifteen publications retrospectively evaluated by historians of geology as the most important for the development of geology between 1935 and 1949, none appears in the list of the most cited geological articles between 1945 and 1954 (Brush 1991, p. 395 and 396).

(60) The question is how scientometric measures can yield valid results concerning the cognitive quality of publications or the scientific achievements of authors—if these measures cannot even serve as a reliable index of use, if the distribution of citations mirrors the structure of power and reward in a discipline more than its cognitive structure and if one understands by achievement the contribution to the expansion or improvement of our knowledge and not the consensus which has been created at a given time. It is easy to predict that the malfunctions of science which have been analyzed above will have an unimpeded impact on measures of citation. That can also be seen positively. Scientometric methods are a good means for investigating the social and political structure of special areas and disciplines, the autopoiesis of a science, to speak metaphorically (Fischer 1993). The result is an image of the self-perception of a discipline, which not only contains cognitive achievements presently perceived as such, but also the institutional mesh, the structure of authority and power, social commitments, economic ties, the spirit of the times, ongoing interpretations—all that in a mixture subject to variation, which can be separated into its components only with great difficulty.

LITERATURE

Armstrong, J. S. (1997), Peer review for journals: Evidence on quality control, fairness, and innovation, in: Science and Engineering Ethics 3, p. 63-84.

Armstrong, J. Sc. (1982), Barriers to scientific communication: The author's formula, in: The Behavioral and Brain Sciences (BBS), Vol. 5, p. 197.

Bloor, D. (1978), Polyhedra and the abominations of Leviticus, in: Brit J Hist Sc 11, p. 245-272.

Breuer, H. (2003), Das Wispern der Mikroben, in: DIE ZEIT No. 40, 25. 9. 2003, p. 36.

Broad, W.; Wade, N. (1984), Betrug und Täuschung in der Wissenschaft, Basel etc.

Brush, S. G. (1991), The most-cited physical-sciences publications in the 1945-1954 Science Citation Index. Part 3: Astronomy and earth sciences, in: Eugene Garfield, Journalogy, Key Words Plus, and Other Essays (Essays of an Information Scientist, Vol. 13, 1990), Philadelphia, p. 389-398.

Caneva, K. L. (1981), What should we do with the monster? Electromagnetism and the psychosociology of knowledge, in: Mendelsohn, E.; Elkana, Y. (ed.), Sciences and Cultures, Dordrecht.

Collins, H. M. (1975), The seven sexes: A study in the sociology of a phenomenon, or the replication of experiments in physics, in: Sociology, 205-24;

Collins, H. M. (1981), Son of seven sexes: The social destruction of a physical phenomenon, in: Social Studies of Science 11, p. 33-64

Douglas, M.; Wildavsky, A. (1982), Risk and Culture. An Essay on the Selection of Technological and Environmental Dangers, Berkeley etc.

Eldrege, J. D. (1997), Identifying peer-reviewed journals in clinical medicine, in: Bull Med Libr Assoc 85(4) October, p. 418-422.

Fischer, K. (1987), Kognitive Grundlagen der Soziologie, Berlin.

Fischer, K. (1988), The functional architecture of adaptive cognitive systems with limited capacity, in: SEMIOTICA. Journal of the International Association of Semiotic Studies, Vol. 68-3/4, p. 191-243.

Fischer, K. (1993), Changing Landscapes of Nuclear Physics. A Scientometric Study, Berlin u.a.

Fischer, K. (2002), Ist Evaluation unvermeidlich innovationshemmend? In: Pipp 2002, p.

109-128.

Fischer, K. (2004a), Spielräume wissenschaftlichen Handelns. Die Grauzone der Wissenschaftspraxis, in: Freiheit und Verantwortung in Forschung, Lehre und Studium. Die ethische Dimension der Wissenschaft, Berlin: Bund Freiheit der Wissenschaft (34. Bildungspolitisches Forum) 2004, p. 41-110.

Fischer, K. (2004b), Soziale und kognitive Aspekte des Peer Review Verfahrens, in: Fischer, K.; Parthey, H. (Hg.), Evaluation wissenschaftlicher Institutionen. (Wissenschaftsforschung Jahrb. 2003), Berlin, p. 23-62.

Fischer, K. (2005), Code, System und Konflikt. Probleme intersystemischer Kommunikation, in: Becker, R.; Orth, E. W. (Hg.), Medien und Kultur. Mediale Weltauffassungen, Würzburg, p. 83-118.

Fischer, K. (2006), Wahrheit, Konsens und Macht, in: Fischer, K.; Parthey, H. (Hg.), Gesellschaftliche Integrität der Forschung (Wissenschaftsforschung Jahrb. 2005), Berlin, p. 9-58.

Frank, G. (2005)., Mentaler Kapitalismus. Eine politische Ökonomie des Geistes, München/Wien.

Fröhlich, G. (2002), Anonyme Kritik. Peer Review auf dem Prüfstand der Wissenschaftsforschung, in: Pipp 2002, p. 129-146.

Garfield, E. (1988), The Articles Most Cited in the SCI, 1961-1982. 9. More Contemporary Classics of Science, in: Toward Scientography (Essays of an Information Scientist, Vol. 9, 1986), Philadelphia, p. 55-64.

Gernert, D. (1999), Erscheinungsformen und Argumentationsmuster dogmatisch fixierten Denkens, in: Ethik und Sozialwissenschaften 10, p. 20-23.

Goodman, B. (1995), Observers fear funding practices may spell the Death of innovative grant proposals, The Scientist, June 1995 http://www.the-scientist.library.)

Hanisch, C. (1997), Eine Expertin der Abwehr, in: DIE ZEIT No. 2, 30.5.1997, p. 34.

Hook, E. B. (ed.) (2002), Prematurity in Scientific Discovery. On Resistance and Neglect, Berkeley u.a.

Horrobin, D. F. (1990), The philosophical basis of peer review and the suppression of innovation, in: JAMA (Journal of the American Medical Association), Vol. 263, No. 10., p. 1438-1441.

Kirby, K.; Houle, F. A., Ethics and the Welfare of the Physics Profession, in: Physics Today, 57, p. 42. http://www.physicstoday.org/vol-57/iss-11/p42.html

Lutterotti, N. v. (2003), Das Schweigen der Forscher", FAZ, Dec. 17, 2003, No. 293, p. N2.

MacRoberts, M. H.; MacRoberts, B. R. (1986), Quantitative measures of communication in science: A Study at the Formal Level, in: Social Studies of Science, Vol. 16, p. 151-172.

Mahoney, M. (1976), Scientists as Subject: The psychological imperative. Cambridge.

Margulis, L. (1999), Die andere Evolution, Berlin/Heidelberg.

Martin, B. (1997), Suppression Stories, Wollongong. http://www.uow.edu.au/arts/sts/bmartin/dissent/documents/

Martin, B. (1992), Scientific Fraud and the Power Structure of Science, in: Prometheus 10, p. 83-98

Martinson, B. C.; Anderson, M.S.; Vries R. de (2005), Scientists behaving badly, in: Nature, Vol. 435/9, June 2005, p. 737.

Merton, R. K. (1997), Foreword to: Garfield, E., Citation Indexing - Its Theory and Application in Science, Technology, and Humanities, New York et. al., p. VII-XI

Moed, H. F. et al. (1985), The use of bibliometric data for the measurement of university research performance, in: Research Policy 14, p. 131-149.

Montada, L. et al. (1995), Die internationale Rezeption der deutschsprachigen Psychologie, in: Psychologische Rundschau 46, p. 186 - 199.

Montori, V. M. et al. (2004), User's guide to detecting misleading claims in clinical research reports, in: BMJ (British Medical Journal), 329, p. 1093-1096.

Neidhardt, F. (1988), Selbststeuerung in der Forschungsförderung. Das Gutachterwesen der DFG, Opladen.

Nissani, M. (1995), The plight of the obscure innovator in science, in: Soc St Sci 25, p. 165-183.

Ofshe, R.; Watters, E. (1996), Die mißbrauchte Erinnerung, München.

Paulus, J. (2004), Die Tricks der Pillendreher", DIE ZEIT, No. 18, April 22, 2004, p. 40.

Pert, C. B. (1999), Moleküle der Gefühle, Reinbek.

Peters, D. P.; Ceci, S.J. (1982), Peer-review practices of psychological journals: The fate of published articles, submitted again, in: BBS (The Behavioral and Brain Sciences), Vol. 5, p. 187-195.

Physik Journal, 4 (2005), Nr. 1, p. 10.

Pipp, E. (Hg.) (2002), Drehscheibe E-Mitteleuropa. Information: Produzenten, Vermittler, Nutzer. Die gemeinsame Zukunft (Biblos-Schriften Band 173), Wien.

Roberts, R. M. (1989), Serendipity. Accidental Discoveries in Science, New York u.a.

Sahner, H. (1979), Veröffentlichte empirische Sozialforschung: Eine Kumulation von Artefakten? Eine Analyse von Periodika, in: ZfS (Zeitschrift für Soziologie), 8, p. 267-278.

Sahner, H., Zur Selektivität von Herausgebern: Eine Input-output-Analyse der Zeitschrift für Soziologie", in: ZfS, 11 (1982), p. 82-98.

Shadish, W. R., et al. (1995), Author judgements about works they cite: Three studies from psychology journals, in: Soc St Sci, Vol. 25, p. 477-498.

Showalter, E. (1997), Hystorien. Hysterische Epidemien im Zeitalter der Medien, Berlin.

Sokal, A.; Bricmont, J. (1999), Eleganter Unsinn. Wie die Denker der Postmoderne die Wissenschaften mißbrauchen, München.

Sommer, T. J. (2001), Suppression of scientific research: Bahramdipity and nulltiple scientific discoveries, in: Science and Engineering Ethics, Vol. 7, No. 1, p. 77-104

Thiel, R. (1944), Ruhm und Leiden der Erfinder, Wien.

Tröger, J. (2005), Die Mühsal der Aufklärung, in: DIE ZEIT No. 52, Dec. 21, 2005, p. 32.

Wali, K. C. (1991), Chandra. A Biography of S. Chandrasekhar, Chicago and London.

Willgerodt, H. (2004), Die Anmaßung von Unwissen, in: Ordo, Vol. 55, Stuttgart.

Ziman, J. (1982), Bias, incompetence, or bad management? In: BBS, Vol. 5, p. 245.

Uncited Classics:

Thomas S. Kuhn, Karl Popper, Paul Feyerabend, and Hans Albert.

Scientific Peer Review

An Analysis of the Peer Review Process from the Perspective of Sociology of Science Theories

Lutz Bornmann

Swiss Federal Institute of Technology Zurich, Switzerland

bornmann@gess.ethz.ch

Abstract: Although a large number of studies have been published up to now on peer review in science, only few of them employ a theory-guided approach. Most of the studies fail to connect the empirical findings to theory. The few available theory-guided empirical studies are found predominantly in the older peer review research, which was strongly influenced by the more traditional sociologists of science, referred to as the Mertonians, or the "North American school" of Robert K. Merton. The aim of the present contribution is to present a theoretical framework that includes also recent theory developments and that can serve as a basis for future empirical research on peer review. To do this, peer review is examined from the points of view of the three most important theoretical directions in the sociology of science: (1) the North American school, (2) social constructivism, and (3) social systems theory.

I. Introduction

In modern science, peer review has become the most important instrument for assessing scientific work (Ziman 2000). Through the peer review process, not only are manuscripts selected for publication but also prizes (like the Nobel Prize) and grants are awarded and jobs allocated (Hemlin and Rasmussen 2006). Beyond these, peer review is also used to evaluate research groups and academic institutions (Hemlin 1996). With universities needing to cut costs in recent years, the trend in research project funding is that researchers can rely less and less on regular research funds from their universities and more and more have to seek external research grants that are allocated through peer review (Guston 2003).

In the typical areas of application of peer review—the review of manuscripts in-

Lutz Bornmann is a postdoctoral researcher at ETH Zurich. Since the late 1990s he has been working on issues in the promotion of young academics and scientists in the sciences and on quality assurance in higher education. For his doctoral dissertation, he evaluated the peer review process of the Boehringer Ingelheim Fonds—a foundation for the promotion of basic research in biomedicine. Up to now, the findings of that evaluation have been published as a monograph and in many journal articles. The present contribution is a modified version of an article that appeared in German in the *Swiss Journal of Sociology* (volume 33, issue 2, pp. 327-347) in 2007. Key words: theory of peer review, sociology of science, "ethos" of science, social constructivism, social systems theory.

tended for publication and grant proposals for research funding—it is the task of the 'reviewers' in the review process, as 'gatekeepers' of science, to recommend for selection the best scientific research under the condition of scarce resources (such as limited space in journals, limited funds) (Hackett and Chubin 2003). Moreover, the reviewers are supposed to uncover errors in scientific papers and recognize scientific misconduct (Smith 2006).

For testing and legitimation of scientific work, the proponents of peer review find that the process is better suited than any other method suggested thus far (see, for example, the suggestions by Roy 1985). Active researchers in the same field of research are considered to be the persons best suited to assess the quality of their colleagues' scholarly work (Eisenhart 2002). Critics of peer review see as a weakness of the process that (1) different reviewers' assessments of one and the same piece of scholarly work hardly agree, (2) reviewers' recommendations show systematic biases in judgment (the judgments are not based on the scientific quality of the work but instead on non-scientific criteria), and (3) there is little connection between peer review judgments and the quality of the reviewed work (on criticism of peer review, see, for example, Eysenck and Eysenck 1992; Ross 1980). It has been said that the only reason for continued use of the peer review process is that there is no clear consensus on a 'better' alternative (Young 2003).

The research on peer review, which has taken up criticism of the peer review process and examined it systematically, deals for the most part with peer review for journals (for an overview, see Campanario 1998a; Campanario 1998b; Overbeke and Wager 2003; Weller 2002) and somewhat less frequently with peer review for research and grant proposals (for an overview, see Bornmann and Daniel 2003; Demicheli and Pietrantonj 2004; Wessely 1998). There is hardly any research in other areas of application of peer review (except, for example, in Bornmann, Mittag, and Daniel 2006; Wissenschaftliche Kommission Niedersachsen 2006). And although up to now a large number of studies have been conducted on peer review—Weller (2002) considered a total of 1,439 studies for the most comprehensive literature overview of the research on manuscript review to date)—very few of the studies were conducted using a theory-guided approach (Hirschauer 2004). As Gläser and Laudel (2006:187) state, "There is a stark discrepancy between the number of empirical peer review studies and the theoretical understanding of the process." The few theory-guided empirical studies that are available are found predominantly in the older research on peer review, which was strongly shaped by Robert K. Merton (1973) and what is called the North American school (see, for example, Cole, Cole, and Simon 1981; Cole and Rubin 1978).

In peer review research, the fact of the North American school losing its dominant position in sociology of science and being superseded by social constructivism in the late 1970s and early 1980s was connected with a transition from a more theory-guided to a more atheoretical, empirical analysis of the peer review process: "After Mertonian sociology of science had been supplanted ... the literature shifts to atheoretical, empiricist approaches by scientists and editors who were mainly interested in the validity and reliability of peer review processes" (Gläser and Laudel 2006:187).

The social constructivist sociology of science dealt with the topic of peer review only to some extent in its ethnomethodological studies (see here the few passages in Knorr-Cetina 1981; Myers 1990; see also Bedeian 2004). According to Gläser (2006), this blind spot is due to the fact that peer review is not a local interaction, observable at one locality, but instead takes place remotely as interaction at a spatial distance. The entire process eludes ethnographic observations. Because the social constructivist and systems theory perspectives are guided by similar methodological premises, social systems theory following Luhmann (1992; 1998) has also hardly dealt with peer review up to now.

As a result of this development, the theories in sociology of science (especially the more recent theories) have remained unconsidered in empirical analysis of the scientific peer review process. The suppositions of the North American school are often viewed as outdated, and a more extensive theoretical analysis of peer review from the perspective of social constructivism and social systems theory has so far failed to appear. Therefore, the goal of the present contribution is to present a theoretical framework for peer review that also considers more recent theory developments and that can serve as a basis for future empirical peer review research.

Peer review will be analyzed from the perspectives of the three most important theoretical directions in sociology of science (see here Bornmann 2008; Cole 1992): (1) the North American school of sociology of science founded by Merton (1938), (2) social constructivism (Knorr-Cetina 1981; Latour and Woolgar 1979), and (3) Luhmann's (1992) social systems theory. Before analyzing peer review from these theoretical perspectives, the most important empirical findings of peer review research will be presented in the section just below. Based on the findings of the empirical studies, conclusions can be drawn regarding the validity of the theoretical assumptions as applied to peer review.

II. EMPIRICAL STUDIES IN PEER REVIEW

The peer review process can be conceptualized formally as a social judgment process of individuals in a small group (for example, one or more reviewers and one or more editors of a disciplinary in-group in manuscript review) (Krampen and Montada 2002). As the peer review process can be said to be of high quality if the judgments are reliable, fair, and valid (Hackett and Chubin 2003), a large part of the studies in peer review research based the empirical analysis on these quality criteria (Daniel 1993/2004).

1. *Reliability of Peer Review*

In peer review, judgments (recommendations and decisions) on one and the same piece of scientific work are called reliable if there is a high level of agreement among the individual judgments. In peer review research the degree of agreement among the judgments of reviewers was mostly investigated in the area of review of manuscripts submitted for journal publication. All in all, the results of the studies show similar results: the coefficients measuring the inter-rater reliability among reviewers fall in the range of 0.2 to 0.4 (see an overview in Cicchetti 1991). As coefficients (kappa and intraclass correlation coefficients) below 0.5 indicate a rather low level of agreement among reviewers, this indicates that judgments in peer review are not very reliable.

Based on these numbers, Eckberg (1991) and Kostoff (1995) point out that differing judgments in peer review are not necessarily a sign of disagreement about the quality of a manuscript but may be instead explained by differing positions (paradigms), judgment criteria, and areas of competency among the reviewers. According to Cole (2000) low correlation coefficients for reviewers' judgments reflect the lack of consensus that is prevalent in all scientific disciplines at the 'research frontier.' Cole says that at the frontiers of research it is usually impossible to make a reliable assessment of scientific work. Hargens and

Herting (1990) criticize that in studies on reviewers' assessments, it is implicitly assumed that reviewers' judgments vary along one latent scientific quality dimension, and that this assumption can hardly be tested by calculating kappa or intraclass correlation coefficients. For this reason Hargens and Herting (1990) calculate row-column (RC) association models (Goodman 1984). Their investigation of manuscript review for five journals shows that at four of the journals, it can be assumed that one quality dimension accounts for the association in the reviewers' judgments. In contrast to the findings of the other studies on the reliability of peer review, Hargens and Herting's (1990) results thus indicate that there is a substantial statistical association among reviewers' judgments (see also Bornmann, Mutz, and Daniel 2007).

Although a high level of agreement among the reviewers in the peer review process is striven after, some scientists see high agreement at the same time as disadvantageous to the review process: "Too much agreement is in fact a sign that the review process is *not* working well, that reviewers are not properly selected for diversity, and that some are redundant" (Bailar 1991:138). According to Marsh and Ball (1991) the selection of reviewers that represent diverse perspectives, use different judgment criteria, and so on, reduces reliability. But validity can be increased considerably, as the decision-makers (such as journal editors or grant program managers) can make their decisions about a manuscript or proposal based on much broader information. For this reason, in the opinion of researchers that see low reliability as helpful for the decision-making process, reviewers should be selected precisely because of their different perspectives, judgment criteria, and so on (Stricker 1991).

2. *Fairness of Peer Review*

Overviews of the peer review research literature (Hojat, Gonnella, and Caelleigh 2003; Owen 1982; Pruthi, Jain, Wahid, Mehra, and Nabi 1997; Ross 1980; Sharp 1990; Wood and Wessely 2003) have named up to 25 different biases that can potentially endanger the fairness of the peer review process. Bias is usually defined as follows: "Bias is any feature of an evaluator's cognitive or attitudinal mind-set that could interfere with an objective evaluation" (Shatz 2004:36). Even if numerous studies have reported a lack of fairness in the peer review process, the research on fairness faces two fundamental problems that make generalization of the findings difficult. For one, the published results on the influence of the individual biases on reviewers' judgments are inconsistent. For instance, numerous studies demonstrated gender bias in the peer review of research and grant proposals (for example, Brouns 2000; Wenners and Wold 1997). However, a similar number of studies were able to identify only a moderate gender effect or none at all (for example, Sandström and Hällsten 2006; Ward and Donnelly 1998) or reported mixed results (for example, Bornmann and Daniel 2005). For another, as there are no experimental studies in the research on peer review, it is impossible to establish whether a particular group of scientists (such as, for example, grant applicants or manuscript authors at less prestigious universities) receive worse reviews due to preferential biases on the part of the reviewers or whether the unfavorable reviews are the consequence of the insufficient quality of the proposals or manuscripts (Daniel 1993/2004).

3. *Predictive Validity of Peer Review*

The goal of peer review of grant applications and manuscripts is usually to select the 'best' from among the work submitted (Smith 2006). The selection function is considered to be a difficult research topic to investigate. According to Jayasinghe, Marsh, and Bond (2001) there exists no uniform definition of what makes a manuscript 'worthy of publication,' or of what makes a research proposal 'worthy of funding.' But as the number of citations of a publication reflects the international impact of the reported research (Daniel 2005; van Raan 2004), it is a common approach in peer review research to evaluate the success of the process on the basis of the citation rate of the reviewed research.

Bornmann and Daniel (2008), Daniel (1993/2004), Lock (1985), Wilson (1978) and Opthof, Furstner, van Geer, and Coronel (2000) examined the predictive validity of peer reviewing of manuscripts based on the citation counts of manuscripts accepted for publication and manuscripts rejected by a journal but then published elsewhere. The results of all of these four studies show unanimously that manuscripts accepted and published by *Angewandte Chemie*, *British Medical Journal*, *Journal of Clinical Investigation*, and *Cardiovascular Research* are clearly more frequently cited than manuscripts that were rejected by these journals and later published elsewhere.

III. Peer Review from the Point of View of Sociology of Science Theories

In the following sections, the theoretical assumptions of the North American school of sociology of science, social constructivism, and social systems theory are presented in a first step. In a second step, the assumptions are applied to the peer review process.

1. *The Point of View of the North American School on Peer Review*

With the "ethos of science" Merton (1973) conceptualized in the tradition of structural functionalism the (positive) norms and values of science that are held to characterize appropriate and correct behavior on the part of scientists: (1) scientific knowledge should be made public knowledge (communism, or communality), (2) the scientist forgoes all forms of personal gain (disinterestedness), (3) knowledge must always be scrutinized (organized skepticism), and (4) knowledge claims should be judged impersonally, independently of their source (universalism). Since the beginnings of modern science, the ethos of science has developed in the interaction of (professional) scientists (Merton 1973). As prescriptive norms (see Coleman 1990) the norms prescribe certain behaviors and create an implicit scientific conscience. As imperatives, the norms are internalized by scientists today through socialization in scientific institutions (Merton 1973).

For Merton (1973) the peer review process is an institutional guarantor that science will be oriented towards the prescriptions and principles of practicing 'good' science—the ethos of science. Only with an institutionalized form of assessment (that is, with a recognized quality assurance system) can science make a claim to being that functional system in society that is responsible *exclusively* for knowledge production and knowledge accumulation (Hansson 2002). Two of the four norms in the ethos of science play a particularly important role in the analysis of the peer review process: (1) the norm of universalism, and (2) the norm of organized skepticism (the norms of communism and disinterestedness have more to do with guaranteeing unrestricted access to research findings):

With the norm of universalism being operative in peer review, the aim is to prevent economic, political, religious, or other non-scientific interests from influencing the process of selecting a scientific paper or a grand proposal and thus the process of knowledge gain (Ziman 2000). If scientists are to assess the work of a colleague in their field through a process of peer review, they are encouraged to base their judgments exclusively on scientific quality criteria and not to draw up any irrelevant criteria (on this, see Cole 1992). The most suitable persons for assessing others' work according to scientific quality criteria are successful colleagues of the scientist whose work is being reviewed (grant applicant or author of a manuscript, see on peer review the findings by Jayasinghe 2003). When assessing the scientific quality of a proposal or manuscript, for the reviewer the correspondence of the research results with the 'laws of nature' plays a decisive role, as reviewers assess as positive only those facts that agree with the 'laws of nature' (Cole 1992). The norm of universalism in peer review would, finally, result in acceptance and communication of knowledge independently of the local and social conditions of knowledge production (in the laboratory) or selection (in peer review) (Ziman 2000).

With the norm of organized skepticism being operative in peer review, it could be guaranteed that new findings have to be put up for debate among experts until, once assessed positively, they are ascribed a recognized degree of credibility and certainty. Through the norm of organized skepticism, only knowledge that is tested thoroughly and authorized by specialist colleagues would flow into the body of knowledge and inform the further knowledge utilization in a specialist field. Thus, as a result of the workings of the norm of organized skepticism, knowledge that is legitimized and thus capable of consensus (Ziman 1991) would ultimately produce a high level of agreement in a specialized area of study. For the followers of the North American school, a high level of consensus in the process of knowledge construction is an important prerequisite to ensure in a field of inquiry the sustainable accumulation of verified knowledge (Mulkay 1969).

According to Merton (1973) norms take the form of rules, prohibitions, and principles that determine what conduct is to be favored and what is permissible. In peer review the rules and principles for good scientific conduct are imposed by means of sanctions. These sanctions are predominantly of an incremental type (on this, see Coleman 1990): Where there is proof of a scientist's intentional misconduct, that scientist is (collectively) 'cut off.' That these sanctions exist is shown by the many recommendations lists concerning conduct that scientific institutions have published for the prevention of and for dealing with scientific misconduct (see, for example, European Association for Chemical and Molecular Sciences 2006). Following Merton (1973) attacks on the integrity of science show that science has to rely on these sanctions so that the process of knowledge gain is not endangered.

A part of the studies in peer review research (see above, Empirical Studies in Peer Review) contains indications that the assumptions of the North American school of sociology of science can indeed adequately describe scientists' conduct in the peer review process: (1) The studies conducted by Hargens and Herting (1990) show that reviewers judge manuscripts according to a uniform scientific quality dimension. This consensus about quality seems to show that the two norms (universalism and organized skepticism) are operative in peer review. (2) Up to now no investigations *having an experimental study design* have been conducted on fairness in peer review that were able to demonstrate without a doubt systematic biases in judgment. Indeed, the conduct of reviewers, editors, and program managers does seem to be guided by the norm of universalism. (3) The studies on predictive validity make it clear that editors and reviewers can assess the scientific impact of manuscripts correctly: Rejected manuscripts published later elsewhere have distinctly lower citation counts than manuscripts accepted for publication. The finding that the review process is oriented towards the scientific quality (or the possible impact) of a piece of research can again be seen as showing that the two norms are operative in peer review.

2. *The Social Constructivist Point of View on Peer Review*

The followers of social constructivism among the sociologists of science reject the theoretical assumptions of the North American school—in particular the existence of the norms that Merton (1973) summed up in the ethos of science—and replace them with their own assumptions (for an overview, see here Mulkay 1979). In the following, the most important assumptions of the social constructivists are described in a first step and applied to peer review in a second step. As we will see, the theoretical assumptions show a recognizable relation to criticism that has been directed at traditional peer review (whose principles are shaped by the North American school) for years (see introduction above).

First assumption: Scientific behavior is not governed by the norms that Merton identified and set out as the ethos of science.

The followers of social constructivism are convinced that science is not steered by norms. They base this conviction on a number of microsociological studies (see, for example, Knorr-Cetina 1981). These studies do not provide indications that (1) scientists try to follow certain norms in their behavior and that (2) the behavior of scientists can be satisfactorily explained by the ethos of science (Knorr-Cetina 1991). Scientific conduct is therefore seen as not conforming to norms. From this assumption of the independency of scientific conduct and normative prescriptions, the followers of social constructivism derive the hypothesis that normatively guided behavior is *not* a necessary prerequisite for the process of knowledge construction and scientific progress.

A number of studies have been conducted in peer review research that examined the behavior of scientists in peer review against the background of the norms of the ethos of science. All of the studies on fairness in peer review (see the section on empirical studies above), for example, tested implicitly or explicitly the norm of universalism in the peer review process. The results of some of these studies indicate that judgments in the peer review process are systematically influenced by non-scientific criteria, such as gender. In one of the best-known studies, the results of which were published in the journal *Nature*, Wenners and Wold (1997) demonstrate for the peer review process at the Swedish Medical Research Council (MRC; Stockholm) that a woman must be approximately 2.5 times more productive scientifically (through publications in high-quality journals) than a man in order to receive the same judgment by the reviewers.

While studies such as that by Wenners and Wold (1997) call into question that scientists' behavior in peer review behavior conforms to norms, the results of a study (Ellison, Rosato, and Outram 2005) presented at the Fifth International Congress on Peer Review and Biomedical Publication (Chicago, USA) in 2005 make it clear, moreover, that scientists are hardly willing to align their behavior to institutionally decreed principles and regulations. The study investigated the extent to which authors publishing articles in the *British Medical Journal* (*BMJ*) followed the journal's ethical guidelines. The researchers found the following: "It is clear that the *BMJ*'s 'guidelines on the use of ethnic, racial, and cultural descriptions in published research' have not been followed" (Ellison, Rosato, and Outram 2005).

Second assumption: Scientists do not own the privilege of participation in the scientific discourse in their field of science.

According to the assumptions of the North American school of sociology of science, scientists acquire specific knowledge and abilities in their field through long years of university studies and research activity that certify them as experts in that field (Collins and Evans 2002). Due to this knowledge and abilities, it is ultimately only experts like this that can conduct the scientific discourse in a field of expertise at an appropriate level. The social constructivists contradict this assumption and deny the experts in a given specialist field the sole privilege of participating in the scientific discourse (see here Fuchs 1996). It might seem that anyone can be an expert (Collins and Evans 2002:238). Also outsiders (persons or scientists outside of the specialism), who may indeed lack technical or mathematical understanding in a field but who are familiar with the contents, theories, and technical terminology, may participate, similar to those possessing expertise, in the scientific discourse.

In an experiment conducted by Harry M. Collins—one of the most influential researchers among the social constructivists—and reported in the journal *Nature* (Giles 2006), it could be shown that an expert's answers to questions in his area of expertise (gravitational wave physics) could not be distinguished from the answers given by an outsider (a social scientist) passing himself off as a physicist: "The experiments show that the linguistic performance of those well socialized in the language of a specialist group is indistinguishable from those with full blown practical socialization but distinguishable from those who are not well socialized" (Collins, Evans, Ribeiro, and Hall 2006:656).

Another study, which was presented in 2005 at the Fifth International Congress on Peer Review and Biomedical Publication (Chicago, USA), called into question the privilege of experts to take part in the peer review process: Bryan, Fletcher, and Kale (2005) examined whether a nonqualified but experienced member of the *BMJ* editorial administrative team makes final decisions on submitted manuscripts similar to the BMJ editors' decisions made within a particular time period based on reviewers' recommendations. The results show that the staff member would make the same decision on 90% of the manuscripts originally rejected by the editors and would accept for publication only 10% of the papers that the editors reject.

The findings by Collins, Evans, Ribeiro and Hall (2006) and Bryan, Fletcher, and Kale (2005) contradict the assumption of the North American school that active scientists within the same field of research are the best suited persons—in accordance with the norm of organized skepticism—to assess the work of colleagues in their fields as to scientific quality. Outsiders (persons or scientists outside of the specialism) that are familiar with the contents, theories, and technical terminology can arrive at similar judgments.

Third assumption: Scientific knowledge does not reflect 'natural reality' but is instead constructed socially and locally.

One of the most important assumptions of social constructivism is that scientific insights are social and local constructions (Knorr-Cetina 1991). Scientific knowledge does not reflect any 'natural reality' but instead is manufactured in the research process under the given local conditions (Knorr-Cetina 1981). These conditions include the cultural and social contexts, and also the scientists' personal interests, social relationships, and biases in judgment. In order to describe the processes of the context-specific construction on knowledge, social constructivists make use of concepts such as social situatedness, contextuality, interpretative flexibility, opportunistic rationality, and local idiosyncrasies (Krohn 2000).

For social constructivists, 'laws of nature' and 'truth' play little role not only in the process of knowledge acquisition but also in the peer review process. Favorable and unfavorable assessments by reviewers, editors, and grant program managers are said to be (exclusively) the result of social processes and local conditions. Those involved in a peer review process make their assessments upon the background of their personal biographies, individual interpretations of specialized knowledge of their subject, the social network, and the local conditions under which they work (Gläser and Laudel 2006).

For Cole (1992) the low level of agreement among reviewers reported by some of the studies on peer review substantiates the social constructivist assumption that it is not the scientific validity of a work but rather local and social conditions that determine the judgments (see above, Reliability of Peer Review): "If consensus can be determined by comparing a contribution with nature or by the application of a set of rational rules to evaluate the validity of a contribution, why do we find little consensus?" (Cole 1992:83). Also the many biases in judgment that are discussed in connection with peer review (see above, Fairness of Peer Review) give rise to doubt as to whether there can be objective review without local and social influences.

The results of the studies on predictive validity (see above, Predictive Validity of Peer Review) show, on the one hand, that manuscripts accepted for publication have a greater impact on subsequent research than rejected manuscripts published later elsewhere. The peer review process thus seems to fulfill its function of selection between 'successful' and 'unsuccessful' research. On the other hand, studies on predictive validity demonstrate that in the peer review process of one and the same manuscript, reviewers arrive at different judgments: Many manuscripts that are rejected by a journal (through a peer review process) are then accepted by another journal (through a peer review process). This finding indicates that manuscript review is *not* only based on generally valid quality criteria that a scientific work can fulfill (acceptance) or not fulfill (rejection); the review (or the outcome of the review) seems also to be dependent upon the local and social conditions under which the peer review process at the individual journals takes place.

Fourth assumption: Scientific work is a social construction of the scientist under review <u>and</u> the reviewers.

For the followers of the North American school, manuscripts are the result of scientists' research work, who by submitting manuscripts to journals seek to secure ownership of their intellectual property with sole rights to recognition (Merton 1973). The followers of social constructivism contradict this assumption. To them, it is not only the author that is responsible for the content of a manuscript (and has rights to the intellectual property); the content of a manuscript is a joint product by the author, reviewer, and editor (on this perspective extended to research project proposals, see Laudel 2006). Authors anticipate the peer review process already while writing (Knorr-Cetina 1981), and in the review process reviewers and editors seek to realize personal interests with regard to the content of manuscripts (Gläser 2006; Gläser and Laudel 2006). The theoretical assumption that in the peer review process a manuscript is socially constructed by both the scientists under review *and* the reviewers was developed by social constructivists based on observations of the research and publication process (Knorr-Cetina 1981) as well as on analyses of manuscripts and reviewers' assessments (Myers 1990).

3. The Social Systems Theory Point of View on Peer Review

Sociologists working within the North American school framework and the social constructivism framework have always tended to reject each other's positions (see here, for example, Knorr-Cetina 1991; Sokal and Bricmont 1999), although some papers proposing a mediating position have been published (for example, Murphy 1994). Krohn (2000) sees great potential in Luhmann's (1992) social systems theory to overcome the crisis of the social studies of science that arose due to the theoretical fragmentation. By considering the perspective of an observer, Luhmann (1992) maintains the difference that the constructivists make between knowledge and object. For Luhmann knowledge is the construction of distinctions, whereby that which is distinguished has no direct correspondence in reality. In contrast to the social constructivist approach (and following the North American school), science is for Luhmann (1992) characterized by a specifically shaped form of communication that is oriented to 'truth.' Through this orientation, science distinguishes itself from its surroundings as a functionally differentiated system and can remain an autonomous system.

In his analysis of science, Luhmann (1992) utilized two components from social systems theory that are said to be fundamentally important in the construction of solid knowledge (see here also Leydesdorff 2001): 1) second-order observation, and 2) the evolution of knowledge. In contrast to the theoretical assumptions of the North American school and the social constructivists, these components are *not* mechanisms that seek to explain the (specific) phenomena in science through the actions of persons in social contexts. Instead, these components can fundamentally describe the course of scientific communication and point out the central importance of review and selection in the process.

According to social systems theory, science is differentiated as a societal subsystem that is an autopoietic system having the function of producing knowledge of high certainty (Luhmann 1992). Autopoietic systems are recursive, operatively closed systems—there are no operations entering the system from the environment and vice versa. The operative element of science is the communication of knowledge that is reproduced in the symbolically generalized medium of 'truth' (Stichweh 1994). Communication can be called specifically scientific, only if it is used to establish a distinction (the binary code) of 'true' and 'not true.' With the attribution of 'true,' it is symbolized in the system that a finding can be held to be generalizable statement of reality.

1) Second-order observation. Luhmann (1992) bases the theory of social systems on a formal concept of observation that goes back to Spencer-Brown (1969). Spencer-Brown (1969) developed the *Laws of Form*, according to which every observation is constructed of two components (indication and distinction), which can explain all possible operations of an observer: The basis of all indications is a distinction drawn by an observer between inside and outside. In this process, there is no complete and consistent system of description available to the observer—that is, "no observation can observe what it cannot observe" (Fuchs 1996:321). Once having drawn the distinction, the observer has a 'blind spot,' seeing only one side of the distinction, which is then discernible only through observation of a higher order. As every observation is fraught with a blind spot, no 'better' description of reality can be expected with every further observation of a higher order than with the observation of a lower order.

Luhmann (1992) applied the formal concept of observation to the process of knowledge production in science. Since the natural world cannot speak for itself (Gieryn 1982:288), in the research process one has to rely on generating knowledge through a first-order observation that can claim to explain a certain natural phenomenon (or social phenomenon). As the observation can only refer exclusively to the phenomenon and not to the observation itself, it can only be called naive realistic and not constructivist (Fuchs 1996): The first-order observer can not forgo the assumption of reality; the first-order observer draws distinctions in an object world (Schmidt 1998). If the observation is formulated as a knowledge claim, it can be coded, through a second-order observation, as 'true' (verified) or 'not true' (not verified).

"On the second level, communication must decide, amongst other issues, which observations are relevant and deserve being noticed by others," states Fuchs (1996:317). Findings (from first-order observation) are as a rule described in manuscripts and submitted to scientific journals for review. Through second-order observation (critical assessment), peers code the findings as 'true' or 'not true' (or verified or not verified). As the attribution of 'true' or 'not true' can be made only by second-order observation (and *not* by first-order observation in an object world), it is therefore only through critical assessment that it can be guaranteed that science produce the substantiated knowledge that society expects of it in fulfillment of its function. Only at the level of second-order observation can the binary code of 'true'/'not true' unfold and science differentiate itself as a system (Luhmann 1992).

2) The evolution of knowledge. For the analysis of science, Luhmann (1992) applied the theory of evolution to the process of the generation of (verified) knowledge, with the main elements of evolutionary theory as variation, selection, and stabilization. Science produces vast amounts of findings with a high degree of *variation*. Here we have the global supply of findings that can contribute towards explanation of natural and social phenomena. The great range of variation of the findings means that more findings are produced in science than are capable of reception. For this reason, in the process of selection a part of the findings are taken up in communication and marked as 'true' (the other part is not taken up and marked as 'not true'). In science, the (critical) selection process is undertaken by the peer review process. Another selection process, downstream of peer review, is the utilization (or the ignoring) of findings and thinking in the further research work in a field of study, which finds expression in citation by colleagues in the field. Through the rejection of manuscripts, research proposals, and so on, in peer review, only a part of the findings is given the possibility to be taken up in the body of knowledge, or to be further utilized, in a field. The other part does not clear the 'hurdle' of peer review and is subject to the fate of being forgotten. With the selection of scientific work, peer review at the same time can fulfill the function of stabilization (of knowledge) in science. According to Luhmann (1992), in the evolutionary process, *stabilization* is the tendency in the system to largely avoid changes in the body of knowledge and to maintain the previous body of knowledge, or to persist in the state of the previous structure. In the end, the only new findings that can get past the stabilizing tendencies in the system as substantiated knowledge are those that stand out sufficiently and prove to be 'true' intersubjectively.

IV. DISCUSSION

In the above, peer review was analyzed from the points of view of the three most important theoretical directions in sociology of science: (1) the North American school, (2) social constructivism, and (3) social systems theory. In the face of the inconsistent findings in peer review research (see above, Empirical Findings), in the analysis of peer review from the points of view of the North American school and social constructivism, empirical results (as a rule from studies having a strict methodology) could be found *in each instance* that could support diametrically opposed theoretical assumptions. The inconsistent findings in peer review research thus reflect the theoretical fragmentation in sociology of science (and vice versa).

The empirical validation of the assumptions of *both* theoretical directions (as they were shown in the sections above) can in the end support those sociology of science contributions that have proposed a mediating position between the two theoretical directions (see, for example, Murphy 1994). As the components from social systems theory (second-order observation and evolution of knowledge) that were utilized in this paper for the analysis of peer review are 'abstract' communication processes (with no direct relation to the level of the conduct of scientists), no empirical results from peer review research could be used to validate these processes.

With the analysis of peer review from the perspectives of the three theoretical directions, a theoretical framework was gained for the empirical investigation of peer review, which can be summarized as follows: For the description of fundamental processes in peer review, second-order observation and the evolution of knowledge can be utilized. In the course of scientific communication, these components can show in what way assessment in peer review and selection by means of peer review can contribute towards fulfilling continuously the specific function of science for society—the generation of *verified* knowledge. Verified knowledge is only then generated, if (1) second-order observations of (not verified) findings are coded as 'true' and 'not true' (assessment in peer review), and (2) under the condition of scarce resources, a part of the knowledge is selected out of a reservoir that is rich in variation (selection by means of peer review).

For the analysis of peer review, in addition to second-order observation and the evolution of knowledge (at the level of communication in science), (1) the norms in the ethos of science and (2) the assumptions of the supporters of social constructivism (at the level of the conduct of scientists) were utilized. The norms of universalism and organized skepticism are held to be imperative for the disciplining of the conduct of scientists. This disciplining can (and should) lend stability to that system of assessment and selection that is continuously generated through second-order observation and the evolution of knowledge. Norms can (and should) lead to 'control' of scientists' conduct according to the principles of 'good' scientific practice, in that they work against particularism (the operative effect of universalism) and guarantee the intersubjectivity of knowledge construction (the operative effect of organized skepticism). The theoretical assumptions of the followers of social constructivism point on the one hand to the necessity to discipline the conduct of scientists by peer review. Adherence to the norms—as the results of the microsociological research and a part of the studies in peer review research show clearly—is not a matter of course; conduct deviating from the norms must be reckoned with. On the other hand, the microsociological research of the social constructivists makes it clear that the local and social conditions under which scientists work can have a decisive influence on the review process: "'Facts' needed long construction and acceptance procedures" (Evetts, Mieg, and Felt 2006:116).

By pointing out scientific conduct deviating from norms and the local and social conditions of the review process, we can see the theoretical assumptions of the social constructivists, on the one hand, as a complement to the assumptions of the North American school that leads to a 'denser' (that is, fuller) description of the conduct of scientists in peer review (see here Collins and Evans 2002; Gilbert and Mulkay 1984). On the other hand, the assumptions of the social constructivists show that certain conditions that are viewed in traditional peer review as necessary for a 'successful' assessment process are subject to flexibility and variability: (1) Assessments do not necessarily have to be undertaken by peers (experts), and (2) not only the reviewed scientist (author or grant applicant) but also the reviewing scientist (reviewer, editor, or grant program manager) contributes in the peer review process towards the coming into being of a scientific work. Recognition for the content of a scientific work is due therefore not only to the author or grant applicant but also to the reviewer.

With the assumptions of the social constructivists, the North American school, and social systems theory, we come to a description of processes of communication and conduct that are fundamental for the conducting of peer review in science, which has become a functionally differentiated subsystem in society for the production of verified knowledge. The assumptions of the three most important theoretical directions in sociology of science are a valuable basis for in-depth, theoretically founded examination of the (future) findings of peer review research.

REFERENCES

Bailar, J. C. 1991. "Reliability, fairness, objectivity, and other inappropriate goals in peer review." *Behavioral and Brain Sciences* 14:137-138.

Bedeian, A. G. 2004. "Peer review and the social construction of knowledge in the management discipline." *Academy of Management Learning and Education* 3:198-216.

Bornmann, L. 2008. "Grundlegende Kommunikationsprozesse und stabilisierte Mechanismen für die Erzeugung gesicherter Erkenntnisse—eine Übersicht über den Theoriestand in der Wissenschaftssoziologie." *Sociologica Internationalis* 45: 177-200.

Bornmann, L. and H.-D. Daniel. 2003. "Begutachtung durch Fachkollegen in der Wissenschaft. Stand der Forschung zur Reliabilität, Fairness und Validität des Peer-Review-Verfahrens." Pp. 211-230 in *Universität auf dem Prüfstand. Konzepte und Befunde der Hochschulforschung*, edited by S. Schwarz and U. Teichler. Frankfurt am Main, Germany: Campus.

Bornmann, L. and H.-D. Daniel. 2008. "Selecting manuscripts for a high impact journal through peer review: a citation analysis of Communications that were accepted by Angewandte Chemie International Edition, or rejected but published elsewhere." *Journal of the American Society for Information Science and Technology* 59: 1841-1852.

Bornmann, L. and H.-D. Daniel. 2005. "Selection of research fellowship recipients by committee peer review. Analysis of reliability, fairness and predictive validity of Board of Trustees' decisions." *Scientometrics* 63:297-320.

Bornmann, L., S. Mittag, and H.-D. Daniel. 2006. "Quality assurance in higher education—meta-evaluation of multi-stage evaluation procedures in Germany." *Higher Education* 52:687-709.

Bornmann, L., R. Mutz, and H.-D. Daniel. 2007. "Row-column (RC) association model applied to grant peer review." *Scientometrics* 73:139-147.

Brouns, M. 2000. "The gendered nature of assessment procedures in scientific research funding: the Dutch case." *Higher Education in Europe* 25:193-199.

Bryan, G., J. Fletcher, and R. Kale. 2005. "Do you need to be an editor to accept or reject research papers sent to the *BMJ*?" in *5th International Congress on Peer Review and Biomedical Publication*, edited by D. Rennie, F. Godlee, A. Flanagin, and J. Smith. Chicago, IL, USA.

Campanario, J. M. 1998a. "Peer review for journals as it stands today - part 1." *Science Communication* 19:181-211.

Campanario, J. M. 1998b. "Peer review for journals as it stands today - part 2." *Science Communication* 19:277-306.

Cicchetti, D. V. 1991. "The reliability of peer review for manuscript and grant submissions: a cross-disciplinary investigation." *Behavioral and Brain Sciences* 14:119-135.

Cole, J. R. 2000. "The role of journals in the growth of scientific knowledge." Pp. 109-142 in *The web of knowledge. A festschrift in honor of Eugene Garfield*, edited by B. Cronin and H. B. Atkins. Medford, NJ, USA: Information Today.

Cole, S. 1992. *Making science. Between nature and society.* Cambridge, MA, USA: Harvard University Press.

Cole, S., J. R. Cole, and G. A. Simon. 1981. "Chance and consensus in peer-review." *Science* 214:881-886.

Cole, S. and L. Rubin. 1978. *Peer Review in the National Science Foundation. Phase one of a study.* Washington, DC, USA: National Academic Press.

Coleman, J. S. 1990. *Foundations of social theory.* Cambridge, MA, USA: Belknap Press of Harvard University Press.

Collins, H., R. Evans, R. Ribeiro, and M. Hall.

2006. "Experiments with interactional expertise." *Studies in History and Philosophy of Science Part A* 37:656-674.

Collins, H. M. and R. Evans. 2002. "The third wave of science studies: studies of expertise and experience." *Social Studies of Science* 32:235-296.

Daniel, H.-D. 1993/2004. *Guardians of science. Fairness and reliability of peer review.* Weinheim, Germany: Wiley-VCH. Published online 16 July 2004, Wiley Interscience, DOI: 10.1002/3527602208.

Daniel, H.-D. 2005. "Publications as a measure of scientific advancement and of scientists' productivity." *Learned Publishing* 18:143-148.

Demicheli, V. and C. Pietrantonj. 2004. "Peer review for improving the quality of grant applications (Cochrane Methodology Review)." in *The Cochrane Library, Issue 1*. Chichester, UK: John Wiley & Sons, Ltd.

Eckberg, D. L. 1991. "When non reliability of reviews indicates solid science." *Behavioral and Brain Sciences* 14:145-146.

Eisenhart, M. 2002. "The paradox of peer review: admitting too much or allowing too little?" *Research in Science Education* 32:241-255.

Ellison, G. T. H., M. Rosato, and S. Outram. 2005. "The impact of editorial guidelines on the classification of race/ ethnicity in the *BMJ*." in *5th International Congress on Peer Review and Biomedical Publication,* edited by D. Rennie, F. Godlee, A. Flanagin, and J. Smith. Chicago, IL, USA. Poster Session Abstract available at http://www.ama-assn.org/public/peer/abstracts.html#class2.

European Association for Chemical and Molecular Sciences. 2006. "Ethical guidelines for publication in journals and reviews." European Association for Chemical and Molecular Sciences (EuCheMS), Brussels, Belgium.

Evetts, J., H. A. Mieg, and U. Felt. 2006. "Professionalization, scientific expertise, and elitism: a sociological perspective." Pp. 105-123 in *The Cambridge Handbook of Expertise and Expert Performance,* edited by K. A. Ericsson, N. Charness, P. J. Feltovich, and R. R. Hoffman. New York, NY, USA: Cambridge University Press.

Eysenck, H. J. and S. B. G. Eysenck. 1992. "Peer review: advice to referees and contributors." *Personality and Individual Differences* 13:393-399.

Fuchs, S. 1996. "The new wars of truth: conflicts over science studies as differential modes of observation." *Social Science Information Sur Les Sciences Sociales* 35:307-326.

Gieryn, T. F. 1982. "Relativist/ constructivist programs in the sociology of science - redundance and retreat." *Social Studies of Science* 12:279-297.

Gilbert, G. N. and M. Mulkay. 1984. *Opening Pandora box: a sociological analysis of scientists' discourse*. Cambridge, UK: Cambridge University Press.

Giles, J. 2006. "Sociologist fools physics judges." *Nature* 442:8-8.

Gläser, J. 2006. *Wissenschaftliche Produktionsgemeinschaften. Die soziale Ordnung der Forschung*. Frankfurt am Main, Germany: Campus.

Gläser, J. and G. Laudel. 2006. "Advantages and dangers of 'remote' peer evaluation." *Research Evaluation* 14:186-198.

Goodman, L. A. 1984. *The analysis of cross-classified data having ordered categories*. Cambridge, MA, USA: Harvard University Press.

Guston, D. H. 2003. "The expanding role of peer review processes in the United States." Pp. 81-97 in *Learning from science and technology policy evaluation. Experiences from the United States and Europe,* edited by P. Shapira and S. Kuhlmann. Cheltenham, UK: Edward Elgar.

Hackett, E. J. and D. E. Chubin. 2003. "Peer review for the 21st century: applications to education research." in *Peer review of education research grant applications. Implications, considerations, and future directions,* edited by National Research Council. Washington, DC, USA.

Hansson, F. 2002. "How to evaluate and select new scientific knowledge? Taking the social dimension seriously in the evaluation of research quality." *VEST* 15:27-52.

Hargens, L. L. and J. R. Herting. 1990. "A new approach to referees assessments of manuscripts." *Social Science Research* 19:1-16.

Hemlin, S. 1996. "Research on research evaluations." *Social Epistemology* 10:209-250.

Hemlin, S. and S. B. Rasmussen. 2006. "The shift in academic quality control." *Science Technology & Human Values* 31:173-198.

Hirschauer, S. 2004. "Peer review research - reviewed - sociological shortcomings of academic evaluation." *Zeitschrift fur Soziologie* 33:62-83.

Hojat, M., J. S. Gonnella, and A. S. Caelleigh. 2003. "Impartial judgment by the "gatekeepers" of science: fallibility and accountability in the peer review process." *Advances in Health Sciences Education* 8:75-96.

Jayasinghe, U. W. 2003. "Peer review in the

assessment and funding of research by the Australian Research Council." University of Western Sydney, Greater Western Sydney, Australia.

Jayasinghe, U. W., H. W. Marsh, and N. Bond. 2001. "Peer review in the funding of research in higher education: the Australian experience." *Educational Evaluation and Policy Analysis* 23:343-346.

Knorr-Cetina, K. 1981. *The manufacture of knowledge: an essay on the constructivist and contextual nature of science*. Oxford, UK: Pergamon Press.

Knorr-Cetina, K. 1991. "Merton sociology of science: the first and the last sociology of science." *Contemporary Sociology* 20:522-526.

Kostoff, R. N. 1995. "Federal, research impact assessment - axioms, approaches, applications." *Scientometrics* 34:163-206.

Krampen, G. and L. Montada. 2002. *Wissenschaftsforschung in der Psychologie*. Hogrefe: Göttingen, Germany.

Krohn, W. 2000. "Wissenschaftssoziologie: zwischen Modernisierungstheorie und Sozialkonstruktivismus auf schwankendem epistemischem Boden." *Soziologische Revue* 23:314-325.

Latour, B. and S. Woolgar. 1979. *Laboratory life: the social construction of scientific facts*. London, UK: Sage.

Laudel, G. 2006. "Conclave in the Tower of Babel: how peers review interdisciplinary research proposals." *Research Evaluation* 15:57-68.

Leydesdorff, L. 2001. *A sociological theory of communication. The self-organization of the knowledge-based society*. Boca Raton, FL, USA: Universal Publishers.

Lock, S. 1985. *A difficult balance: editorial peer review in medicine*. Philadelphia, PA, USA: ISI Press.

Luhmann, N. 1992. *Die Wissenschaft der Gesellschaft*. Frankfurt am Main, Germany: Suhrkamp.

Luhmann, N. 1998. *Die Gesellschaft der Gesellschaft*. Frankfurt am Main, Germany: Suhrkamp.

Marsh, H. W. and S. Ball. 1991. "Reflections on the peer review process." *Behavioral and Brain Sciences* 14:157-157.

Merton, R. K. 1938. *Science, technology and society in 17th century England*. New York, NY, USA: Harper & Row.

. 1973. "The sociology of science: theoretical and empirical investigations." Chicago, IL, USA: University of Chicago Press.

Mulkay, M. 1969. "Some aspects of cultural growth in natural sciences." *Social Research* 36:22-52.

. 1979. *Science and the sociology of knowledge*. London, UK: George Allen and Unwin.

Murphy, R. 1994. "The sociological construction of science without nature." *Sociology - the Journal of the British Sociological Association* 28:957-974.

Myers, G. 1990. *Writing biology: texts in the social construction of scientific knowledge*. Madison, WI, USA: The University of Wisconsin Press.

Opthof, T., F. Furstner, M. van Geer, and R. Coronel. 2000. "Regrets or no regrets? No regrets! The fate of rejected manuscripts." *Cardiovascular Research* 45:255-258.

Overbeke, J. and E. Wager. 2003. "The state of the evidence: what we know and what we don't know about journal peer review." Pp. 45-61 in *Peer review in health sciences*, edited by F. Godlee and T. Jefferson. London, UK: BMJ Books.

Owen, R. 1982. "Reader bias." *Journal of the American Medical Association* 247:2533-2534.

Pruthi, S., A. Jain, A. Wahid, K. Mehra, and S. A. Nabi. 1997. "Scientific community and peer review system - a case study of a central government funding scheme in India." *Journal of Scientific & Industrial Research* 56:398-407.

Ross, P. F. 1980. *The sciences' self-management: manuscript refereeing, peer review, and goals in science*. Massachusetts, MA, USA: The Ross Company, Todd Pond.

Roy, R. 1985. "Funding science the real defects of peer-review and an alternative to it." *Science Technology & Human Values* 52:73-81.

Sandström, U. and M. Hällsten. 2006. "Productivism in grant peer review: Wenners and Wold revisited." Pp. 121-123 in *Ninth International Conference on Science and Technology Indicators*, edited by W. Glänzel. Leuven, Belgium.

Schmidt, S. J. 1998. *Die Zähmung des Blicks. Konstruktivismus - Empirie - Wissenschaft*. Frankfurt am Main, Germany: Suhrkamp.

Sharp, D. W. 1990. "What can and should be done to reduce publication bias - the perspective of an editor." *Journal of the American Medical Association* 263:1390-1391.

Shatz, D. 2004. *Peer review: a critical inquiry*. Lanham, MD, USA: Rowman & Littlefield.

Smith, R. 2006. "Peer review: a flawed process at the heart of science and journals." *Journal of the Royal Society of Medicine* 99:178-182.

Sokal, A. and J. Bricmont. 1999. *Eleganter Unsinn: wie die Denker der Postmoderne die Wissenschaften missbrauchen*. Munich, Germany: Beck.

Spencer-Brown, G. 1969. *Laws of Form*. New York, NY, USA: Dutton.

Stichweh, R. 1994. *Wissenschaft, Universität, Professionen: soziologische Analysen*. Frankfurt am Main, Germany: Suhrkamp.

Stricker, L. J. 1991. "Disagreement among journal reviewers: no cause for undue alarm." *Behavioral and Brain Sciences* 14:163-164.

van Raan, A. F. J. 2004. "Measuring science. Capita selecta of current main issues." Pp. 19-50 in *Handbook of quantitative science and technology research. The use of publication and patent statistics in studies of S&T systems*, edited by H. F. Moed, W. Glänzel, and U. Schmoch. Dordrecht, The Netherlands: Kluwer Academic Publishers.

Ward, J. E. and N. Donnelly. 1998. "Is there gender bias in research fellowships awarded by the NHMRC?" *Medical Journal of Australia* 169:623-624.

Weller, A. C. 2002. *Editorial peer review: its strengths and weaknesses*. Medford, NJ, USA: Information Today, Inc.

Wenners, C. and A. Wold. 1997. "Nepotism and sexism in peer-review." *Nature* 387:341-343.

Wessely, S. 1998. "Peer review of grant applications: what do we know?" *Lancet* 352:301-305.

Wilson, J. D. 1978. "Peer review and publication." *Journal of Clinical Investigation* 61:1697-1701.

Wissenschaftliche Kommission Niedersachsen. 2006. "Forschungsevaluation an niedersächsischen Hochschulen und Forschungseinrichtungen. Bewertung des Evaluationsverfahrens." Wissenschaftliche Kommission Niedersachsen, Hannover, Germany.

Wood, F. Q. and S. Wessely. 2003. "Peer review of grant applications: a systematic review." Pp. 14-44 in *Peer review in health sciences*, edited by F. Godlee and T. Jefferson. London, UK: BMJ Books.

Young, S. N. 2003. "Peer review of manuscripts: theory and practice." *Journal of Psychiatry & Neuroscience* 28:327-330.

Ziman, J. 1991. *Reliable knowledge: an exploration of the grounds for belief in science*. Cambridge, UK: Cambridge University Press.

Ziman, J. 1991. 2000. *Real science. What it is, and what it means*. Cambridge, UK: Cambridge University Press.

Studying Ourselves as Scholar-Teachers in the Age of HIV and AIDS in Southern Africa

Mathabo Khau & Kathleen Pithouse

University of KwaZulu-Natal, South Africa • McGill University, Canada

khaumsc@gmail.com • kpithouse@gmail.com

Abstract: In southern Africa, critically high levels of HIV and AIDS have brought about an expectation that all teachers should integrate HIV- and AIDS-related education into their subject areas. Research suggests that such integration can be supported by teacher education that takes account of the vital mediating role that teachers' experiences and perspectives play in HIV- and AIDS-related education. In this essay, we reflect on our experiences of learning and teaching in a graduate course for teachers called "Studying ourselves as scholar-teachers in the age of HIV and AIDS in southern Africa" which took place at a university in South Africa. Kathleen (at that time a novice teacher educator and PhD student) designed and taught the course and Mathabo (at that time a school teacher and Masters student) was one of course participants. Our discussion explores the nature and value of the learning that took place during the course and draws attention to the complexity and challenges involved in bringing self-awareness and self-questioning to educational practice and inquiry.

INTRODUCTION

In southern Africa, critically high levels of HIV and AIDS, particularly among the youth (see UNAIDS, 2007), have led to a growing expectation that all teachers should integrate HIV- and AIDS-related education into their subject areas (see, for example, Department of Education South Africa, 2000; Ministry of Education Lesotho, 2005). Research (for instance, Baxen & Breidlid, 2004; Simbayi & Skinner, 2005) and our own experience as teachers and education scholars suggests to us that

Mathabo Khau is a Ph.D. (Gender and Education) student in the Faculty of Education, University of KwaZulu-Natal. Her Masters dissertation was titled "Understanding adolescent sexuality through the memories of four Basotho women teachers: An auto/biographical study," which looked into how women teachers' own experiences of adolescent sexuality influence their teaching of sexuality and HIV and AIDS education in Lesotho classrooms. Her interests are on the intersections of religion, culture, gender, and education in the understandings of sexual health issues, sexuality education, and the formation of sexual identities. Kathleen Pithouse is a postdoctoral scholar in the Faculty of Education, McGill University. She recently completed a Ph.D. in the Faculty of Education, University of KwaZulu-Natal, in which she explored narrative self-study as a research methodology and as a pedagogic approach in teacher education. Kathleen is currently co-editing a trans-disciplinary book that explores connections between self-study and social inquiry, knowledge, and action.

such integration can be supported by teacher education that incorporates teachers' experiences and viewpoints into pedagogic processes and that acknowledges teachers as intellectuals and significant producers of knowledge. In this essay, we reflect on our experiences of learning and teaching in a graduate course for teachers called, "Studying ourselves as scholar-teachers in the age of HIV and AIDS in southern Africa." This course was part of a Masters level module on Health, Sexuality, and HIV/AIDS in Education offered in 2006 in the Gender and Education program at the Faculty of Education, University of KwaZulu-Natal, South Africa. Kathleen (at that time a novice teacher educator and Ph.D. student) designed and taught the course and Mathabo (at that time a school-teacher and Masters student) was one of nine experienced teachers from the southern African countries of Lesotho and South Africa who participated in the course.

KATHLEEN'S INTENTIONS IN DESIGNING THE COURSE

In my recent doctoral research, I undertook a self-study (see, among others, Loughran, Hamilton, LaBoskey, & Russell, 2004; Mitchell, Weber, & O'Reilly-Scallon, 2005) of my experience as a novice teacher educator (see Pithouse, 2007). As a component of this self-study, I designed and taught a Masters level course on HIV and AIDS in education. The learning activities that I designed for the course were aimed at providing opportunities and support for the students to make connections between their own experiences and concerns as teachers in the age of HIV and AIDS in southern Africa and a range of relevant literature and public issues. My understanding was that, as experienced teachers, the students would bring diverse professional and personal experiences to our class and that interaction with their own and each other's lived experience, as well as with academic material, could engender new knowledge and understanding about HIV and AIDS in education. I also wanted to create openings for the students to identify and inquire into individual areas of interest within the broad topic of HIV and AIDS in education. In this way, I hoped to facilitate the students' intellectual and emotional engagement with the complex and challenging topic of HIV and AIDS in education and to encourage them to envisage possibilities for taking action through their work as teachers and education scholars.

When planning the course, I decided to make it a requirement for the students to keep course journals. As part of my self-study process, I also kept a journal in which I recorded and reflected on my experience of teaching the course. I aimed to encourage the students to use their journals as informal, unthreatening spaces to muse on and articulate ideas and questions, keep reading logs, write drafts of their assignments, and express their thoughts and feelings about the experience of the course (see Richardson, 2003; Ballantyne & Packer, 1995).

The opening task for the course was a mini literature review, which required small groups of students to select one of a number of themed collections of texts I had compiled and then to work together to read and discuss some texts from their chosen collection. I devised a number of questions to guide the reading and discussion. These questions were designed to encourage the students to identify key public issues and scholarly conversations in the area of HIV and AIDS in education and to consider these in relation to their own experiences, concerns, and viewpoints.

A subsequent activity required the students to write and then present and discuss brief "lived-experience descriptions" (van Manen, 1990, pp. 64-65) of experiences they had had that related in some way to the topic of HIV and AIDS in education and that they found "emotionally as well as intellectually interesting" (Conle, 2000b, p. 190). I decided to present a lived-experience description of my own as a prompt for the task. Writing this description of an incident that still evoked strong, unresolved feelings in me gave me a sense of how emotionally challenging it could be to bring self-awareness to the study of HIV and AIDS in education and of how important it would be to cultivate an atmosphere of support and mutual trust in the class.

Because I was concerned that revisiting HIV and AIDS-related experiences—which intersect with emotionally loaded issues of sickness, death, sexuality, and often, sexual violence—could feel quite overwhelming, I designed an activity in which the students and I would read and discuss Rager's (2005) article on "self-care" in qualitative research and then negotiate guidelines for 'group-care' within the class. I also arranged for the campus student counselor to attend one of our classes so that the students could meet her and find out how to access her support services if need be.

Another learning activity involved constructing an individual and public timeline of the age of HIV and AIDS in southern Africa. My intention was for the students to map out their own experiences in relation to public HIV and AIDS-related issues and happenings. As with the lived-experience descriptions, I offered my own timeline as a pedagogic resource. I linked the timeline activity with a slideshow on public events, issues, and debates in the area of HIV and AIDS in South Africa between 1982 and 2006. When developing the slideshow, I made a point of including controversial and thought-provoking issues and happenings that I anticipated would stimulate critical debate and interest in this area. One example of such an issue was South African President Thabo Mbeki's highly contentious questioning of orthodox scientific views about HIV and AIDS.

The main coursework task required the students to write scholarly autobiographical essays in which they would map the development of their interest in HIV and AIDS in education. In the essay guidelines, I encouraged the students to locate their areas of interest in relation to their own experiences and contexts as well as to relevant literature and public issues. I also asked the students to consider how they might translate their concerns into some action (no matter how small) in their professional and/or academic contexts. My aim in devising this essay task was to give the students a framework within which to review and consolidate what they had learned through their thinking, reading, and discussion during the course. I envisioned critical and creative narratives that would move inward, outward, backward, and forward to connect the students' scholar-teacher selves to broader issues and happenings, and to realistic possibilities for taking action in their professional and academic communities (see Clandinin & Connelly, 2000; Nash, 2004; Ritchie & Wilson, 2000).

MATHABO'S EXPERIENCE OF THE COURSE

This course was designed such that students would undertake a mini-research project on their personal/professional/academic interest in the area of HIV and AIDS in education. This involved making journal entry reflections on their interest, drawing up a timeline of the development of this interest, and writing an autobiography of their interest. The interactive nature of the module enabled the students to engage actively with each other and with a range of literature related to HIV and AIDS in education. As educators tasked with the duty of teaching the youth about their sexuality and HIV and AIDS, we all had to face our own experiences and our own selves as teachers, parents, spouses, siblings, and researchers. We had to be able to work out our identities in relation to the complex and sensitive subject of HIV and AIDS and come to terms with how 'who we are' influences us within the classroom.

The Story I Shared During the Course

I first heard of HIV and AIDS in 1986, when the first patient was identified in Lesotho. Then the predominant discourse was that HIV and AIDS infects only homosexual people. At this time, for me, HIV and AIDS was a very foreign disease and like many others in my community and country, I believed that it would never infect me because I was heterosexual.

The reality of HIV and AIDS came closer to home for me when one day in April 1997, I had a fight with my husband and I decided to spend the night at my mother's house. The following morning when I got to my house, my husband refused to let me in. I could sense that there was something wrong. When he finally decided to let me in, he forced me to have sex with him right there and then in the lounge. I told him that I was not ready but he pretended not to hear me. He managed to penetrate me and continue with the sexual act as I lay there like a corpse, mortified by what was happening to me. When he was through, I wanted to go to the bathroom to clean myself of the act. He tried to stop me but I managed to go through.

That is when I saw her, another woman lying on my bed with nothing on. I could not believe my eyes. I knew her. I became nauseous, my vision blurred and my knees could not hold me. I had to hold on to the door to steady myself. The bedroom smelled like a cheap motel room with stale cigarette smoke and empty beer bottles all over. I felt so dirty and cheap. I wished I could disappear or erase what had just happened to me in the lounge. Why did he force me to have sex with him while he had obviously spent the whole night with this other woman? He could not tell me why he had done what he did, and at that moment, I knew that my marriage was over.

I started wondering whether he had used condoms with her because he never used them with me. I asked myself why I had never had the courage to demand that he use condoms. "Am I HIV positive?" "Am I going to die and leave my daughter?" All these questions were running through my mind as I left my house for good, with uncontrollable tears streaming down my face. I was now staring the threat of HIV and AIDS in the eye. It was no longer a foreign disease. It was right here in my family. I knew that with my husband's track record, I had a very good chance of being HIV positive. I wanted to get tested but I was too scared. I kept silent about the sexual and other abuses I had endured in my marriage because I thought people would say that I deserved it. I finally got tested after my divorce came through. I was HIV negative. After three more negative tests over a period of nine months, I decided that I was in the clear.

In 2001, my father was brought home to my mother's house after 23 years of separation. He had been staying with other women during this separation time and he never even knew where my mother was staying or how we as his children were surviving. My mother never considered divorce because she was afraid of being 'cut off' from the Catholic Church. When my father became fatally ill, the woman he was staying with told him that she could not take care of him and that he should find his wife. My younger brother, who had always been in contact with my father, decided to bring him to my mother.

I was very angry with my brother and mother for taking care of my father after so many years during which he never bothered about anyone else but himself. When I saw my father where he was sleeping, I was shocked. He was a corpse, barely breathing. He was very dark and emaciated. His mouth was covered with thrush and he was wearing adult nappies. I told my mother that the man had AIDS, but she just dismissed me because she had been made to believe that he was only suffering from pneumonia. She was taking care of him without any precautions. She was blind to the other symptoms that were so obviously evident. I was asking myself why my mother felt she had to take care of a man who never loved her, who had always been violent to her, who had spent all his youthful years and his money with other women. It made me feel sick looking at how she was playing the dutiful wife. She took care of him until she buried him. During the same year, four of my friends, aged between 27 and 30 years, died of AIDS-related illnesses.

I was more devastated by the deaths of my friends than by my father's death. I felt guilty that I was still alive and they had died. I started asking myself what I could have done or what I can do to stop such senseless deaths. From then on, I took every chance I could to further my knowledge on HIV and AIDS.

My Interaction with the Course

When I enrolled for the course, I was just making up the required number of modules for a Masters degree. However, my interest was sparked during the literature review activity at the start of the course when I read about sexual violence against women and children and how this links to HIV and AIDS (see, for example, Pettifor, Measham, Rees, & Padian, 2004; Dunkle et al., 2004). All the memories of sexual violence I had kept locked inside came flooding back. I could see myself back then when my own husband sexually abused me, and all the emotion and passion I had for changing the status quo for other women resurfaced. During the course, the sharing of stories of the students' remembered experiences and the subsequent discussions allowed us to learn from each other and to be able to face our fears, demons, challenges, and misgivings in relation to issues of HIV and AIDS.

My personal involvement in the course helped me come to terms with my experiences and to forge an understanding of how I was teaching my own learners in relation to sexuality and HIV and AIDS. I was able to face the fears and demons from my well-hidden past. The course led to a reflective engagement among the teachers in the class concerning how remembered experiences have influenced our professional lives, particularly in relation to our classroom interaction with learners during HIV- and AIDS-related lessons. We started asking ourselves questions such as, "How can I improve my teaching?" According to Allender and Allender (2006, p. 15), "Unless we are conscientiously aware of what is driving our choices of behavior in the classroom, we are all too likely to revert to the ways of the teachers who taught us." They argue that self-awareness is "essential to breaking with nonfunctional teaching behaviors in an effort to discover personal strengths as a classroom teacher." A key part of our formation as teachers is our own experiences as learners. Much of what we do and do not do is in reaction to and influenced by those experiences. Our past "experiences create hidden personal narratives about education, school, and schooling that have a profound and sometimes intractable impact on the way we teach our students" (Samaras, Hicks, & Berger, 2004, p. 908). The importance of providing opportunities and support for teachers to develop self-awareness in relation to sexuality and HIV and AIDS is highlighted by Baxen and Breidlid (2004), who draw attention to the need to acknowledge and explore teachers' lives and experiences as a key mediating factor in HIV and AIDS education.

Using Autobiography

Cole and Knowles (2001) point out that autobiography is a structured account of life written by and about oneself. As I embarked on writing my autobiographical essay for the course, I became aware of the challenges and complexity of using autobiographical writing within an academic setting. These involve a perceived lack of objectivity, lack of scientific evidence, and a reliance on memory. I remember very well the struggles we all went through trying to write the first draft of our autobiographical essays and how we felt it was against the conventional way of doing academic writing. In addition, an autobiographical approach involves the writer in self-examination and scrutiny whose potential effect is exposure to emotional threat. The danger of emotional harm was offset by discussing how to take care of ourselves as researchers especially when dealing with emotionally taxing subjects. Rager (2005) outlines several strategies for self-care that were useful for our class. Since the process of writing our autobiographies and sharing our stories touched the whole person and not just the intellect, "there needed to be a sense of trust, security and mutual respect" (Conle, Louden, & Mildon, 1998) among the members of the group. Understanding ourselves shaped the way we interacted with each other. Acknowledging insecurities, fears, and weaknesses in 'self' helped us view the world and others in a different, more empathetic light.

At first, it was very difficult for me to put myself in the centre of my inquiry. I was uncomfortable with starting with the self and making myself vulnerable. I found myself having to share with my fellow students stories that are close to my very being and who I have become. While these stories were very sensitive in nature, I thought that baring my experiences to this small group of friends and acquaintances would help them better understand who I was and where I was coming from. For me this was a very therapeutic experience because the spaces created in this environment allowed me to let go and cry, releasing all my pent-up emotions from years gone by. I believe that being able to deal with who we are as teachers in relation to our own gendered experiences and sexual identities is necessary if we are to teach effectively on HIV and AIDS. Who we are as people shapes who we are as teachers and this influences our interactions with learners in our classrooms. Who we are as people is shaped by different actors in our lives. Acknowledging the influence of all these social actors in our lives allows us to engage ourselves in reflection on how they have shaped us.

So What?

Being part of the "Studying ourselves as scholar-teachers in the age of HIV and AIDS in southern Africa" course has made me better understand my position as a woman science teacher in relation to issues of sexuality and HIV and AIDS in Lesotho classrooms. The position I occupy is surrounded with conflicting social constructions of masculinity and femininity. As a Mosotho woman, I am expected to be submissive, unquestioning, and sexually innocent. However, as a science teacher, I am expected to be knowledgeable, powerful, and able to make decisions in any situation. These positions that I occupy have an important bearing on my teacher self and how I handle boys and girls in my lessons. It is important for my professional development to find a balance or some common ground between being a woman and a science teacher so that my learners can benefit from our interactions. Walkerdine (1990) draws our attention to the "impossible fiction" of being a woman teacher in today's society. She points to the tensions and contradictions that are inherent in an identity that asserts power and status and commands respect (teacher) at the same time that it speaks of subordination, marginalization, and repression (woman). In support, Munro explains that "to be a woman is to lack authority, knowledge, and power. To be a teacher is to have authority, knowledge, and power" (1998, p. 1). Looking at the contradictions in what constitutes a good Mosotho woman and a good science teacher, I believe the only way forward is to disturb and challenge the assumptions that my society has regarding these two seemingly conflicting positions. I believe that social constructions of masculinity and femininity in Lesotho can be contested and reconstructed. I would like to see women in Lesotho treated as autonomous beings, with knowledge and opinions, and with a right to enjoy sex. I think that this could help to make women teachers more comfortable in teaching about HIV and AIDS. Maybe such sexual enlightenment would change the power dynamics in sexual partnerships, such that women become empowered to decide on what they really want.

My engagement with the course was also very helpful in shaping my research topic for my Masters thesis. The stories we explored contributed to my reflection on the power dynamics that are characteristic of sexual relationships, and the role of sexuality education in reshaping the landscape of sexual decision-making and HIV infections. I started thinking of how I was taught as an adolescent girl and how I am teaching adolescent boys and girls on issues of sexuality and HIV and AIDS in my classrooms. This led to a reflection on my practice and, in my Masters research, I involved other teachers in reflecting together on our memories of adolescent sexual experiences and their influence on our teaching (see Motalingoane-Khau, 2007).

KATHLEEN'S REFLECTION ON THE COURSE

For me, Mathabo's response to the course underlines the educative value of offering opportunities and support for teachers to work together to make sense of the complex and challenging area of HIV and AIDS in the light of their own experiences and concerns as well as pertinent literature and public issues. During the course, Mathabo and her fellow students began to look more closely and carefully at their own contexts, viewpoints, and professional practices in relation to HIV and AIDS in education. They also began to think in new ways about a variety of HIV- and AIDS-related public issues. The storytelling aspects of the course allowed the students to look through each other's eyes to gain new perspectives on HIV- and AIDS-related experiences and issues (cf. Conle, 1996, 2000a; Conle, Li, & Tan, 2002). Mathabo's reflection highlights how the students learned with and from each other and how their learning was supported and facilitated by their relationships and interaction with each other (see, among others, Hollingsworth, 1992; Palmer, n.d.). In addition, perhaps most importantly, the course encouraged the students to identify and explore aspects of HIV and AIDS in education that they genuinely cared about and to imagine and share possibilities for making choices and taking action in response to those issues in their professional and scholarly practice.

As well as identifying benefits of studying the 'teacher self' in relation to HIV and AIDS in education, Mathabo also raises some queries and concerns about this pedagogic approach. Teaching this course helped me to understand more about the complexity of bringing self-awareness and self-questioning to teacher education. I came to see that self-study could actually make a teacher's professional life more difficult because colleagues or parents might feel threatened by new ideas or ways of working. I also became more aware of tensions between individual teachers' intentions and deep-rooted patterns of school and community life. I realized that these tensions might not necessarily be resolved in the university classroom, but that teacher educators and fellow students could offer encouragement and support for teachers to acknowledge and grapple with them.

Part of my pedagogic approach to the course involved giving broad outlines for the main coursework assignment and requiring the students to discover their own foci for their essays. I anticipated that my inquiry-orientated approach would help the students to find intrinsic motivation for and satisfaction in learning (see Kleiser, 2004; LaBoskey, 2004). I also hoped that this approach would make the learning in our class more relevant to the students' lived experience and to their professional and academic work. In my view, Mathabo's interaction with the course illustrates how positioning students as agents of their own learning can stimulate and enhance their learning. Nonetheless, during the course, I became conscious of how many students can find this positioning unfamiliar and can feel unsure and anxious about what it requires from them. I realized the importance of monitoring and supporting the growth of students' agency through class interaction, appropriate learning activities, and ongoing guidance and assistance. In addition, I came to see that, as I worked to support and challenge the students to develop more self-reliance, I needed to make sure that I was being as clear and explicit as possible about why I was taking this particular pedagogic approach and that I was inviting the students to interrogate this approach with me. Furthermore, I was surprised and somewhat disconcerted to find that, like several of the students, I too experienced anxiety about not having a wholly pre-determined assignment 'question.' Although I had deliberately chosen to work with the students to develop individual foci for their essays, the tentative, emergent nature of the process that we went through did make me feel unsure and 'out of control' at times. Nevertheless, I came to see how it was possible for a teacher-educator to find the way together with students rather than to prescribe the way, and that although this might involve relinquishing some 'control,' it did not necessarily amount to an abdication of responsibility.

Mathabo's discussion of the course also draws attention to the emotional complexity of bringing self-awareness into the classroom, particularly when dealing with the emotionally loaded topic of HIV and AIDS. While I had foreseen this to some degree, and had made some provision for it, I do not think that I had fully anticipated the emotional weight that some of the learning activities would carry. Additionally, I had not expected to be so personally affected by this emotional intensity. Consequently, I had not realized how stressful I might find it to facilitate sessions in which I would need to be constantly alert and ready to make immediate decisions about when and how to intervene in group processes in order to provide support where necessary and to minimize potential for harm.

Through my experience of this course, I have become more conscious of the emotional responsibility that I take on when I encourage students to bringing awareness of self to an area such as HIV and AIDS, which is permeated with experiences of suffering and loss. I find that responsibility quite daunting. However, I understand from Mathabo and her fellow students' responses to the course that even though they found some of the learning activities to be emotionally painful at times, they did not experience them as harmful, but rather as valuable and rare learning experiences (cf. Conle, 1996; Elbaz-Luwisch, 2002). Moreover, their responses emphasized the atmosphere of encouragement, understanding, and trust that developed in the class as a central factor in helping them to deal with emotionally challenging learning.

LOOKING FORWARD

Our shared reflection on this course has emphasized for us that starting with the self and making one's self visible and thus vulnerable in educational practice and inquiry can be a demanding experience. We are mindful of the complexity and responsibilities involved in bringing self-awareness and self-questioning to teaching and learning, particularly in relation to sensitive and emotionally taxing topics such as HIV and AIDS. Nonetheless, because our reflections have also highlighted the educative potential of linking the development of self-knowledge to the study of significant educational and social issues, we are committed to the ongoing exploration of ways in which to work skillfully and constructively with self-study and autobiography in our practice as scholar-teachers.

REFERENCES

Allender, J., & Allender, D. (2006). How did our early education determine who we are as teachers? In L. M. Fitzgerald, M. L. Heston & D. L. Tidwell (Eds.), *Collaboration and community: Pushing boundaries through self-study. Proceedings of the sixth international conference on self-study of teacher education practices, Herstmonceux Castle, East Sussex, England, July 30 - August 3* (pp. 14-17). Cedar Falls, IA: University of Northern Iowa.

Ballantyne, R., & Packer, J. (1995). The role of student journals in facilitating reflection at doctoral level. *Studies in Continuing Education, 17*(1&2), 29-45.

Baxen, J., & Breidlid, A. (2004). Researching HIV/AIDS and education in Sub-Saharan Africa: Examining the gaps and challenges. *Journal of Education, 34*, 9-29.

Clandinin, D. J., & Connelly, F. M. (2000). *Narrative inquiry: Experience and story in qualitative research*. San Francisco, CA: Jossey-Bass Publishers.

Cole, A. L., & Knowles, J. G. (2001). *Lives in context: The art of life history research*. Walnut Creek: AltaMira Press.

Conle, C. (2000a). Narrative inquiry: Research tool and medium for professional development. *European Journal of Teacher Education, 23*(1), 49-63.

Conle, C. (2000b). Thesis as narrative or 'What is the inquiry in narrative inquiry?' *Curriculum Inquiry, 30*(2), 189-214.

Conle, C., Li, X., & Tan, J. (2002). Connecting vicarious experience to practice. *Curriculum Inquiry, 32*(4), 429-452.

Conle, C., Louden, W., & Mildon, D. A. (1998). Tensions and intentions in group inquiry: A joint self-study. In M. L Hamilton, S. Pinnegar, T. Russell, J. Loughran, & V. Laboskey (Eds.), *Reconceptualising teaching practice: Self-study in teacher education* (pp. 178-193). London: Falmer Press.

Department of Education South Africa. (2000). *The HIV/AIDS emergency: Guidelines for educators*. Pretoria: Government Press.

Dunkle, K. L., Jewkes, R. K., Brown, H. C., Gray, G. E., McIntyre, J. A., & Harlow, S. D. (2004). Gender-based violence, relationship power, and risk of HIV infection in women attending antenatal clinics in South Africa. *The Lancet, 363*(9419), 1415-1421.

Elbaz-Luwisch, F. (2002). Writing as inquiry: Storying the teaching self in writing workshops. *Curriculum Inquiry, 32*(4), 403-428.

Hollingsworth, S. (1992). Learning to teach through collaborative conversation: A feminist approach. *American Educational Research Journal, 29*(2), 373-404.

Kleiser, E. (2004). My investigation into the use of portfolios as a teaching, learning, and assessment tool in my higher education classes. In D. L. Tidwell, L. M. Fitzgerald, & M. L. Heston (Eds.), *Journeys of hope: Risking self-study in a diverse world. Proceedings of the fifth international conference on self-study of teacher education practices, Herstmonceux Castle, East Sussex, England, June 27 - July 1* (pp. 162-165). Cedar Falls, IA: University of Northern Iowa.

LaBoskey, V. K. (2004). The methodology of self-study and its theoretical underpinnings. In J. J. Loughran, M. L. Hamilton, V. K. LaBoskey, & T. Russell (Eds.), *International handbook of self-study of teaching and teacher education practices* (Vol. 2, pp. 817-869). Dordrecht; Boston; London: Kluwer Academic Publishers.

Loughran, J. J., Hamilton, M. L., LaBoskey, V. K., & Russell, T. L. (Eds.). (2004). *International handbook of self-study of teaching and teacher education practices*. Dordrecht; Boston; London: Kluwer Academic Publishers.

Motalingoane-Khau, M. (2007). *Understanding adolescent sexuality in the memories of four female Basotho teachers: An auto/biographical study*. Unpublished MEd thesis, University of KwaZulu-Natal, Durban.

Ministry of Education Lesotho. (2005). *Education sector strategic plan*. Maseru:UNESCO.

Mitchell, C., Weber, S., & O'Reilly-Scanlon, K. (Eds.). (2005). *Just who do we think we are? Methodologies for autobiography and self-study in teaching*. New York: RoutledgeFalmer.

Munro, P. (1998). *Subject to fiction*. Buckingham: Open University Press.

Nash, R. J. (2004). *Liberating scholarly writing: The power of personal narrative*. New York: Teachers College Press.

Palmer, P. J. (n.d.). Good talk about good teaching: Improving teaching through conversation and community. Retrieved September 20, 2006, from http://www.mcli.dist.maricopa.edu/events/afc99/articles.html

Pettifor, A. E., Measham, D. M., Rees, H. V., & Padian, N. S. (2004, November). Sexual power and HIV risk, South Africa: Unequal sexual power reduces condom use in South Africa. *Emerging Infectious Diseases*. Retrieved May 12, 2005, from http://www.cdc.gov/ncidod/EID/vol10no11/04-0252.htm

Pithouse, K. (2007). *Learning through teaching: A narrative self-study of a novice teacher educator*. Unpublished PhD thesis, University of KwaZulu-Natal, Durban.

Rager, K. (2005). Self-care and the qualitative researcher: When collecting data can break your heart. *Educational Researcher, 34*(4), 23-27.

Richardson, L. (2003). Writing: A method of inquiry. In N. K. Denzin, & Y. S. Lincoln (Eds.), *Collecting and interpreting qualitative materials* (Second ed., pp. 499-541). Thousand Oaks, California: Sage Publications.

Ritchie, J. S., & Wilson, D. E. (2000). *Teacher narrative as critical inquiry: Rewriting the script*. New York: Teachers College Press.

Samaras, A. P., Hicks, M. A., & Berger, J. G. (2004). Self-study through personal history. In J. J. Loughran, M. L. Hamilton, V. K. LaBoskey & T. Russell (Eds.), *International handbook of self-study of teaching and teacher education practices* (Vol. 2, pp. 905-942). Dordrecht; Boston; London: Kluwer Academic Publishers.

Simbayi, L., & Skinner, D. (Eds.). (2005). *Workplace policies in education: A review focusing on HIV/AIDS*. Cape Town: HSRC Press.

Walkerdine, V. (1990). *Schoolgirl fictions*. New York: Verso.

UNAIDS. (2007). *Report on the global AIDS epidemic*. Retrieved January 21, 2008, from http://www.unaids.org/en/KnowledgeCentre/HIVData/EpiUpdate/EpiUpdArchive/2007/2007default.asp

van Manen, M. (1990). *Researching lived experience: Human science for an action sensitive pedagogy*. Albany, N.Y.: State University of New York Press.

Grappling with Global-Personal and Victim-Culprit Tensions

Reflections on Teaching Globalization Courses

Satoshi Ikeda

Concordia University, Montréal, Canada

sikeda@alcor.concordia.ca

Abstract: My globalization courses are the place where I learn from the students. Each student examines her/his involvement in globalization as victim and culprit, and devises actions to counteract corporate domination and neoliberal globalization. From classroom discussion and reports, I am often surprised by the deep insights students develop about themselves and globalization. I have succeeded at least to turn the table around between the student and the instructor. Personal examination of individual involvement in globalization helps the student grasp the macro process of globalization as personally relevant historical and social process. They start seeing themselves relationally and being the connected parts of the macro structure. I hope that they take this world-systemic understanding of the whole and parts beyond my class, or continue to live through the global-personal tension. I am also impressed by the students' understanding of the complicated and often-contradictory nature of their involvement in globalization. The perspective of victim-culprit forces them to see their being not as singular unitary existence but as a collection of fragmented and contradictory 'selves.' I see myself as a contradictory being. My critical view on the corporation is accompanied by a middle class lifestyle that is quite dependent on goods and services supplied by the corporation. My pension fund is invested in military contractors and oil companies. I am struggling to come to terms with my own multiple and contradictory existence. I think that my pedagogical praxis is my way of pursuing utopystics (Wallerstein 1998; Tamdgidi 2007, forthcoming) by inviting students to form their communities so that they could overcome victim-culprit schizophrenia imposed on them by neoliberal globalization and the corporation.

INTRODUCTION

This essay is a reflection on my experience in teaching sociology courses on globalization.

I was hired by the Department of Sociology of the University of Alberta ten years ago to create globalization courses at the graduate and undergraduate levels. Trained in world-system studies at SUNY-

Satoshi Ikeda is Associate Professor in the Department of Sociology and Anthropology, Concordia University, Montréal, Canada. He is a Canada Research Chair (political sociology of global futures), studying emerging post-corporate and post-global alternatives that are sustainable socially, economically, and ecologically. The projects he is working on include organic/natural farming and distribution through direct marketing and community supported agriculture networks, local production and consumption networks, local trading and exchange systems and local currencies. Through research and university instruction, he hopes to promote post-global and post-corporate movements for equitable, sustainable, and just futures.

Binghamton, I thought I could introduce to the students the useful tools to critically examine globalization. My pedagogical objective for the undergraduate courses, however, quickly turned to juxtaposing globalization against 'personal' for a couple of reasons.

First, I was frustrated with the division of roles in classroom between the instructor and the students as the provider and receiver of knowledge, respectively. This role assignment usually makes students passive participants in classroom while the instructor is expected to turn into an entertainer. I was not impressed with my ability to entertain, and I believed that the students could play the role of entertainer with their personal examination of their roles in globalization.

Secondly, I was frustrated with the complacency of the privileged middle class students from the Global North. Critical examination of globalization would involve questioning their lifestyles and privileges they take for granted. Also, they may be victim of globalization due to corporate infringements on environment, labor rights, and sovereignty particularly in the Global South,[1] even though the university students in Canada may be primarily the beneficiary of corporate activities. The best way to make them realize their involvement is to critically examine their personal involvement in globalization. These frustrations made me develop globalization courses that juxtapose a large-scale social process and students' personal involvement, and that examine contradictory and complicated ways we interact with globalization both as victim and culprit. I am using the term 'global-personal tension' to describe the challenge of putting personal against global, and the term 'victim-culprit tension' to describe the exercise of examining our multiple and often contradictory involvements in globalization.

The lineage of my pedagogical concerns involving 'personal' in the world-system studies started with the exciting undergraduate courses my good friend Mohammad H. (Behrooz) Tamdgidi taught at SUNY-Binghamton and then at SUNY-Oneonta in the 1990s. He asked his students to apply the world-system approach to the understanding of their personal development, thus providing opportunity to critically reflect upon themselves and those who were influencing them at various life stages. At the end of the Spring 1997 semester at SUNY-Binghamton, he published a selection of his students' essays in a book titled *"I" in the World-System: Stories from an Odd Sociology Class.* The volume was a rare accomplishment in an undergraduate course since most of the term papers in university/college courses remain personal. Making student papers open to the public empowers the students (and instructor) since it provides a forum to continue engaging in learning/teaching experience. Even though the book initially met a challenge when the parents of a student contributor demanded the university to stop its publication because they were embarrassed by the critical assessment they received from their child, the university solidly stood on the side of the instructor and the student's freedom of expression and the book was published. While I do not have the capability to publish student essays, I learned from Tamdgidi's efforts the importance of putting the student at the center of learning experience.

Tamdgidi criticized the world-system studies for their Euro-centric, economic, and macro bias, and he suggested alternatives that build on his reinterpretation and

[1] The term "Global South" is used here to indicate the countries and regions that are negatively affected by globalization. They are the so-called periphery in the world-system approach, together with the semiperipheral countries that experienced economic decline under globalization. The "Global North" designates the countries in the core zone, and they overlap with the OECD (Organization of Economic Cooperation and Development) countries.

hybrid introduction of Marx, Gurdjieff, and Mannheim (Tamdgidi 2003). The two tensions I was grappling in my globalization courses were among the primary scholastic concerns he has been engaging in since his graduate student days. The global-personal tension resonates with his critique of world-system studies that place primacy on the 'macro' while ignoring the 'micro', or personal level, leading to a shortcoming in imagining emancipation from within (Tamdgidi 2004a; 2005; 2006b; 2007; 2008; forthcoming). The victim-culprit tension resonates with Gurdjieff's emphases on the paradigmatic significance of human inner division and multiple selfhood which Tamdgidi insists as the key when we imagine realistic historical alternatives to the existing world-system (Tamdgidi 2004b; 2006a; forthcoming).

Reading his work gave me a chance to reflect upon my experiences as an instructor of globalization courses, and that led me to this essay where I intend to introduce my approach to learning/teaching globalization by focusing on the global-personal and victim-culprit tensions. I hope that this essay could stimulate the imagination of the instructors and students toward fruitful and critical learning/teaching experience.

GLOBAL IS PERSONAL, AND PERSONAL IS GLOBAL

Globalization is a world historical process manifesting in the 1980s where the 'national' restriction on the cross-border flow of goods, services, money, and information was greatly removed. The neoliberal ideologues, those who believe in small government and limited public intervention in the market (or regulation over corporate activities), want us to believe that globalization was a natural phenomenon, or the only possible outcome of the failed Keynesian policy in the 1970s. From the political economy perspective, however, globalization was an unexpected consequence of the failed economic policies based on utterly unrealistic economic model that were imposed by the US government, the World Bank, and the International Monetary Fund (IMF). In the process of globalization, the US regained dominant economic position using finance as the leverage while the major corporations became dominant economic players that were largely free from regulations on working condition, product safety, and ecological sustainability (Ikeda 2004, Ellwood 2001). As a result of unrestricted corporate activities, the working condition for the majority of workers both in factories and on farm deteriorated together with the declining wages. The earth's ecological system suffered from land, sea, and air pollution. Globalization, however, created opportunities for the privileged few, including the middle class students in Canada.

When I asked students "Do you think you will work for a foreign company or Canadian company operating abroad?" the answer was yes by a big margin (80%). Those who aspired to work for non-profit or non-government sectors had working abroad as part of their future plan. Permanently staying in Canada to earn a living was a future vision shared only by a few. When I asked them "How do you benefit from globalization?" students pointed out the varieties of consumer goods including food that are produced abroad. They enjoyed "ethnic food" available thanks to the spread of various diaspora communities and the increased communication with people all over the world using computers and telephones. Many students said they enjoyed music and movies made in various countries in the world that are now available through cyberspace. When I asked students "Are there any way you are victim of globalization?" those who were aware of global warming and the poor working condition in the factories in the Global South pointed out their indirect 'victimhood', but the majority of the students did not see themselves as victims. I then turned their attention to their relationship with the corporation.[2] I asked them "Who tells you what brand name in cloths is cool? What is the brand of automobile do you want to have?" these questions invited students to cast doubt about their 'independence' in their consumer choice, leading them to the realization that they are dependent on the corporation for the choice they have over consumption items and the cultural meaning attached to them. The question could be escalated to: "Do you think you are a victim of corporate domination?"

Our lives in the age of globalization are dependent on the corporation, a legal entity that behaves rather pathologically (Bakan 2005). Unleashed from the various forms of regulation, profit-driven corporations create our wants and needs that are both essential and superfluous, even to the degree of determining the meaning of our lives. As I invite students to discuss various aspects of our engagement with the corporation, students start reflecting on their values and often-uncritical acceptance of the lifestyle prepared by the corporation. By buying into the corporate-promoted lifestyle, the consumers become part of the global structure of exploitation of the people and nature in the Global South.[3] Furthermore, the involvement of the residents of the Global North is much deeper and more complicated than their immediate involvement through consumption. The young generation knows that the corporation provides high income earning opportunity. Because most students are not yet working as full-time workers, however, they are often unaware of the corporate control of their savings, investment, and post-retirement preparation. Once upon a time, children were the assets that provided livelihood to the elders. Today the middle class workers have to save money to prepare for retirement, and the principal investment venue is the corporate stocks. The future prospect of the workers is closely tied to the continued growth of the corporation. The lives of the middle class corporate employees are dependent on the corporation for income earning activities, consumption, and investment to the degree that they could not survive without the corporation.

When I ask the students "Do you know under what condition the products and food you purchase from the corporation are manufactured and prepared?" some students are aware of the sweatshop conditions in the Global South. News on the chemical contamination of pet and human food makes them aware of the potential safety problem of the foods we import. I introduce materials to inform students of corporate offense against workers and environment involving the production of their favorite consumption items. Then I ask "Are you culprits of corporate offense against the workers and environment in the Global South?" This gives an opportunity for students to reflect critically on the consequence of their consumption. By purchasing corporate-supplied goods and services, we are endorsing the policy of the US, World Bank, and the IMF and become the part of the problem of global inequality, worker exploitation, environmental degradation, and resource depletion.

Our personal lives are the source of global problem when we continue to dependent on the corporation. The fact that Canadians cannot vote in the U.S. presidential

---

[2] "The corporation" designates the major corporations, or the corporate sector, that became the dominant economic actors under globalization. Joel Bakan (2005) has analyzed the corporation critically for their irresponsible nature under globalization pushed by neoliberal ideology.

[3] Corporate exploitation is not limited to the Global South. Even in the Global North, neoliberal globalization escalated ecological degradation and worker exploitation. Food security has been compromised as indicated by repeated e-coli contamination of meat and vegetable prepared by large-scale corporate operation.

election and that Canada's share in the World Bank and the IMF is not big enough to challenge US domination do not exonerate the Canadian citizens for the global problem caused by global corporate rule as long as the Canadians continue to support the corporation with their employment, consumption, and investment. I have been successful in sharing this realization that our personal lives are intricately connected to global processes that create inequality, poverty, and environmental degradation. Macro historical process (the global), therefore, is closely connected to micro level processes (the personal).

MULTIPLE AND CONTRADICTORY INVOLVEMENT IN GLOBALIZATION

When I first arrived at the University of Alberta, I was asked to teach an undergraduate course on inequality and social stratification. When I read the calendar description of the course, I was at loss. The course was a Canada content course when in fact I did not have any knowledge on Canadian history or society. I knew roughly the shape of the Canadian border with the U.S., but I had no idea about where and how Canada ends at its northern border. The undergraduate director at the department told me that I could change the content into globalization, and I tried to discuss global inequality and stratification based on nationality, ethnicity, language, etc.

The course turned out to be a boring one, particularly because the content was far away from students' interest and they often lacked basic knowledge to engage in discussion. I decided to stick to the calendar description, but decided to turn the table of teacher-student relationship around. Since the students are the resident/citizen of Canada who have longer experience of living in this country than their newly arrived foreign instructor, they could be the teacher. The discussion format based on weekly reading assignment proved to be an effective tool to stimulate lively debate.

As I acquired knowledge on the history and society of Canada, I realized that most Canadian students hold a positive image that Canada is a country with gender equality and ethnic tolerance without extreme poverty. The literature, however, informed me otherwise. Then my objective in the course changed to challenging such naïve notions about Canada by introducing readings and videos that point out persisting gender bias, deep and historical legacy of discrimination against aboriginals and newcomers that still continue today, and persistent income class division overlapping with gender and ethnic inequality. The motto of my inequality and social stratification class became "make Canadian students feel uncomfortable for being Canadians." The similar motto applies to my globalization courses: "make Canadian students feel uncomfortable for being the victim and culprit of globalization."

In midterm and term reports, I ask students to perform critical examination of their involvement in globalization both as victim and culprit. As compared to the assignments in inequality and social stratification course where I ask students to examine their involvement in and contribution to discrimination based on gender, age, income class, ethnicity, language capability, etc., the assignments to examine their involvement in globalization may appear easy. But our participation in globalization via employment, consumption, and investment is often accepted uncritically while discrimination of any sorts are frowned upon or outright condemned these days. Therefore, examining our involvement in global inequality, environmental degradation, and resource depletion requires more careful and imaginative thinking.

Being Canadian for many students brings a sense of pride, and they usually express displeasure toward the U.S. policies in the world. It, therefore, is a challenge for them to realize that they are, or will be, part of the privileged corporate employee/consumer/investor in the Global North and they are the culprits of global problems. I have not had a student who defended the corporation and globalization perhaps because my students are sociology majors and minors who share the consciousness concerning social problems and a keen sense of social justice. But because of their sensitivity toward social justice, it becomes difficult at times for them to face up to the reality of their offense against the workers and environment in the Global South. Many students realize that they are indeed the victims of corporate hegemony over material culture, lifestyle, and the values. I had many excellent papers that accomplished deep and critical reflection of their involvement both as culprit and victim of globalization. The success of the course, however, would be evaluated if the course empowered the students toward personal and social transformation for the better.

ACTION IS THE BEST TEACHER

Except for a few more mature students, most of my students are in their late teens and early twenties. They are the survivors of peer pressure to conform to the consumption norms set by the corporations. Class discussion on their personal brand addiction, for instance, becomes a lively exchange of what is cool and what is not. It is shocking for the students to learn about the sweatshops and unthinkably low wages paid to the young female workers in the Global South who manufacture the brand name shoes and cloths. The students face a dilemma between brand addiction and socially responsible behavior.

The key to overcome this dilemma is action/activism against global injustice. In the third year course on globalization, I introduced the action/activism component. The midterm project is a group project where students find and participate in anti-globalization and anti-corporate activist groups off-campus. The midterm report introduces the organization and its activities and assesses the effectiveness and impact of their action. For the term project, I ask students to devise their own anti-globalization action, implement it, and report the result. By activism, I mean 'work on others to change their behavior.' Simply changing the student's behavior, like eliminating brand name cloths from their closet, is not enough to be an acceptable term project. To my surprise, most students enjoy these assignments and come up with quite inventive activism.

One student, for instance, collected from numerous people pledges not to purchase from a global hamburger franchise for one year. After one year, he handed the certificate of the completion of their pledge to those who stood by the pledge. Another student attempted water tasting so as to inform people that brand name bottled water does not necessarily taste better than tap water. She had fun challenging people's brand-name addiction. Yet another student challenged his friends to give up their favorite brand name cloth. He brought expensive shirts and jerseys and tore them in front of his classmates as a symbol of the emancipation from corporate branding. Some classmate criticized him that the cloth could have served those in need. But I saluted him for his courage to talk to his close friends and to make them give up brand goods.

How is it possible to create alternatives to the existing world-system? From my research and instructional experience, I came to a conclusion at this point that the target of our individual activism is the corporation. How can we become free from corporate domination in our lives? I asked students "What are the things you do that do not involve goods and services supplied by the corporation?" "Is there anything you produce for your own consumption?" The examples given were limited, such as growing herbs, buying food from local farmers, cooking own meals instead of eating out even though the materials may be purchased from supermarket. When I asked students "How can you reduce dependency on the corporation?" the immediate answers included efforts to reduce the amount of spending by switching from brand name to no-brand name and to brew their own 'fair trade' coffee at home instead of buying it from the global coffee franchise.

But the expected impact of these actions is small. I introduce the examples of Community Supported Agriculture (CSA) where urban consumers pay in advance in spring to local organic farmers and receive weekly food box.[4] Other examples are the Local Exchange and Trading System (LETS) and Local Currency.[5] The LETS is a community-based network for exchanging goods and services and Local Currency is the currency accepted only in the community. These movements stimulate local economic activities outside the realm of corporate activities even though corporate supplied goods such as automobile may be necessary. Promoting local production for local consumption is a trend, and I hope that students could become involved more in these activities.

Realistically it is not possible for consumers of the Global North to achieve self- or community-sufficiency in their daily lives unless they are willing to live without advanced technology embodied in latest corporate products. There are so many black boxes around us which we have no expertise to produce or repair. Expanding alternative economies outside the corporate realm, however, will encourage the corporation to take social and environmental responsibility. There is no major corporation today that does not pay at least lip service to corporate social responsibility. This is a result of anti-corporate activism that targeted notable global corporations that used sweatshops and engaged in ecologically damaging practices. It is possible to pressure corporations with consumers' purchasing power, but it would be the existence of post-corporate alternatives that would truly motivate the corporation toward socially acceptable behavior.

Is it possible to challenge corporate domination in income earning activities? I asked students "Is it possible to find employment and or conduct income earning activities that do not depend on the corporation?" The alternatives students suggested included lawyers, medical professionals, teachers, and government employees. I suggested looking into the emerging post-corporate sector called social enterprise. A social enterprise is motivated to serve the society, in contrast to the corporation that is profit-driven, although it has to generate income to cover costs.[6] For the students, the non-profit and non-governmental sector also appeals as the meaningful place of work. One possible pitfall of this sector is that their operational fund often comes from the corporation or the government, and there is no guarantee that the fund will keep coming. Overall, students were excited to know that there are various employment opportunities outside the corporation. Income earning activities are essential, but we do not need to sell our soul to the corporation to do so. Earning income while engaging in activism is an ideal way to apply the theory and engage in praxis simultaneously.

Is it possible to challenge corporate

---

[4] See http://www.localharvest.org/csa/ and http://www.nal.usda.gov/afsic/pubs/csa/csa.shtml for US examples on CSA.

[5] For information on Local Currency, see http://www.ithacahours.org/ and http://www.communitycurrency.org/resources.html, and for the LETS, see http://www.gmlets.u-net.com/ and http://www.letslinkuk.org/.

[6] See http://www.hbs.edu/socialenterprise/ and http://www.socialenterprise.org.uk/.

domination in post-retirement preparation? This is an area that is most challenging because of the lack of alternatives. Those who are concerned about unacceptable corporate behavior could invest in the Sustainable and Responsible Fund (SRF), Socially Responsible Investment (SRI), Eco Fund, etc. These funds enable the investor to take into account concerns over social, ecological, health, development, etc., in addition to the financial concerns. By selecting the stocks of the corporation that behave acceptably, these funds may be able to pressure corporations to conform to extra-financial social concerns.[7] Investment in these funds may sooth the consciousness of the middle class investors, but such investment supports the corporation and its drive for profit, which in turn will benefit the large-scale upper class investors. Truly alternative investment should serve the community and society, not the interest of corporate managers and the upper class.

Community investment was once performed by the credit union. Even though neoliberal globalization forced credit unions to conform to the global financial standard, they continue to service individuals and communities. It is possible to establish investment institutions that take into account the community needs and long-term sustainability in ecology, society, and economy. In Quebec, cooperative Desjardins provides an alternative to profit-motivated banks.[8] It is possible to nurture credit unions that serve the interest of the community so that even our investment becomes post-corporate activism. I have not yet challenged the students to device alternative investment networks, but it would be exciting to ask the youth to develop intergenerational investment mechanisms so that they could prepare for a sustainable future.

[7] See http://www.socialinvest.org/ and http://www.socialinvestments.com/.

[8] See http://www.desjardins.com/fr/bienvenue.jsp.

CONCLUSION

My globalization courses are the place where I learn from the students. Each student examines her/his involvement in globalization as victim and culprit, and devises actions to counteract corporate domination and neoliberal globalization. From classroom discussion and reports, I am often surprised by the deep insights students develop about themselves and globalization. I have succeeded at least to turn the table around between the student and the instructor. Personal examination of individual involvement in globalization helps the student grasp the macro process of globalization as personally relevant historical and social process. They start seeing themselves relationally and being the connected parts of the macro structure. I hope that they take this world-systemic understanding of the whole and parts beyond my class, or continue to live through the global-personal tension.

I am also impressed by the students' understanding of the complicated and often-contradictory nature of their involvement in globalization. The perspective of victim-culprit forces them to see their being not as singular unitary existence but as a collection of fragmented and contradictory 'selves.' I see myself as a contradictory being. My critical view on the corporation is accompanied by a middle class lifestyle that is quite dependent on goods and services supplied by the corporation. My pension fund is invested in military contractors and oil companies. I am struggling to come to terms with my own multiple and contradictory existence. I think that my pedagogical praxis is my way of pursuing utopystics (Wallerstein 1998; Tamdgidi 2007, forthcoming) by inviting students to form their communities so that they could overcome victim-culprit schizophrenia imposed on them by neoliberal globalization and the corporation.

REFERENCES

Bakan, Joel. (2005). *The Corporation: The Pathological Pursuit of Profit and Power*. NY: Free Press.

Ellwood, Wayne. (2001). *No-Nonsense Guide to Globalization*. Toronto: New International Publications.

Ikeda, Satoshi. (2004). "Japan and the Changing Regime of Accumulation: A World-System Study of Japan's Trajectory from Miracle to Debacle." *Journal of World-System Research*, x, 2, summer, 363-394.

Tamdgidi, Mohammad H., ed. (1997). *"I" in the World-System: Stories from an Odd Sociology Class (Selected Student Writings: Soc. 280Z-Sociology of Knowledge: Mysticism, Science, and Utopia)*. Binghamton, NY: Crumbling Façades Press (revised/reprinted in 2004 by the Okcir Press, Medford, MA).

Tamdgidi, Mohammad H. (2003). "Mysticism and Utopia: Towards the Sociology of Self-Knowledge and Human Architecture (A Study in Marx, Gurdjieff, and Mannheim), Ph.D. Dissertation, SUNY-Binghamton. Ann Arbor: University Microfilms, Inc.

Tamdgidi, Mohammad H. (2004a). "Rethinking Sociology: Self, Knowledge, Practice, and Dialectics in Transitions to Quantum Social Science." *The Discourse of Sociological Practice*, Vol. 6, Issue 1(Spring): 61-81.

Tamdgidi, Mohammad H. (2004b). "Freire Meets Gurdjieff and Rumi: Toward the Pedagogy of the Oppressed and Oppressive Selves." *The Discourse of Sociological Practice*, Vol. 6, Issue 2(Fall): 165-185.

Tamdgidi, Mohammad H. (2005). "Orientalist and Liberating Discourses of East-West Difference: Revisiting Edward Said and the Rubaiyat of Omar Khayyam." *The Discourse of Sociological Practice*, Vol. 7, Issues 1&2(Spring/Fall): 187-201.

Tamdgidi, Mohammad H. (2006a). "Toward a Dialectical Conception of Imperiality: The Transitory (Heuristic) Nature of the Primacy of Analyses of Economies in World-Historical Social Science." *Review*, XXIX, 3, 2006.

Tamdgidi, Mohammad H. (2006b). "Middle Eastern Insights into Anzaldúa's Utopystic and Quantal Socio-logical Imagination: Toward New Agenda" *Human Architecture: Journal of the Sociology of Self-Knowledge*, Vol. IV, Special Issue (Summer): in press.

Tamdgidi, Mohammad H. (2007). *Advancing Utopistics: The Three Component Parts and Errors of Marxism*. Boulder, Colorado: Paradigm Publishers.

Tamdgidi, Mohammad H., (2008). "From Utopistics to Utopystics: Integrative Reflections on Potential Contributions of Mysticism to World-Systems Analyses and Praxes of Historical Alternatives." In *Islam and the Orientalist World-System*, co-edited by Khaldoun Samman and Mazhar Al-Zo'by. Boulder, CO: Paradigm Publishers.

Tamdgidi, Mohammad H. (Forthcoming). "Utopystics and the Asiatic Modes of Liberation: Gurdjieffian Contributions to the Sociological Imaginations of Inner and Global World-Systems." In *Asian Regionality in the Twenty-First Century*, edited by Ganesh Trishur. Boulder, CO: Paradigm Publishers.

Wallerstein, Immanuel. 1998. *Utopistics: Or, Historical Choices of the Twenty-First Century*. New York: New Press.

"I Arrived Late to This Book"

Teaching Sociology with Julie Dash's *Daughters of the Dust*, the 'Novel'

Karen M. Gagne

University of Wisconsin at Platteville

gagnek@uwplatt.edu

Abstract: Cultural productions like Julie Dash's trilogy *Daughters of the Dust* need to play a more central role in the teaching of anti-colonial sociology. Representing the human condition and rethinking the purpose of doing sociological research, these works offer an important corrective to Western social science and historiography. As non-linear Diasporic texts, they require that author and reader alike actively engage in their culture, learn its symbols and participate in the struggle for human freedom. Investigating the linked past, present and future of all Africans, including their persistent resistance and revolt against the more than 500 years of enslavement and colonization, these works represent invaluable tools for social change.

I arrived late to this book. It is more than ten years old and I feel as if I just missed a friend who had waited for me as long as they could, but had to leave.

—"Ysis," September 2004 reviewer of *Daughters of the Dust*, at Amazon.com

I finished the book in less than three days. This book gave me a closure I never knew I needed. In reading this book I was able to give my ancestors a proper burial and set all our souls free.

—Felicity Jackson, October 2006 reviewer of *Daughters of the Dust*, at Barnesandnoble.com

Dr. Karen M. Gagne joins the sociology faculty at the University of Wisconsin at Platteville in August 2008. She recently completed her Ph.D. in sociology at Binghamton University. Her dissertation was entitled "Fiction as a Guerilla Activity: Towards a New Science of the Human." Gagne is the author of "On the Obsolescence of the Disciplines: Frantz Fanon and Sylvia Wynter Propose a New Mode of Being Human," (*Human Architecture: Journal of the Sociology of Self-Knowledge*, Summer 2007); "Fighting Amnesia as a Guerilla Activity: Poetics for a New Mode of Being Human," (*Human Architecture: Journal of the Sociology of Self-Knowledge*, Summer 2006); "Falling in Love with Indians: the Metaphysics of Becoming America" (*CR: The New Centennial Review*, Fall 2003); and co-author of "On Coloniality and Condemnation: A Roundtable Discussion" (*Proud Flesh: The New Afrikan Journal of Culture, Politics, and Consciousness* http://www.proudfleshjournal.com, 2003). The author thanks Greg Thomas for encouraging her to write about *Daughters of the Dust* and for his editorial assistance in writing this article.

I picked up the hardcover edition in a discount bin at my university bookstore in 1998. It was on sale for $5.00. It was a clean copy, in good shape, and had a lovely picture on the jacket. I envisioned I would enjoy reading it when I got around to reading a novel—post-Ph.D. I placed it on the shelf with others of similar classification. I'm glad I didn't wait *that* long.

When I thought I was taking a break from my "sociology" work, the cover of *Daughters of the Dust: A Novel* again caught my eye; so I took it off the shelf and read it. It was hard to read, with some language barriers, but I still couldn't put it down. Each page demanded another. I tried to learn the language and keep the names of the African gods straight. It turns out, reading *Daughters of the Dust* proved to be an unforgettable journey toward rethinking the meaning of doing sociological research in the first place.

Soon after I finished reading the book, I found a copy of the film by the same title. I was so thrilled that I would be able to see some of the scenes portrayed on the screen—especially those Saturday nights at Carrie Mae Johnson's backwoods juke joint, when Amelia and Elizabeth visit the place where Haagar (Amelia's grandmother) came from, and that most unforgettable scene when Lucy Peazant finds the human skull and leg bones with chains still on them while plowing in the field she intends to buy. Little did I know at the time that the book and the film were not one and the same; the book actually begins where the film ends. I didn't find this out until toward the end of the film when I began to recognize the characters from the book. I started over and watched the film and then reread the book.

A few years later when I was teaching a class called "Images of the Family," I assigned all three works by the same title: the film (1991), the novel (1997), and the book about the making of the film entitled *Daughters of the Dust: The Making of an African American Women's Film* (Dash 1992). What a trilogy! Dash makes an invaluable contribution toward the understanding of the lives of Africans in the New World. She makes plain the connection between the Gullah/Geechee people of the Sea Coast Islands and all Africans in the Diaspora, those who moved North and those who live in the South—in the Caribbean and the Americas. In addition to consulting her own family for their histories, Dash did over ten years of research to prepare for this cultural production at the Schomburg Center for Research in Black Culture in Harlem; at the National Archives, the Library of Congress, the Smithsonian, in Washington, DC; and at the Penn Center on St. Helena Island, off the Coast of South Carolina (Dash 1992: 5). Sandra Grayson, in her analysis of the film, argues that when juxtaposed with others such as Haile Gerima's *Sankofa* (1993) and Kasi Lemmons's *Eve's Bayou* (1997) Dash's *Daughters of the Dust* functions as a "continuous narrative about enslaved Africans and their descendents in North America, vignettes that represent experiences spanning from the antebellum period through the 20th century" (Grayson 2000: 2). Grounded in historical research, Dash uses spoken word and cultural symbols to present factual information that is "infused with an imaginative construction creating a mythic memory" (2).

However, while the film *Daughters of the Dust* has received considerable attention in academic circles,[1] the novel has received *no* attention at all. Before it becomes lost to the dusty shelves of used bookstores (for at Barnesandnoble.com it is only available used) in spite of the positive praise by a handful of brief Barnes and Noble and Amazon online reviewers, I would like to call attention to the powerful role this book

[1] There are fifty-seven articles that analyze the film, as listed in "African American Filmmakers, African American Films: A Bibliography of Materials in the UC Berkeley Library" (Handman).

can play in teaching sociology, representing the human condition, in healing a community, and in "making the struggle irresistible"—as Toni Cade Bambara suggests it is the job of a writer to do.

By way of *Daughters of the Dust,* Dash makes a critical intervention against the Western social sciences and historiography. However, the novel is not *only* a continuation of the story of the Peazants, the family portrayed in the film about to move north to the mainland. It does indeed pick up the story where the film leaves off and give insight into the characters we met in the film. *Daughters of the Dust,* the novel, is further a powerful cultural history all by itself. Through its "nonlinear, multilayered unfolding" (1992: xiii), the story connects African gods (Ogun, Elegba, Oya, Osun, and Yemoja) and African ancestors to those Africans living in the present. It shows the importance of storytelling, regardless of the form or method one uses to tell stories, in carrying out the struggle of African peoples, as each new generation is connected to each other and to their ancestors by keeping collective memories alive.

Dash, like other Black women writers before her,[2] disrupts and rewrites "history." She offers a counter-narrative to the conventional and canonical stories of hegemonic discourse. Mae Gwendolyn Henderson argues that Black women writers have encoded oppression as a discursive dilemma. In other words, these writers have consistently raised the issue of their relationship to power and discourse, and negotiate this dilemma by "entering into both 'testimonial' and 'competitive' discourses with their readers" (Henderson, cited in Brondum 154). Regarding the film *Daughters of the Dust,* Lene Brondum argues that Dash speaks to Black women in an "intimate voice" by focusing on the stories of the Peazant women and by "mythologizing their past and their present." Brondum adds that this is not merely a personal discussion. Dash "also enters into a public and political dialogue with the hegemonic systems" by subverting the traditional standards of historiography (154). In fact, Dash does two things. Firstly, she makes a film entirely about Black women's lives. This act alone blows up "History." Dash, in other words, rewrites the whole story (of "History") by writing Black women into the "canon" (154). Secondly, Dash writes a nonlinear "composite history" of the Sea Islands by representing this history in a nonchronological manner—in which historical facts are "selected for representation according to whether they effectively reveal something about the 'essence' of a culture or a historical period, rather than according to 'objective,' historical demands for chronological accuracy" (155).

Writing about Paule Marshall's *Praisesong for the Widow* (1983), Abena Busia (1989) argues that it is crucial that we not look at such texts by Black women writers as "abstractions." They must be understood as concrete aspects of the lives of Africans, "where our meaning—our story—becomes what we can read about ourselves and our lives" (198). Reading texts like these is an active exercise in self-identification whereby one recognizes both life experiences and historic transformations that lead to celebration and a coming together "attainable only through an understanding and acceptance of the demands of the past, which are transformed into a gift for the future" (98). The impact of *Praisesong for the Widow,* for example, on Busia herself was a profound one:

> In many respects, the reason I am so powerfully attached to this novel is because when reading it for the first time, I experienced many powerful moments of recognition and could therefore feel the lyrical

---

[2] Zora Neale Hurston, Paule Marshall, Erna Brodber, Toni Morrison, Audre Lorde, and Toni Cade Bambara are but a few.

> joy Avatara feels when she recognizes again elements from out of her childhood, which come also out of mine. The moment I first read the account of the Ring Shout, for instance, I recognized the outlines of the dance: I, like, Avey, had been a visitor to my home, watching on the sidelines while only the old people took part. The descriptions of the durbar in Ghana were so familiar, as were Lebert Joseph's ceremonies for absent children. Those supplications to the ancestors, performed for wandering children such as myself who seem to take a long time finding their way home again, have been done to ensure my safe return or to celebrate my safe arrival. My personal reading of this work then was in the first instance an "autobiographical" reading, which reminded me that all exiles can be transformed into homecomings. (240, n. 5)

Avatara's private history of material acquisition and cultural dispossession is also the history of all Africans as a result of the enslavement-colonization process for 500 years.

Diaspora texts always begin and end with Africa. These works carry out the symbolic reversal of that Middle Passage for the writer *and the reader*, as well as for the people in the stories. Storytelling, Busia argues, "incorporates a wealth of nonwritten cultural forms that must be reinvested with meaning." Re-validation of these cultural forms, for the individual and for the collective, happens within the text and is part of the lifelong process of "African Reconnection" (197). The writer and the reader must engage in the same active process, as the protagonists, of recognizing and reassembling cultural signs; they must also actively take this journey and thereby reverse the Middle Passage in effect (196).

Through the journey of her protagonist Avatara, Marshall takes her readers on a journey both backwards and forward, toward a new "self-recognition and healing." This is, according to Busia, vital for the reconnecting of the scattered peoples of African trauma that is "constantly repeated anew in the lives of each new generation" (197). The trauma is reproduced by the conditions under which the people have to live, conditions which dictate that they sever from their African cultural roots. However, Marshall and Busia maintain that this is a sacrificial price much too high to pay.[3] To understand the magnitude of the sacrifice, it is necessary for Marshall's readers to take the journey along with Avey Johnson. Still, Marshall requires first that her readers have some knowledge of "diaspora literacy" (Clark 1990) so that all can see again the "fragments that make up the whole." In other words, readers must have an ability to read a variety of cultural signs of the lives of "Africa's children at home and in the New World" (197).

The journey Avey/Avatara takes in *Praisesong* is the same journey that Amelia has to take in *Daughters of the Dust*. In both

[3] This is suggested by both James Baldwin (1985: xiii) and Jacques Depelchin (2005: 154). Depelchin, citing Baldwin, argues that in order to understand the source of this trauma, one needs to remember. However, there are overwhelming forces against this remembering. Baldwin writes, "It has something to do with the fact that no one wishes to be plunged, head down, into the torrent of what he does not remember and does not wish to remember. It has something to do with the fact that we all came here as candidates for the slaughter of innocents. It has something to do with the fact that all survivors, however they accommodate or fail to remember it, bear the inexorable guilt of the survivor. It has something to do, in my own case, with having once been a Black child in a White Country." Depelchin adds, "The pull to conform to the views of the dominant history would be irresistible, but for the fact of having seen the evidence. The witness cannot deny its existence even if everything and everyone has conspired to claim they cannot see it" (Depelchin 154).

cases, this journey begins with a boat crossing, as I shall describe. Those who read these texts must also complete the journey–or "cross over." This is to say, again, Marshall's and Dash's readers must also "reverse the Middle Passage" (Benton 2000), if not literally by boat, and learn to read the cultural signs to become whole again.

Marshall's *Praisesong* inspired Dash's idea for *Daughters of the Dust*. The central story in *Daughters* about Ibo Landing is borrowed from Marshall (Benton 2000: 227; Brondum 1999: 159; Dash 1992: 141; Marshall 1983: 37-39). In giving Ibo Landing the central place in her text on Dawtah Island, Dash puts forth her stance like Marshall "regarding the great American afflictions, amnesia and disconnectedness" (Dash 1992, xi-xiii; Bambara 1996, 109). Both Marshall and Dash link the histories of all Africans in the Diaspora and their persistent resistance and revolt against their capture to that moment when the Ibos saw what was ahead of them when they got off the slave ships (and the many while still on the ships). Rejecting what would be in store for them on those Carolina Sea Islands, they headed back toward Africa. The name "Ibo Landing" in the film and novel recalls stories still told on these islands *and* in the Caribbean. Toni Cade Bambara notes that this place, the site for the film, is called "the secret isle" because the land is both bloody and blessed:

> A port of entry for the European slaving ships, the Carolina Sea Islands (Port Royal County) were where captured Africans were "seasoned" for servitude. Even after the trade was outlawed, traffickers used the dense and marshy area to hide forbidden cargo. But the difficult terrain was also a haven for both self-emancipated Africans and indigenous peoples, just as the Florida Everglades and the Louisiana bayous were for the Seminoles and Africans, and for the Filipinos conscripted by the French to fight proxy wars (French and Indian wars). (Bambara, 1996: 94-95)

The last ship on the U.S. records to have "imported" Africans to the United States is the slave ship *The Wanderer*. This ship transported four hundred Africans to Jekyll Island, Georgia, in 1858, fifty-one years after the importation of Africans as slaves had been banned in 1807 (Brondum 1999: 153). The story of this ship is told to the young Peazants (and Dash's audience) by Nana, Bilal, and Eula in the film. Even Viola and Yellow Mary dismiss Mr. Snead's "official history" which counters the existence of this "too late date" ship. This important story, though, is retold to a new generation (and again to Dash's readers) in the book. Bilal[4] Muhammad arrived on this ship with the same Ibo peoples from whom the Landing received its name. In the novel, there is another ship called *The Sorcerer*. Evidence of *this* ship is found engraved on iron shackles that are accidentally dug up, still attached to human skulls and leg bones. The ship, Miz Emma Julia tells us, got its name because "it run so fast folks not sure dey see it. It come an go jus like dat!" (Dash 1997: 234). This ship was constantly escaping the watch of the Yankees who, just before the War, in the waters between Beaufort and Charleston, looked for illegal transporting of new captives (234).

The film *Daughters of the Dust* represents a powerful and critical commentary on ethnographic film. This is suggested by the characters of Viola Peazant, a modern Jesus-saved woman who previously escaped her "heathenness" by leaving "this God-forsaken Island," and her hired photographer Mr. Snead (hooks 1992). They come to the Island to record the movement

[4] In the novel, Muhammad goes by the name Paymore.

of Viola's family from Island toward the "progress" that awaits them in the cities up North. Dash's commentary is palpable in the novel in the person of Amelia Varnes. The story of Amelia is a biting critique of the fields of ethnographic film and anthropology, their rise in the university, and the funding of these "sciences" by "benefactors" in the private sector who have interests in "developing" the land for their own profit.

Amelia is one of the first Black female anthropologists, studying at Brooklyn College. A graduate student in 1926, she would have been a contemporary of Zora Neale Hurston. Amelia has received word that the Foundation of Brooklyn College has provided her with a grant to complete her field research, which is to study "her people" of the Carolina Sea Coast Islands. Amelia's benefactor wants to remain anonymous, but it is revealed to be someone who made a fortune dealing in land, railroads and other "investments." Her advisor, Professor Colby adds, "Actually I suspect he has an interest in the land down there; he's been talking about another summer home" (31).

Amelia is sent equipped with a 16mm, hand-cranked Akeley camera. In exchange for her testing out the old camera, her benefactor will pay for all the film and processing. The images, her professor suggests, will "add to the weight of her research."

Six months later, when Amelia finally returns to New York to complete her thesis, she is a completely different person. Fully intending to remain "objective" and "outside" her subjects of study when she began the project, Amelia finds out that this is not only impossible, but increasingly undesirable. She sees the need to protect her family from encroachment of the kind of researchers of the type that she was—especially those sponsored by land-hungry "philanthropists." As she attempts to edit the film footage taken by her young cousin Ben she is reluctant to share these images of family and friends with strangers:

> Contrary to everything that she had started out to do, she did not want their culture, their ways, how they talked to each other described in academic terms that would leave out the spirit and overlook the common heritage and stories that bound them so closely. She had winced when she read her thesis proposal with its distant perspective and cold language that attempted to describe people whose lives were filled with wonder, vibrancy, and an uncomplicated honesty. Life for them was hard, a struggle from one moment to the next, but it was filled with the natural pleasures that she was now just beginning to understand and appreciate. (283-4)

She carefully edits and provides only glimpses of the scenery and non-intrusive or more "impersonal" moments—although she is sure this is not going to sit well with her committee, or her benefactor. She does show a film at the end of a grueling defense meeting, where she is grilled about her thesis.

One committee member who has purportedly "completed the formative work on Negro communities, and…is not familiar with these people," poses the question, "What is it about these colored people that makes them different from any other colored people?" Amelia restates what she has been taught to be the requirements of a "culture," including a shared history. This interjection produces an uproar from Professor Anderson, who states, "They are simply stories. There is little there of historic merit. They have long since lost their ties to where they came from!" Professor Colby interjects with "Surely you're not going to dismiss the importance of oral tradition!" Anderson snaps, "Of course not!"

But, he continues,

> I'm just not sure that what these people have to say amounts to a history. Quite frankly, what you have described seems to me to be some kind of crude imitation of the old plantation culture, a patois derived from broken English and the flurried accents of Irish and Huguenot planters, all of which is the obvious product of European influence. Surely you're not suggesting that in the land of secession and large plantations, there can be anything else? (289)

Amelia struggles to maintain her composure. Professor Rehnquist pipes in:

> Mere vestiges, if anything, Miss Varnes. Not what would be acceptable in a scientific sense. Why, it is an established fact accepted by your own colored sociologists, that the colored people, bereft of a history, culture, traditions, have long since adopted the ways of the dominant culture, shedding all ties with the African continent. (1997 289-90)

This exchange goes back and forth as Amelia struggled to defend her position, with Professor Colby's support, including her assertion that the people maintained a distinct *language* and not merely a *dialect* or simple form of English. When the others were not able to convince Colby or Amelia, they turned to more stylistic problems with her thesis:

"The writing...," Professor Anderson insisted. "There seems to be an inordinate amount of adjectives and adverbs. It's not as clean of personal...ah...feelings as it could be." It is at this point that Amelia shows them her film:

> She had just completed it the night before, carefully, protectively choosing what she wanted to share with them and what she would share with her family when she and her mother returned. From the footage of Elizabeth and the children sprinting down the dunes, to the men cleaning the fish on the docks, to the churchgoers shouting in church, there was only the sound of the projector running as the images flashed on the screen. She had put together a pictorial story that ran through the lives of the people, never intruding, but showing their everyday realities and capturing their grace. She had chosen the footprints on the beach as the final sequence. (291)

Her committee is silenced. She passes with flying colors. A few days later, her advisor, breathless with exciting news, tells Amelia that she has been asked to present her paper at the yearly conference of the American Anthropological Society, which would guarantee that her work be published in "at least two of the most preeminent journals in the country." He adds, "You have no idea of what you have started!" The committee was thrilled to hear about the film.

It is at this point that Amelia truly shows what she has learned in the last six months when she most emphatically rejects the path of fame and upward mobility that would be hers. She declines the invitation, first with "I can't present it" and then firmly, "I won't present it." When she is asked why she says:

> "I'm scared...scared about the kind of change this is going to bring... What's going to happen to my family and my friends when everyone starts flooding in, asking about the details of their lives that are just natural to them? I know it's selfish,

> but they are a proud people, an independent kind of people. They've survived slavery, crop disasters, hurricanes, and floods, and I don't want them to lose what brought them through…The first class I took with you, you taught us about respecting what we were to study…but you also taught us about change and how as anthropologists we were not to introduce change. I can't help but worry that I've already brought change, and I'm afraid that if this paper gets published, it will bring about the worst kind of change." (294)

Dash's biting critique of this kind of research is illustrated in the passage that follows:

> "Miss Varnes, change is going to come to the Island," Professor Colby warned her. "If not through you, it will be someone else. You can't stop that. Why, your benefactor is sitting down with planners as we speak talking about what he's going to build down there in the next few years."
>
> "I know I can't stop change, but Lord knows, I don't have to be an agent of it."
>
> "But what of all your work?" Professor Colby argued. "It's wonderful! You deserve the acclaim it will bring."
>
> Amelia considered this and then smiled slightly. "I didn't write it for that, and I don't need that! I got so much more out of it than that."
>
> "It's a vital piece of work." He pressed her again. "It needs to be shared with others."
>
> Amelia stood to leave and extended her hand. "I understand what you're saying, but I don't want it published." They gazed at each other until he nodded with respect and understanding, then handed her the thesis.
>
> "You've done excellent work. When you're ready to take it further, you know how to contact me."
>
> She looked around the room and then turned to open the door. Professor Colby stopped her. "Will we have the pleasure of your company at graduation?"
>
> When she turned to him, he was startled at the bright smile that lit her face. "No, I'm going back to the Island. My mother and I, we're going home." (294-5)

What happens on her six-month field trip to this gifted anthropologist, between the time she arrives at the train station in Charleston, South Carolina—dressed in a "city-smart suit and fashionably bobbed hair" and when she returns to New York to finish her thesis? What would make Amelia give up such a promising career, only to return with her mother to that Island for good? As I suggested above, her real journey, just like Avey Johnson in *Praisesong for the Widow* (a journey that we, the readers, must also take), requires a purging and a cleansing, assisted by a boat ride to the Island.

Amelia manages to keep herself "together," through the choppy waters, suppressing the nausea and dizziness that swelled in her stomach (69). The water gets even rougher, the closer they get to their destination. Her cousin Elizabeth Peazant helps her get through:

> Amelia could sense the change in the mood as the men fought the current to get the boat close enough to the pier. They stopped singing and listened carefully for Willis George's commands as they pushed and waited, using the strength of the current to move closer. Amelia gasped when a wave slapped the side of the boat, spraying everyone with water. Elizabeth threw her arm around her and held her close as the men heaved together, sending the boat directly into the current. Fear shot through Amelia as she felt the current catch the boat and push it just past the pier. A shout from Willis George and the men heaved again, popping the boat out of the current. As they reached the calmer waters, the men grinned at each other, their white teeth flashing in the dark and at Willis George, nodding in satisfaction. They then rowed for pier at Ibo Landing. (71)

As she reaches the edge of the pier, Amelia "bent double." Elizabeth holds her head until she stops heaving and then she helps Amelia to sit. Amelia, like Avey Johnson in *Praisesong for the Widow,* sleeps for two days straight, rises on the third day. Eli Peazant notices this and says, "You sure she all right. She aint moved since her lay down. Her sleep like she dead." Eula Peazant understands, "She jus dat bone weary. Dat long trip and rough crossin. From what [Amelia's mother] Myown wrote, I spect dere was no peace to be had in dat house when her made known what her was gonna do" (72).

Getting the people of the island to tell their stories requires that Amelia shed her "baggage," layer by layer. Eliciting the trust of her relatives and their neighbors, especially when lugging the huge camera around, is a big feat. She eagerly gives over the filming to her eager cousin Ben, who right away has a knack and a desire for the job. Still, not all of her relatives are so thrilled with her presence. This includes her cousin Lucy, Elizabeth's younger sister. Lucy dismisses Amelia immediately and doesn't speak a word to her, even though the two share a bed during Amelia's stay.

At the beginning of the film, there is a magnificent shot of a young Nana Peazant's hands, as she holds some of the soil. A still photograph of this shot is reproduced in book on the making of the film. The elder Nana, throughout the film, reminds her departing family that the ancestors are in this soil; to leave it is to leave them behind. She begs them to stay, and if they cannot stay, to make sure to take a part of their ancestors with them, to keep the connection. She warns against leaving this soil and tells of those that came before them. In the novel, Lucy is the one woman who carries that appreciation for the soil with her. She works the land, speaks to it, and loves it everyday. Not even she, however, is prepared for the *literal* presence of those ancestors in that soil.

What happens to Lucy in the story is one of the most self-transforming moments, so much so it is to be anticipated with much trepidation each time the story is revisited or read. Lucy Peazant is the second daughter to Eli and Eula Peazant.[5] She is someone who from early childhood has been committed to this life and community on the Island; she has been an unflinching resister of the "new ways" and never considers such ways as a desirable alternative to their life. Her father had attempted to send her to the mainland for schooling, but

[5] Eli and Eula Peazant, who we get to know in the film, have six children. Elizabeth, the oldest, was the "unborn child" in the film, sent by the ancestors. Elizabeth, who went away for her education, has recently come back to be the Island's only teacher. She lives in her great, great grandmother Nana's house deep in the forest. Then there is Lucy, Ben, twins Henry and James, and Rebecca, the youngest.

after one term she refused to go back. Since then she has been working the soil at the old Wilkerson place, the "piece of land that meant so much to Lucy" (224), which she and her Charlie were saving up to buy. It had once been a large plantation where one "could still see the fire-scorched foundation stones of the Big House on the far edge of the pasture" (224). Trinity Wilkerson was the cook for the whites who owned the plantation and she got her "forty acres" after the War. She took some of the richest soil around. The land around this plot has long since been abandoned. All the Wilkersons except for Ol Trent have since left the place for good and Lucy has not only been working the land, but is saving money to buy it.

The relationship between Lucy and Amelia has remained distant. She has barely spoken a word to her cousin. Still, Lucy agrees to let Amelia help her plow the fields—mostly to give her a real taste of hard work. Amelia is determined not to let Lucy see her falter. Lucy fails to warn Amelia how hard it would be to get their mule Homer to move. Amelia doesn't make it very long. Lucy finally makes her sit down. Amelia sits and watches in amazement Lucy's command of Homer. Amelia turns to daydreaming about her mother and her life back in New York. She is sorry to have left her mother alone to deal with her father and grandmother. She is snapped out of her daydreaming by Lucy's piercing screams.

Lucy begins to run from the field. Amelia figures it is a snake, but Lucy is running in "wild patterns, waving and screaming." Pointing to where Homer stood, Lucy is bent over, clutching her stomach. "It evil. It evil back dere!" (227). Amelia goes to the last spot Lucy has plowed. This is what happened next:

> She stared at the spot, then kneeled down and looked at an object that the plow had uncovered. She felt her heart begin to pound as she brushed the dirt from a pair of rusted shackles, a chain running from them into the ground. Despite her misgivings, she grabbed a jagged-edged stick that lay in the furrow and began to dig. As she continued burrowing into the earth, her eyes filled with tears as she recognized a human leg bone. Dropping the stick, she jerked away and walked around to the front of the plow and saw, at last, what had sent Lucy spinning. A piece of skull was impaled on the front blade; a jawbone with several teeth lay in the scattered earth. She backed away and ran over to Lucy, who was curled on the ground, crying. She pulled Lucy up, slipped around her waist, and half-carried her home. (227)

In total, there are three sets of bones found. Eli states, "Three of dem, tied together like dogs." He adds, "We keep looking tomorrow" (230). Lucy is unable to leave her bed for days afterward. For years Lucy has been set on that buying that meadow. Her sister Elizabeth worries that this discovery will crush her dreams.

The discovery of these human remains brings the folks from all over the Island. Word spreads like wildfire and the men, women, and children—old and young—make their way to the old Wilkerson place. Miz Emma Julia, one of Nana's oldest and closest of friends, arrives on a horse-drawn cart. Upon seeing Eula, she immediately asks, "How Lucy?" Eula replies, "Her in a bad way." Miz Emma Julia says, "She come across dat evil buried dere in dat field! ...It up to we to make it right! Us got to claim our own, take dem back from de buckra!"

"You sound like my Nana," Eula nods. Miz Emma Julia snaps, "Us all of de same spirit!" When Miz Emma Julia is brought close to the bones now wrapped in quilts, she rubs the shackles and sees a faint imprint on the worn iron. She asks Elizabeth to tell her "what it say," and she responds, "The Sorcerer." Miz Emma Julia then tells everyone the story.

The whole Island comes together to prepare the bones for a proper burial. This takes place over several days. They construct an altar and prepare the burial ground. The women have missed this ritual; it disappeared from the days when Nana and the others passed on. Everyone tries to get Lucy to join them but she refuses to move from her bed.

But, Lucy does come to the final ceremony. As they began, Miz Emma Julia orders the bones, wrapped in the quilts, to be brought to her. She lifts her arms and begins to chant so that the gods of "de many peoples of Africa" can hear her call. She asks them to hear the children calling them, the stolen children from "de lands of Ibo, Yoruba, Kissee, Dahomey, Angola, Gambia, Whydah!"

The ancestors answer with a sudden strong gust of wind that parts the clouds and reveals the moon. Everyone feels their presence. Some fall to their knees. Lucy shrinks "as the wind seem[s] to circle her, moving along the edge of the group." Amelia feels the "soft caress of the breeze on the back of her neck, her body relaxing." Elizabeth holds her arms out and "let[s] the wind pass through her body." Then the wind stops as suddenly as it starts.

Lucy is beckoned by Miz Emma Julia to carry out the ceremony, "Come, chile, you brought dem to us. You gotta help dem go home." As if pulled by force, she reluctantly and slowly moves forward. "You a lil gal. You gonna need a helper to send dem home." This is a turning point in the relationship between Lucy and Amelia. Lucy looks to Amelia for this help and Amelia takes her place beside her cousin (240). Here is the passage in full:

> "Take de quilts an open dem up." Lucy took one end and Amelia the other and gently spread the quilt. They moved to the second quilt and opened it, Lucy turning her head from the bones that lay in the quilt. Miz Emma Julia held out her hand. "Spread out de shackles and de chain as dey lay in de ground." Amelia and Lucy picked up the pieces and strung them out best they could. Miz Emma Julia spoke, "Now, spread out de bones so we can see de old ones." Lucy looked as if she was about to flee. Caught by Miz Emma Julia's gaze, she began to sob. Eula pressed her hands to her mouth to keep from crying out to her daughter. Eli put his arm around Eula, holding her close.
>
> "Go on, gal. Let dem saltwater tears wash down and cleanse dem. Aint nobody cry for dem for years. Dey lay out in dat field for nobody know how long. Dat why dat field do rich wit de earth. Our elders give it dey life blood. Dey give to we what was took from dem." She gestured for Amelia to spread the bones. Amelia leaned down, hands trembling, and reached for a long piece of bone. When her shaking fingers touched it, she felt a jolt and pulled back. Miz Emma Julia nodded encouragingly. "Dey reachin out to you. Take what dey got to offer." Amelia wanted to refuse, then saw the fear in Lucy's eyes. Heart pounding and breathing heavily, she picked up the bone. She thought it was her imagination, the tingling that started in her fingers and began to spread up her arm. But as it spread over her body, she felt it in every pore as sweat broke out and her skin seemed to flush with blood. She felt the waves of fear, pain, and despair wash over her. The others flinched and moaned with her, while Lucy hid

> her face. Amelia slowly lowered her bone, her arms trembling, her clothes soaking. Miz Emma Julia placed Amelia's coat on her shoulders and stroked her head.
>
> The old woman then reached down and picked up a skull, crossing to Lucy, who cried out and shrank from her. Relentlessly, she grabbed Lucy's hand, struggling with her, and thrust the skull into it. She hissed at her, "Feel deir pain, gal! Feel deir hurt! Only when you feel de pain do de healin begin!
>
> Lucy shook as she stared at the skull. Just as she was about to let it roll off her palm, her head snapped back as the force hit her. She jerked as if her body were receiving the blows, her head lolling and her arms flinging in every direction. Painfully she closed her fingers around the skull and struggled to bring it closer to her. Only when she clasped it to her chest did the force seem to take pity on her. Her head hung low; her body was limp. (240-1)

In words similar to Nana Peazant in the film, Miz Emma Julia reminds everyone,

> "It take a strong people, snatch from de cradle, de wood, de village, put on de boat, and took cross de big water to land dey never know...It take a strong people to keep dey all about dem, to hold on de old ways, to keep de lies true, to know who dey be! It take a strong people to work from day clean to day over to clear de lan, build de house, plow de field, make de indigo, sow de rice, pick de cotton, all to de good of de buckra. Some of we forget how strong dem people was, us look past de old ways, put aside what dey was telling we bout de right way to live. An now dey come back to we, de ancients who seed dis earth wit deir tears, sweat, an blood!" (241-2)

When Miz Emma Julia finishes speaking she holds up a skull and a large piece of bone. Her whole body shakes as the forces move through her. Those nearby duck to avoid the spikes of electricity that emit from her body. They cry out in fear, but Miz Emma Julia holds her head back and laughs. She then says, "Come, chilren, dey telling you to rise up! Free de souls! ... We done took de pain an de sadness! All dat lef is de healin!" (242)

Who among Dash's readers cannot join in this ceremony? Be right there to feel the tingling in the fingers that spread up the arm, to feel the response of those called upon by Miz Emma Julia, and feel the fear, pain and despair when Lucy holds the skull and receives full force of the blows? Who cannot feel the healing begin, as everyone rises and begins to dance in celebration around the quilt that holds the bones?

The Peazants and their neighbors are joined in this story and in this journey through their lives' unbelievable moments. Readers will meet Elizabeth, Eula and Eli, Carrie Mae, Toady, and Yellow Mary (her whole story will be told), Iona and Julien Last Child, and their children. We will meet Willis George, Sallie Lee, and Sugarnun. Readers will understand why Haagar (from the film), Amelia's grandmother, was in such a hurry to leave where she came from. Finally, readers will learn why Ol Trent spends his days walking from one end of the Island to the other, combing the place for treasures and yelling out bible passages.

The rest of the story should be saved for each new reader. Hopefully there will be enough interest to generate a new printing. This will happen when texts like

Dash's are read not only as "novels" but as the cultural and social ("*sociological*") productions that they are. The trilogy of *Daughters of the Dust*—the film text, the "making" of the film text, and the final written "next generation" text—are *all* to be read as part of a continuous telling about the reality of enslaved Africans and their descendants in North America.

Dash has provided some of the most powerful works I have encountered, and has made me change my conception of the fields of "history" and "sociology," for example. Therefore, I find it a shame that we will all have to scramble for the remaining used and library copies in order to provide this resource to a new generation of cultural workers/readers, until the book is back in print.

REFERENCES

Baldwin, James. 1985. *The Evidence of Things Not Seen*. New York: Holt, Rinehart and Winston.

Benton, Jacqueline. 2000. "Grace Nichols' *I Is a Long Memoried Woman* and Julie Jash's *Daughters of the Dust*: Reversing the Middle Passage." Pp. 221-232 in *Black Women Writiers Across Cultures: Analysis of Their Contributions*, edited by Valentine Udoh James. Lanham, MD: International Scholars Publications.

Brondum, Lene. 1999. "'The Persistence of Tradition': The Retelling of Sea Islands Culture in Works by Julie Dash, Gloria Naylor, and Paule Marshall." Pp. 153-163 in *Black Imagination and the Middle Passage*, edited by Maria Diedrich, Henry Louis Gates, Jr., and Carl Pedersen. New York: Oxford University Press.

Busia, Abena P. 1989. "What is Your Nation? Reconnecting Africa and Her Diaspora through Paula Marshall's *Praisesong for a Widow*." Pp. 196-211, 239-240 in *Changing Our Own Words: Essays on Criticism, Theory, and Writing by Black Women*, edited by Cheryl A. Wall. New Brunswick: Rutgers University Press.

Dash, Julie. 1991. *Daughters of the Dust*.

Dash, Julie. 1992. *Daughters of the Dust: The Making of an African American Women's Film*. New York: The New Press.

Dash, Julie. 1997. *Daughters of the Dust*. New York: Penguin Group.

Depelchin, Jacques. 2005. *Silences in African History: Between the Syndromes of Discovery and Abolition*. Dar es Salaam: Mkuki Na Nyota Publishers.

Gerima, Haile. 1993. *Sankofa*.

Grayson, Sandra M. 2000. *Symbolizing the Past: Reading Sankofa, Daughters of the Dust, & Eve's Bayou as Histories*. Lanham, MD: University Press of America.

Handman, Gary. 2007. African American Filmmakers, African American Films: A Bibliography of Materials in the UC Berkeley Library in Media Resources Center. http:www.lib.berkeley.edu/MRC/africanambib2.html

hooks, bell. 1992. "Dialogue Between bell hooks and Julie Dash." Pp. 27-67 in *Daughters of the Dust: The Making of an African American Woman's Film*, edited by Julie Dash. New York: The New Press.

Lemmons, Kasi. 1997. "Eve's Bayou."

Marshall, Paule. 1983. *Praisesong for the Widow*. New York: G. P. Putnam's Sons.

"it's just a dream, just a dream"

The WWII Japanese-American Internment in the U.S. in a Sociological Imagination Class Exercise

Thomas J. Mason, Kathleen M. Powers & Emmett Schaefer

University of Massachusetts Boston

thomas.mason001@umb.edu • kpowers621@aol.com • emmett.schaefer@umb.edu

Abstract: This is an account of a "Race and Power in the US" course exercise conducted at UMass Boston. In a unit on the Japanese-American internment during WWII and the movement for reparations in the 1970s and 1980s that culminated in 1988 when President Reagan signed the Civil Liberties Act into law, the students are asked to put themselves in the shoes of a Japanese-American and to write a letter to a family member. The purpose of the assignment is to help foster in the students sensitivity (to paraphrase C. Wright Mills) to the interplay between socio-historical contexts and the individuals responding to them. The article includes a summary of the sociological imagination exercise as reported by the instructor, and two imaginative letters written by students.

Emmett Schaefer, Lecturer of Sociology, UMass Boston:

In a course I teach at UMass Boston entitled, "Race and Power in the US," we do a unit on the Japanese-American internment during WWII and the movement for reparations in the 1970s and 1980s that culminated in 1988 when President Reagan signed the Civil Liberties Act into law. In the second of two assignments for this unit I ask the students to put themselves in the shoes of a Japanese-American and to write a letter to a family member. The purpose of the assignment is to help foster in the students sensitivity to (to paraphrase C. Wright Mills) the interplay between socio-historical contexts and the individuals responding to them. The assignment reads as follows:

Discussion: After WWII and following the closing of the Japanese-American concentration camps, a heavy silence descended concerning every aspect of this experience. The silence was as deafening inside the Japanese-American community as outside. A combination of factors resulted in renewed discussion of the experience within Japanese-American families and the community in the late-1960s and 1970s. These discussions, and the reparations movement that emerged out of them, were accompanied by

Undergraduate students Thomas J. Mason (Sociology) and Kathleen M. Powers (English) enrolled in a Soc. 211G course on "Race and Power In the U.S.." conducted by Dr. Emmett Schaefer in Fall 2005 and Fall 2007 at UMass Boston respectively. Schaefer is a Lecturer of Sociology at UMass Boston whose research interests include Race and Ethnic Relations, Men's Studies, and Gender Studies. Schaefer is a co-editor (with Cooper Thompson and Harry Brod) of *White Men Challenging Racism: 35 Personal Stories* (Duke University Press, 2003).

intense emotions within Japanese-American families and the community. In this paper you will explore some of these emotions as a way of grasping the interplay between socio-historical contexts and the individuals responding to them.

Assignment: The paper is a role play. You are to take the part of a family member in a Japanese-American family and write a letter to another family member. You need to make three choices to get started. First, are you an Issei, a Nisei or a Sansei [first, second or third generation Japanese immigrant to the US]? Second, are you writing to an Issei, Nisei or Sansei? Third, when are you writing? I want you to pick a time beginning in the late-1960s, so that you can connect the "present" with the past, but it can be during the early years of the Asian-American movement (late-1960s), during the spread of the Reparations' Movement in the 1970s, or after the Wartime Relocation Commission hearings in the early-1980s.

Pick a time and a perspective that you think you have some understanding of, put your self in that perspective, and address another family member. The letter should address aspects of the internment experience, as well as "contemporary" revelations and experiences. For example, you could choose to be a Nisei parent explaining to a Sansei daughter something about the internment, but mostly about why you never talked about it, and why you're opposed to having it aired publicly now.

Be sure to review all relevant course material.

Criteria for evaluation:

1. Level of sophistication of understanding you demonstrate for the tensions within Japanese-American families and the community accumulating over the course of the internment experience and developing during the post-war decades.
2. Level of sophistication you demonstrate for the connections between these tensions and the socio-historical contexts within which they existed.
3. Your ability to convincingly put yourself in another's shoes and take on that person's perspective.

CREATED BY THOMAS MASON, UNDERGRADUATE STUDENT, UMASS BOSTON:

Dear Brother,

It has been years since we last saw each other, and for that matter even talked. It seems so odd to be writing this letter. To think after a little more than twenty years of short ill-fated "Hello's" and "How are you's?" I have finally built up the strength to contact you. Don't take my words wrong brother, in no way am I striking out at you for not contacting me. If anything, I see it as my fault. I'm the older brother and I should have been the one to explain. For that I am sorry.

I understand the tension that grew between us and quickly separated us in that cold January night back in 1943. Not only separated us as individuals, but as a family. Mother had nothing after father was sent away to the camp in Granada, CO. I realize you felt that I was leaving both of you behind and that, there was some malcontent I had. Being called a potential threat, yet still offering to fight for America. I was not in any way ashamed to be Japanese, this was not my reasoning at all, and I know at the time I never gave you my reasons. I just told you and mother that I had to go, shut you out for the rest of my short time in the barracks and then, I was gone.

My regrets are tremendous, and there have been many nights when I sit up and think of my actions and if I actually did the right thing. Joining the 442nd division was not a betrayal of the family in my eyes, and I know if father had been there he would have been taken aback by the thought of me joining. The fact is I did not agree to join the military service for THEM. But yet, I did it for you and mother. Understanding that you felt I had been selfish, I am sorry. Yet you need to understand, that in no way was this a selfish fleeing from our captive world behind the fences. Nor was I attempting to separate myself from you and mother. I did it in the hopes that they would see that we were not of danger, hoping you and mom would be released faster. I thought about it long and hard, and in theory I wondered whether I had made the right choice. Fight for our country, the one that forced us to live in that desolate place, in hopes of early release.

As you know my theory was very wrong, and the loss of mom to the flu that long winter was very painful. I know you feel angry at me for leaving, and I understand your resentment. You being the younger one of us, I can understand how hard it was to first to be separated from our father, then face the death of our mother 8 months after I had left. It saddened me to hear about her death from you, and even though I wrote back and explained how wary it had made me, there was nothing I could do to escape my decision. I had been drafted, and couldn't return "home" till my leave was over.

It hurt to read the letter you wrote, saying that you were ashamed to be my brother. I want you to know now that I was a little ashamed too. Not for the fact that I had done what I felt was going to help you and mom out in the long run, but because I had let you down.

There is just so much to talk about, and it bothers me most of all that suddenly hearing my kids ask "Dad, were you at the camps?", was what finally drove me to write to you. Not that I didn't want to reach you in some way, I just never knew what to say. Above all things I felt that most of what happened to our family, was better off in the past. If that was true I guess I wouldn't be stuck up at night thinking about your life, and how you are doing.

Speaking of which, how is Danny? He must be getting big, around ten or eleven? Wow, its pathetic that I don't even know my own nephew's age. Also scary is how time flies by. Both Susan and Michael are getting up there. Susan will be going into college next year. She's a real smart girl, much smarter than her old man (although that wouldn't take much.)

Anyway, both of the kids keep asking me about the camps. I have no idea what to tell them. So what do I say? Nothing. Not a thing. I wasn't there nearly as long as you, because of the army, but I know how it was when I was there and I'm guessing it didn't change much. It's hard enough talking to you through this letter, let alone talk to my kids about something I feel they could never fully understand. My wife and I avoid talks about the camps if we can, but sometimes they manage to get a little out of us.

How could they understand why we don't talk? How could they even come to grasp the fact that those camps essentially ruined our foundation as a family and split us up? It just seems so petty to tell them about it. I simply tell them never to take anything for granted, with a forced smile.

Have your kids asked about what it was like in the camps? Surely, Danny is too young to ask yet, but have you shared anything with him about the "experience?" I remember when you were just a kid, before we went into the camps. You used to beg me to go play football with you in the backyard, and I'd always tell you to "get lost." You were such a pest!

The camps. They were like broken homes. It disgusts me to even think about them now. It hurts to occasionally hear your voice on the phone, to hear that angst in your voice, talking to me as if I am a stranger. We lost our old house, and everything in it, all for the "protection" of this country. Never did I once think that we would grow so far apart, that this country would divide us. That isn't to say I can not stand this country, and I'm sure the same goes for you, because it is all I know. It's all we've ever known.

I knew it was never mom's and dad's fault we went to the camps, I knew it was inevitable. At the time we were Japanese. The enemy. Still I kept hoping for the war to end. Everyday on our missions you and mom's well being is all I thought about. But I felt regret when I continued to receive letters from you; knowing my attempt to help out proved futile.

The war was ravenous on the 442nd division, and we suffered many losses. I remember training was rigorous and I got called far worse than a "Jap." Drill Sergeants didn't like the thought of us having our own division, and it showed. It showed by the way they talked to us, treated us, pushed us around and by the way still some retorted to us as nothing more than foreign traitors.

To think that even after you serve your country, all someone can say is "You were one of the good ones." I always used to nod my head, give a firm "Yes Sir," and continue on my way. All I could think about was all the other "good ones" we had lived with at camp. But they were never good, you weren't good unless you fought for the country that took your life away. I guess being American just wasn't good enough?

I remember the camps, even though I left earlier than you, I still do. The awful food, the hot days, and colds nights. I remember mom trying to keep us happy (and probably herself), by saying father was ok and that soon enough I and you would be arguing at home again. It's odd enough but some of the best memories I have of you are those of bickering. Any memory is good now, especially since there has been no room for new ones.

I have nightmares all the time, about agents storming into the house and taking me away. It's unsettling waking up almost every night to that. Just keep telling myself "it's just a dream, just a dream." and slowly fall back asleep. Then again being put into the camps seemed like a big enough dream as it was. Do you remember those trains they took us there in? Jesus, did they smell awful by the end. It's funny but I was somehow happy to see desert when it stopped. Anything was better than that horrid odor.

This is so strange to be mentioning any of this to you; haven't been so open with these things, ever, not even with my wife. My thoughts feel so jumbled and I have so many mixed emotions. Was all of what they did to us in vain? Where we really that dangerous as a group of people? If so, why did they bother to release us at all? Also why the hell did they allow us to join their army?

There is a look inside my head. It scares me, and has me biting at my nails every time. That is why I avoid confrontation with my children. They get away with so much compared to what mom and dad let us get away with. This family isn't breaking up, I simply won't allow it. Just avoid the talks of the camps, and avoid as much confrontation as possible.

You're my brother and I will always love you, even if you still resent me. If you have read this far, I know you care a little. Why? Well, because these thoughts were probably just as hard for you to read as they were for me to write about. Our family had no chance in that camp, in those situations. We were split in months. You were too young to go to war, I was just of age. You and mom told me not to go, I had other thoughts. They, THEY had other thoughts.

I miss you, and hope to hear from you soon. It would just be really nice. Regardless, I'm sorry, but in many ways I think they had doomed us from the start. The separations didn't help. Regardless of any animosity, I still care about you and hope your life is going good. This letter was not easy to write. Especially because speaking about all of this for the first time, I have so much more to say. Yet, it has left me mentally and emotionally drained. I just hope you take the time to read it.

Love,

Your Brother

CREATED BY KATHLEEN POWERS, UNDERGRADUATE STUDENT, UMASS BOSTON:

October 24, 1968

Dear Obaasan,

Yes, my dear Nana, I addressed you as my Obaasan, a Japanese nickname for grandmother. Of course I never heard that nickname from you or my mother. I had to ask the old man in the nursing home where my mother works what the word was because I don't know much of anything about being Japanese. She always tells us we are American. It's as if we never had anything to do with being Japanese at all. That is why I am writing you this letter. I want to know more about my heritage, but my mother always puts me off. She says it is foolish to look to the past for answers. She is always pushing me to press forward, to become more educated, and that my education will give me any answers in life that I really need. Well, I do pay attention in school, and I am not getting any answers to my questions, and I thought you could help.

Recently I was helping my mother clean out the attic in our house. I came across a box of pictures. In it was a picture of my mom when she was just a little girl. She was standing next to you. You both look so sad in the picture, kind of lost. I asked where it was taken and she said it was at a camp in Wyoming. I told her I thought she lived in Oregon when she was little, and that she probably looked sad because she had to go to camp so far away. She had tears in her eyes and told me we had done enough cleaning for the day. I sure am glad I never went to camp at a place like that. It did not look like a fun place at all.

I went over to my friend Kathleen Brennan's afterwards and told her and her mother the story of my mom and me in the attic. Her mom said you are a peculiar lot, your family. She said she didn't understand why someone Japanese would want to name their two girls Margaret and Sheila and I thought that was strange because there are a lot of girls with names like that in our neighborhood. She said she was Irish and very aware of her heritage and that there was a lot of good about it and that there was a lot of bad about it and that it was important to be aware of both.

Mrs. Brennan did not go to school very long, but I always get the feeling when she goes on and on about stuff that she is very smart because she gets this excited look in her eyes and I figure what she is saying is very important. She said that she always thought that calling places like my mom stayed at a camp was ludicrous and I said I agreed even though I had no idea what ludicrous meant and I had to go home and look it up in a dictionary after. That only made me more confused when I found out what it meant because I certainly don't want my mom to ever send me to a camp that would be ludicrous. I was going to ask her more about it, but she and my father were so quiet at dinner and looked so sad I decided not to bring it up again.

My sister called from college and I told her the story and she said I should ask my teacher about the camps. Well that was a stupid idea because I asked the teacher about it and she decided to give the class a lesson on Pearl Harbor. I didn't learn anything about the camp at all and afterwards the boys in the class called me bad names and asked me if my grandfather dropped planes on Pearl Harbor for the Japanese. I said he wasn't even in Pearl Harbor for God's sake. He lived on a farm in Oregon and left Japan when he was very young. Then the boys said he was a Jap spy and I said they were stupid, but who knows maybe he was because a lot of the kids in the class won't talk to me now.

My sister came home for a break from college a couple of days later and my mother had a fit. My sister said it was good I was asking questions and that it was never too early to be socially conscious and my mother said that rocking the boat will only lead to trouble. My father had his head in his hands the whole time my mother and sister argued which really surprised me because he usually only has to look at all of us and we know when to stop talking. I knew something big was going on but I just didn't understand it. My sister told my parents she is learning so much about life at some meetings she goes to at school, so much more than she was learning in class. My father got all red when she said that. She talked a lot about some hero she had named Martin Luther King and that he died fighting for all oppressed people, and my father said we are not oppressed, we are nothing like those people, and she said yes we are and there was a lot of yelling and red faces and I cannot believe all of this started because I asked about some camp.

Before my sister went back to school she said she was very proud of me for wanting to know more about my heritage and I guess that means being Japanese because ever since that lecture on Pearl Harbor the boys at school say I am really not American no matter what my parents say. Before she left she gave me a part of a speech that her hero Martin Luther King wrote. She said I wouldn't understand it now, but I should keep it because Mr. King had some very important things to say and someday I will know exactly what it means. She said you and grandpa and mom and dad all know what it means but there isn't any point in talking to you about it because you have all decided to stick your heads in the sand. She gets so crazy sometimes but I get the feeling this time she might be telling the truth because everyone is acting so different to me.

I have always been able to talk to you about anything and I don't feel like waiting to figure out what this says. It is very complicated but I can tell it means a lot. I know how my sister is. She always treats me like a little kid, so I know when she gave me something like this it must be super important. The speech says, "Being a Negro in America means trying to smile when you want to cry. It means trying to hold onto physical life and psychological death. It means the pain of watching your children grow up with clouds of inferiority in their mental skies. It means having their legs cut off, and then being condemned for being a cripple." I know he is talking about being a Negro and my dad says we are not like them, but my sister says we are. My dad always says how educated my sister is, so I think she might be right. What does all of this mean Obaasan?

Love,

Margaret

October 26, 1968

Dearest Margaret,

You are such a bright and inquisitive child! It is hard to believe you are only twelve. Part of me says because you have asked you have a right to know the truth, but then the other part says you are just a child and have a right to be protected from some of the harsh realities of life. Know this Margaret. Despite what your sister in her idealistic way wants you to believe that the world is changing for the better, it is still a white man's world. As you have already guessed, the camp you saw your mother at in the picture is unlike any you may have imagined. We were thrust into a world, your mother, grandfather, and I, simply because we were of Japanese descent, not because we did a single thing wrong. That is the place you saw in the picture. We chose not to tell you about it because it brings back many painful memories, and when we left there we just wanted to put it all behind us and simply just fit in. I owe you now an explanation of it all, but it cannot fit in a letter. I invite you to come to my house next week and I will explain to you as much as a child's heart can understand.

Sincerely,

Your Beloved Obaasan

Exiles and Home

Leila Farsakh

University of Massachusetts Boston

leila.farsakh@umb.edu

Abstract: I have been away from home for over 20 years. But what is home in my case? An Arab woman born to a Palestinian father and an Italian mother, married to a German man and mother to an American daughter, I had traveled a long way. I asked my mother once whether *"Hon casa?"* (Is this home?). I was 2 years old then, and we had just arrived at my grandfather's house in Italy after a long journey from Jordan. I was just starting to talk, but could only do so by mixing the two languages I was born with: "Hon" is "here" in Arabic and "casa" is "home" in Italian. Ever since, I have lived in many houses. I carried many worlds, but kept looking for a home, *the* home. My exile has been layered, a tapestry woven through the course of a diasporic life that could only be completed and reconciled through an embrace of the continual invention of the self. My exile was grounded in the collective sense of exile of the Palestinian people. It was shaped by the unique personal experience of a girl growing amid the reverberation of the 1967 debacle that hit the Arab world. The dialectic between the collective and the personal shaped my alienation as much as my ways to come to term with the person I became.

Exile is predicated on the existence of love for and a real bond with one's native place; the universal truth of exile is not that one has lost that love or home, but that inherent in each is an unexpected, unwelcome loss.

—Edward Said, *Culture and Imperialism* (1993)

"Give me a face for exile," he asked, looking at me as I tried to find refuge in my tears. "Exile has no face," I retorted. "It is sadness, a loss, a nostalgia that cannot be put in words. Exile can only acquire a face once it is overcome!"

I have been away from home for over 20 years. But what is home in my case? An Arab woman born to a Palestinian father and an Italian mother, married to a German man and mother to an American daughter, I had traveled a long way. I asked my mother once whether *"Hon casa?"* (Is this home?). I was 2 years old then, and we had just arrived at my grandfather's house in Italy after a long journey from Jordan. I was just starting to talk, but could only do so by mixing the two languages I was born with: "Hon" is "here" in Arabic and "casa" is "home" in Italian. Ever since, I have lived in many houses. I carried many worlds, but kept looking for a home, *the* home.

Professor Farsakh holds a Ph.D. from the University of London (2003), and an MPhil from the University of Cambridge in the UK (1990). She has published on questions related to Palestinian labor migration, the Oslo Peace Process, and international migration in a wide range of journals including *Middle East Journal, European Journal of Development Research, Journal of Palestine Studies* and *Le Monde Diplomatique.* Her book, *Palestinian Labor Migration to Israel: Labor, Land and Occupation,* was published by Routledge Press in Fall 2005.

ORIGIN OF EXILE

The story of exile started long before. My Palestinian grandmother, who lived in our household up until I was three, was exiled from her native town, Saffuria, north of Nazareth, when she was only 13 years old. She was then being wedded to a man 30 years older than her, and moved to the West Bank in 1926. She was exiled a second time when her town was eradicated by the Hagana (a jewish underground organization) in 1948, with the creation of the state of Israel. She never saw her siblings again, as they became refugees dispersed between Lebanon and Syria. She only reconnected with some of her nephews, who became Israeli Arabs, after 1967. That war reconnected the land, but reaffirmed the exile; her sons had to leave because of the Israeli occupation and many were forbidden from returning to their place of birth.

My grandmother raised four boys and a girl whom she pushed out of the parochialism of village life by insisting that they become enlightened professionals. However, they were forced to be dispersed across the globe, and she suffered for not having them around her, especially in her ailing years. She spent the last years of her life traveling from one country to another to see them, rather than having them visit her and re-establish the home. It was her deepest sorrow that none of her boys witnessed her last breaths, told her farewell, in her, theirs, native land. Only her daughter stood by her.

Yet, for my grandmother, daughters could never be as good, as validating, as boys. Sons, not daughters, gave my grandmother a sense of place, a prominence in the exiled community she came to live among; if her sons had remained she would have perhaps been able to cope better with her exile, been less sad, been comforted with a sense of power, of the status that could compensate, maybe, for the loss of home. Although only her daughter dealt with her daily, never really left her, catered for her needs, comforted her in her pain and sorrow, it was the sons my grandmother longed for. Women might well be those who cater for the home; but men are the ones validating the sense of home, its contours and its content, if not its form.

The exile from Palestine and from womanhood was two exiles that inhabited me, probably ever since I was born. I was in my mother's womb on June 6th, 1967, when the six day war erupted between Israel and its Arab neighbors. My father had wanted me to be a boy, all the more as he became a refugee himself. He was already known by his relatives and friends as Abou-el-Abid, the father of first son, not first daughter. Like many of his third world generation of the 1950s and 1960s, my father had recently returned from being a student in the West, to build his homeland. He returned to work on a pan-Arab engineering irrigation project on the Jordan River. The 1967 war stopped the project, and with that his plan to go home. He saw his relatives come fleeing from Palestine, which they escaped as Israeli troops invaded the West Bank. They became exiled in Kuwait and other places only to be expelled again in 1990. Like them, he too took his family and wandered from place to place, always with the determination to return.

Before I was six years old, I lived in 3 different countries, witnessed 3 wars, and lived in 4 different houses. The 6 days war in 1967 moved me from Irbid, in north Jordan, to Amman, before Black September in 1970 forced us to Kuwait. In 1973, during the Ramadan or Yom Kippur war, we moved to Dubai, the pearl of the Gulf. By then I had three younger siblings, and a mother on the edge of a nervous breakdown, if not fully in it. But my mother did not consider herself exiled, for she was not forced to leave because of a war or dispossession. Exile does not come with freedom, or with the sense of choice. Those who choose to leave their homes cannot be exiled, she argued.

My mother rather willingly chose to leave her native land to marry the man she loved. She departed from Trieste, a city that was a haven for exiles such as James Joyce, Rilke, and others. Trieste was a city that combined a central European tradition within an Italy that was open to many different ethnic groups and cultures. My mother carried with her a curiosity and a tolerance for others that had longed served her in the Arab world in which she lived. She did not feel alienated, she argued, because she had a clear project for her life and found the partner with whom to attain it. Her project for life was to become a mother who would raise independent and responsible children. She found in my father the man who enabled her to achieve her plan. That was at least what she maintained. She also thought that she married him because he would protect her, and complete her, but she soon realized that these were fantasies. That is what she told me when I asked her at age 12 why she married my father. What mattered, though, was her belief that it was her duty to follow her husband and she surely did. She also understood his urge to return to his homeland. As a child my mother saw exiled people, be they Slovians and Italians, longing to return home, to her native Trieste, the city she never left until she was 25 years old. My mother maintained that she did not feel out of place in the different Arab world she came to live in, for she was at peace with herself, with her choices.

And yet, I felt, before I could articulate, that it took time before she found her home, her place, and how ephemeral this sense of created home felt at times. I remember my mother crying, depressed, and complaining about her kids. I remember her having no Arab friend, only one Italian companion, like her married to an Arab. I remember her telling me that I was of help to her in the midst of her loneliness and despair. My mother, however much she lived for her kids and her husband, however much her life was confined to the kitchen and to the role of a taxi driver taking her children from one activity to another, however much she spent years trying and succeeding in understanding the different world of her husband, was also an exile, if only for a while, if only in her own style, in a way that was void of any patriotism or sense of nationalist dispossession that her husband carried with him.

I grew up with these two different exiles. I strove, without always knowing it, to overcome them by reconciling the two worlds that made me. On the one hand, there was the Arab identity, a Palestinian fate and trajectory, a vocal proud but broken narrative entrenched in a rich Arab nationalistic tradition, angry at an European colonialism that was not only the cause of displacement but also of negation. On the other hand, there was my mother, an exile from an admired but suspicious West, occulted but deeply profound, connected to a tradition that came from Trieste carrying with it a cosmopolitanism 'a la recherché du temps perdu.'

My exile, though, like everyone's, is unique. It was hard to be an Arab girl born to an Italian mother in a world that was still male dominated. It was hard to become a Palestinian woman in a world of exiled people.

EXILE AND LANGUAGE

Exile carries with it a sense of banishment, an irremediable loss, loneliness, a feeling of being out of place. Loss was a central element in my history, in the story of my father, in the reality of my grandmother, in the negation of my mother. The challenge for me was to define which loss I experienced and what exile I am living, one that might well descend from that of my parents but is still distinct from them.

Is it the exile from the nation, the land, or the exile from the self? Is it the exile from the parents or the exile from the outside world? Is it the exile from the language, and in this case which language? Italian, the first language I heard and learned, that of my mother, but which I speak with an accent? Or is it Arabic, the language with which I first engaged the outside world, but which was replaced by English, a language that I mastered but with which I never really feel at home?

"L'exile de la langue," which Jacques Hasoun wrote about, was real to me. Language as a sense of displacement often mirrored my sense of exile in this world. I often felt I was living within a Tower of Babel, as I learned to master five languages, but I only felt at home in Arabic. It was at times unbearable to hear these languages resonate in my mind, to witness how they created segmented worlds, in me as much as in the outside world. It was hard to feel and realize that these languages could not meet, that they could not converge, for if they did, my language would become incomprehensible.

Italian proved to be the language of the womb, of my mother, of warmth. It remained the language with which I relate to my mother. It did not carry with it a sense of loneliness, although it did give me a sense of constraint, of limitation. Arabic was the language that allowed me to separate from my mother, to become a separate identity from her. It was tied to my father, but not confined to him. It was the language of my siblings, of the outside world I connected with and of the world of which I was made to be proud of. It also represents for me the language of desire, of poetry, of life. That is at least what I felt, what I feel today.

Yet, English became, after I left home, the language of the intellect, where I learned to be articulate, scientific, in which I acquired my university degrees, and with which I related to the outside world ever since I became an adult. It also became the language I shared with the partner I married, and of the new home I eventually built. I still do not know if it is the language I choose, or the one I just had to take in order to communicate with the Babel outside me, with the external world I ended up living in since I was 18. Yet, it still remained a foreign language, not really mine, one in which I feel somewhat exiled, despite its being today the language in which I best express myself.

My exile has been layered, a tapestry woven through the course of a diasporic life that could only be completed and reconciled through an embrace of the continual invention of the self. My exile was grounded in the collective sense of exile of the Palestinian people. It was shaped by the unique personal experience of a girl growing amid the reverberation of the 1967 debacle that hit the Arab world. The dialectic between the collective and the personal shaped my alienation as much as my ways to come to term with the person I was becoming.

EXILE FROM PALESTINE

Growing up Palestinian was alienating as much as inescapable. Exile from Palestine was not simply the banishment of my father from his land and from his mother. It wasn't only about being away from my grandmother and cousins. It was above all a deep feeling that I should have been born somewhere else, should have had my life over there, in Palestine where my father wanted us to be, not wherein my childhood ended up unfolding.

My father moved to the Gulf but never settled, although he lived there 25 years. His stay was temporary, a station on the way to return home once the land was liberated. He saved money to remit home, but he also continued his activist work for Palestine, an activism that predated my birth.

Like many of his generation, he witnessed the end of colonisation in the Arab World in the 1950s as well as the creation of the State of Israel on Palestine. My father believed in political action and was involved in its various forms, from student activism to political party involvement. With the debacle of 1967 he joined one of the political parties of the Palestinian liberation movement and became active in its diplomatic core. Given that my father studied in Europe and married a European, he mastered a number of foreign languages that made him, together with others, an emissary for the Palestinian cause. Those were the days of the height of the Vietnam War and mass opposition to it, as well as of the 1968 revolts. It was the time when members of liberation movements were not insurgents but members of a legitimate resistance struggle, the struggle for self-determination. My father travelled to Europe, taking me with him when I was only a child. His mission was to explain the cause, foster links with Palestinian youth leadership, and to sow the seeds of European support for the Palestinian struggle. I was a baby then, carrying a broche with the PLO emblem on it. I, with that emblem, became a code by which he and his other Diaspora compatriots could identify themselves to one another. It seemed that I was (pre-)destined to carry on the torch of the struggle, at least as far my father saw it.

Palestine became inevitably the promised land, and I became the Jew of the Jew, as one of Elias Khoury's characters described us in *Bab el Shams (the Sun's Door)*. I lived a diasporic life in the Arab/Persian Gulf, among Palestinians dreaming of returning home. I felt disempowered as much as exiled during the 13 years I lived in Dubai, one of the better cities of the Gulf, one which, however, still did not give me or my parents permanent residence. I grew up in a neighborhood populated by exiled Levantines: Lebanese, Palestinian and Syrian people. My school was run by Lebanese; my friends were Palestinians. Although Dubai buzzed with different ethnic groups, who were unofficially segregated along occupational categories, my life remained confined to a Levantine world.

I did not have local friends, or friends who were Europeans or Asians, although I saw them in the street, shopped with them in the stores, frequented them along the beach. These divisions were self created as much by each ethnic group that came to make money and return home rather than mingle with others, as by the city's policies. Dubai was an international city but not cosmopolitan. It attracted and accepted every immigrant who wanted to work and mind his/her own affairs, but did not seek to melt them into any new identity. Each community was welcomed to create its own schools, businesses, sports clubs and restaurants, not to engage in local or regional politics or question the fully unprotected free market system. Dubai was a place to make money that can be remitted home, not a place to replace one's original home.

Thus, in my world of an exiled teenager, it was Palestine and exiled Levantines who continued to matter, especially as the Israeli invasion of Lebanon erupted in 1982. I was 15 years old at the time and I saw for myself what war and dispossession meant. I managed then to participate in my first conscious political act. I participated in a demonstration, one led by women and children, to denounce the massacres of Sabra and Shatila. It was the only one allowed in a country that forbade any form of political participation. It had no impact though, and it gave me no sense of empowerment. The press was forbidden from reporting on it. The fact that it was organized by women gave it a "humanitarian" rather than political touch that needed to go unnoticed. That first experience of political activism accentuated the sense of exile, both as a Palestinian and as a woman.

Palestine remained the real thing, the only place where the struggle mattered, the only place where home was possible. The longer my life in Dubai continued, the deeper was my sense of dispossession and the more I felt the need to own Palestine. Palestine was mystical to me. My father's story of his childhood in his village, my grandmother's yearly visit telling me about the Israeli atrocities and about the Palestinian's resistance, the unfolding of the Palestinian tragedy with the war in Lebanon, and the exile of the PLO to Tunis, were among the things that created this sense of mysticism, as well as of the deep sense of collective tragedy. But I also needed to own Palestine—to make it mine, even if it remained inevitably linked to my father and the whole community to which I belong. Palestine became real to me when I visited the West Bank, with my mother and siblings, whenever the Israeli government gave us permission to be there. I constructed my own relationship to Palestine, by creating new friendships, discovering my aunts and my cousins, and by being in its landscape, among its olive groves. Palestine gave me, paradoxically, a feeling of freedom and authenticity that I could not feel in my life in the Gulf. It gave me a sense of empowerment. In Palestine people who resisted the Israeli occupation had a clear sense of their lives. They were on solid ground—however much this ground was being encroached upon by Israeli settlements. They were there to stay, like the resilient olive trees, never to give up, never to leave.

My attempt to own Palestine and overcome the collective exile I inherited was an attempt to materialize the dream of return. But this return had to be prepared for and ultimately pass through another exile, that of leaving the Arab world to go and study in the West. My decision to move to the West and study political science was undertaken with an eye to return home, where I wanted to put my acquired knowledge at the service of my people. My father provided a role model for me in this respect, but above all I also wanted to prove that an Arab woman, not a man, can fellow, if not outgrow, his footsteps. Thus Palestine became alive in my student activism. I took upon myself to explain the cause to others, through students' demonstrations, lectures, and collaborative projects to organize westerners to visit the land and see the effects of Israeli occupation. I also had to define the Palestine to which I wanted to return. I was brought up to believe that the only solution for the Israeli-Palestinian conflict is a secular democratic state in which Jews, Christians and Arabs will all live together. That was the motto of the PLO at least until 1974 and officially until 1988. But by 1987 the Intifada erupted against Israeli occupation in the West Bank and Gaza Strip and in 1988 the PLO declared the establishment of the state of Palestine based on UN resolution 242. Arafat by then had accepted the idea of partition as the only solution for Palestine, calling for the establishment of a Palestinian state in the West Bank and Gaza Strip. That was where my father came from, where my cousins were living, were my idyllic, really wonderful summers were spent.

The dream of return materialized in 1998 when I was being offered the job of my life in Palestine. I was hired for a policy advisory post in a Palestinian think tank which would help build the infrastructure of an independent state. It came after spending 13 years in Europe and after marrying a European man. I returned to the West Bank for a year, at the height of the Oslo peace process, when there was so much optimism for peace and independence. I was hoping to find the answer to the end of my exile, personal as well as collective.

But the reality proved to be far off the promises and expectations of Oslo. The situation on the ground deteriorated rather than improved, as Israel failed to implement the signed agreements and Palestinian leaders were more concerned with consolidating their power than with building the country's institutions. The Oslo peace process had divided the West Bank into three major zones, fragmented from one another by over 150 Israeli settlements and 250 miles of bypass roads. Israel had been using the checkpoint and permit policies that locked people up in their enclaves whenever it perceived a threat of a suicide attack. Rather than getting closer to freedom, the Palestinians saw their movement become more restricted and their horizon more constrained. Over 76 new settlements outposts were created during the Oslo years and the settler population doubled to over 370,000 by 2000 and 460,000 by 2008. The Palestinian economic situation deteriorated, as per capita income fell by 30% between 1992 and 1996 and again by over 50% between 2000 and 2006. Unemployment soared to over 25% and poverty to over 60% by 2007. My job, for its good intentions, proved incapable of improving a daily reality that was worsening. Like the rest of the country, the scope for action remained determined by what Israel decided was compatible with its security, not with the Palestinian's right to independence. Despite all the good work and intentions of many returnees and residents to obtain Palestinian independence, economic and social development, they remained dispossessed of control over their destiny.

Yet, the people continued with their resilience. They continued with their lives and with their determination to build institutions. Universities continued their work, new houses were being built, NGOs and financial institutions expanded their outreach and services, and Palestinian workers continued to work on Israeli settlements. The Palestinian state was dead before it was born and separation between Israelis and Palestinians was no longer possible outside an apartheid model. It became clear to me that exile from Palestine can only be overcome through a new project of statehood. Palestinian exile could not be conquered by return, rather by embracing a universal inclusive humanist agenda that I still needed to define and elaborate together with others.

EXILE AND MY MOTHER

Exile from Palestine tormented me but it also defined me. However, it did not resolve the other exile that inhabited me, the exile from my mother.

Growing up as Palestinian was a given, unquestionable, and un-debated. I sought to redefine it for myself and separate what was collectively inherited from what I consciously choose to be. What was not officially recognized, or fully accepted, was the fact that I was also the daughter of an Italian woman. I felt that for my father the fact that my mother was Italian was *accidental* to the fact that I was Arab. It was not a central, albeit specific, element in explaining the *kind* of Arab I became. So while I was aware of an Italian/western heritage, I did not seek to dwell on it or claim it. I refused to think or talk about what is Italian in me, although I always spoke Italian with my mother, visited Trieste annually, and fondly loved my maternal grandfather who meant the world to me. I felt segmented whenever I went to Italy, as if the month I spent there annually had nothing to do with my life back in the Arab world. I wanted, like any child and later a teenager, to conform to my surrounding in the Arab world. Yet, I remained different from my friends because I was the daughter of the foreigner. At school, as well as back among my larger family and in my father's village, I was considered an anomaly for I spoke Arabic well, behaved respectfully as expected from Arab girls, and was obedient, *despite* the fact that my mother was Italian. I felt insulted at such comments and felt I needed to prove more than others that I was the real Arab, the real Palestinian.

The exile of my mother was hard to define, let alone deal with. At times, I am not too sure if that sense of estrangement was more a product of me feeling her distant from me, me exiled from her, or the result of her being exiled from her world. My mother symbolized the West, a world perceived in clichés in the Middle East, even if it was admired. Yet, she did not correspond to the stereotypes of the cold, sexual, decadent, and oppressive West. She cared about her home, dedicated her life to it, and did not want much to change the world. She was not an emancipated, free, career woman and yet she was independent in a paradoxical way. She was not scared of challenging my father's opinion even if she followed him wherever he went. She was not possessive of her kids, did not expect us to live with her, although she lived for them. She appeared to be able to be alone, to be an individual and not to need people around her. At times, I could never really know what she thought, she was unreachable, away in her own thoughts, untouchable behind the cigarette smoke that put an aura around her, as if that cigarette represented for her a "room 'of' her own."

My mother's exile was hard to understand, for she negated it while I felt it. It was not spoken of, unlike the exile of my father. It was unadmitted by my mother for she maintained that she chose to leave her home. It made me want to empathize with her, but also made me feel that there was a world to her I could not share, one in which I did not know my place. My mother did not overtly share her Italy. She took it upon herself to bring up Arab children. She was not nostalgic for Italy and she never talked about what it meant to leave her home, although she transmitted her Italy in her food, her habits, and her language. I remember listening to one of Verdi's operas when I was 14 years old when I discovered by pure chance that my mother was able to repeat all the lyrics. When I asked her how come she knew them, she laughed and said that these songs were for her what dabkeh and dal'ona are for me.[1] I realized then all what she left. When I was 15 years old she took me once on a tour to explore Italy and get to know her historical heritage in Rome and Florence. She considered I was old enough to understand and appreciate her cultural heritage. I was also well grounded in my Arabic heritage not to feel shaken by western influence!!! The West could only come *after* my being an Arab, an Arab produced by an Italian mother.

And yet my mother's world, that Western culture as much as her example of womanhood, was in me as much as around me. I could not escape the West, nor did I want to. I wanted to serve my country but also to know about the West. I did not want to be like my mother, I did not want to occult my culture, but I was fascinated by her sense of self, of what I call her independence. I was far more rebellious than my mother and far more ambitious.

My years as a teenager oscillated between Dubai and London where I went to study. Again I felt a sense of fractured identity as I was trying to reconcile the East and the West in me. I had the hardest struggle with confronting my femininity, both in terms of its content and in terms of my response to it. The West was alienating, with all its clichés about the Arabs, with all its individuality that forced me to discover and face mine. I felt alienated by the gaze of the "other" that often made of me the exotic Arab, who was either underdeveloped or a terrorist, a repressed woman or a fantasized sensual object. I actually wanted to dissociate from being a woman. I was afraid of being an object of desire. I did not know how to handle my femininity in the male world I encountered in the university as well as in my professional life—in the West as well as in Palestine. It was easier to be asexual, untouchable, an ungendered Arab who defended the Palestinian cause.

But I could not escape my gendered

[1] Dabkeh and Dal'ona are traditional Palestinians dance and songs.

identity, even when I was trying to be ungendered. I fell in love when I was 14 with a Westerner who then became my husband. That encounter and marriage helped me confront my femininity with some ease, to accept to be desired and to desire. It also helped me shape my gender identity away from the example of my mother and father. But that was not easy. Before marriage, our relationship was based on a sense of equality. When we got married, I wanted to prove that an Arab woman, not man, can marry a foreigner without losing her identity. I was still hoping that my husband would follow me to Palestine, thereby reversing the story of my parents, the traditional gender role I saw in my family. But that proved impossible, for my husband refused to re-enact the inverse of a story that was not his. He was not going to follow anybody and expected himself as much as me to be equals.

Yet, sorting out what equality meant in a gendered multi-cultural relations proved to be very hard. I lived in Europe as part of my effort to "build my resume," but I missed home. I struggled to find my place in the Western world of my partner, with the career that he was defining for himself. I was alienated in the language we spoke for it was not Arabic. Unlike my mother, my project was not to have children and follow the husband. I had my work, I continued to engage on issues related to the Arab world, to go periodically to Palestine often without my partner. But I was again fractured; the Arab world was something I lived alone, the way my mother lived her Italy alone. At the same time I remained an Arab in an alienating West of which I was a part. I owed to it my intellectual development, my years of struggle to adulthood, my love story, my yearning for a home. My marriage did not escape the torment of exile and self-definition, of depression at times and of despair. It also gave me a sense of home, a redefinition I was not willing to renounce.

COMING TO TERMS WITH EXILE

I continued oscillating between Palestine and my husband who remained in Europe, until an opportunity came for my husband to go to Boston for work. This third major migration in my life gave me an opportunity to come to terms with my exiles, to reinvent myself. I do not know if the place mattered, or if the state of mind in which I was in mattered more in enabling me to make Boston my home. In a way, Boston was a neutral place for me and my husband. It was neither Europe nor the Middle East. It was the U.S., the country of immigrants for people like us who were becoming eternal immigrants. It came above all at a turning point in my life. I had left Europe to work a year in Palestine, but my return home, for all its value in reconnecting me with my language and family, also had its limitations. While it gave me a sense of grounding, it also set limits to my horizon, and did not prevent me from feeling segmented. I became aware that Palestine was an important pillar in my life, but not the only one. Palestine as a quest for return proved its limitation. I wanted to carry Palestine along with me since living in it did not provide all the answers to my angst.

The decision to come to Boston, though, was somewhat ironical, for I never thought of going to the United States. The U.S. after all was the bastion of imperialism. In the Arab world, America, as we call it, was the cause of lots of evil. It did nothing to resolve the Arab-Israeli conflict and if anything proved more and more to be the univocal supporter of the Jewish State. The European part in me looked pretty down on the American culture that was considered dull and shallow compared to continental humanism and culture. I never wanted to study in the U.S., although most of my friends from school aspired to go there. I never had an American dream for I did not seek to make money, nor was I seeking a refuge from my own world. Moreover, the job offered was given to my husband, not to me. I was following my husband, not he me.

Boston, however, represented an opportunity to create my own space while building on my history. It allowed the reconciliation of the personal and the collective sense of exile, the private and the public sense of loss. My husband told me at the time that we do not need to worry much about the move that it will only be for few years, that we can always go back if we do not like it. I laughed for I did not know where 'back' was. I simply told him that I was not going to Boston with the idea that it will be a temporary stay. Out of Boston I was going to create a home, not a house.

And indeed this is what I strove to do. What I meant by home was to create an environment in which I felt I could express my various dimensions, rather than occult one for the sake of the other, or live one at the expense of the other. I needed to express my Arabic, Palestinian dimensions as much as be with my German husband. I needed to speak Arabic as much as communicate in English with the world around me. I wanted to create a family as much as remain connected to mine. Home for me is accepting to oscillate between worlds, to be here and there, to accept the tension between my various dimensions, rather than seek to reconcile them.

My first step toward creating a home was to find my cultural and political space. At first, I sought the Arab community. Although a hyphenated group, the Arab-American could relate to my sense of music, my nostalgia for home. They provided a possibility to speak my language again, if only for a bit, if only with an accent, if only with an American culture that I was to discover. I also found exiled Arabs like me, who were not Arab Americans, who shared with me the eagerness to speak our language, the ability to make the same cultural references. I was soon able to make a group of friends that came from similar backgrounds, as my own; people who lived in various countries, who combined two cultures, one Arab, the other American or another. Sometimes they even shared with me a similar trajectory of exile, having either lived in the Gulf, or having married somebody from a different country, or/and having struggled with their womanhood each in her way. I wanted to create the surrounding of a large family of like minded people.

I also sought to recreate the Middle East in Boston. I joined Arabic reading groups, became fussy about traditions that I wanted to re-establish where I now lived. I started celebrating the feasts of back home, endeavored to host large dinner parties that grouped all the exiled. I looked for Arabic cultural activities, and sought to promote Arabic dancing and literature. I soon realized that I cannot recreate the Middle East here however much I wanted, for I would be creating a false copy of something that is elsewhere. I risked locking myself in an exiled static world that lived off of nostalgia rather than of new ways of expressing one's culture.

Boston allowed me to create a sense of empowerment out of my exile. I became vocally and organizationally active for the Palestinians cause. I increasingly did that through my work, be it my thesis, my book and later my job in academia, or through grassroots work with Arabs as much as with Jewish, often lesbian, women. Whether through writing, speaking or rallying, I attempted to influence the U.S. public. I was actively engaged in talking about the cause, explaining it to a public that was not given the means to know its causes and its tragedy. At first it puzzled me, and maybe irritated me, that the major supporters for the Palestinian cause in the U.S. were Jews, and in context I found it in Boston lesbian women. I soon realized that the fear of any accusation of anti-Semitism, so diabolized in the U.S., makes American Jews the best spokesperson for the Palestinian cause. I also was aware that marginal, outcast, groups were naturally going to join together. The membership of these groups underlined the impending moral obligation of questioning power and hegemony in all its facets, as well as the universality of the quest for justice and respect. Activism for Palestine made me aware of the depth of activism in America, something that I would not have known to exist were I just to stick to the image of the U.S. transmitted through the world.

Another means by which I sought to overcome my exile, was through my evolution concerning the best solution to the Israeli-Palestinian conflict. Both as a result of my research as well as my activism, I became increasingly aware that the two state solution cannot be a viable solution. The fragmentation of Palestinian land through Israeli occupation eroded the physical territorial infrastructure of a potential Palestinian state. My activism, particularly with Jews, showed the limitations of nationalist struggle and the inevitability of an inclusive open strategy for freedom. The one state solution became evidently for me the only solution, the only humanist solution. Never before had anybody in my family pursued a dialogue with Jews, let alone define together a strategy of independence. Boston, however, made that look so evident and so emancipatory.

Yet, what probably most helped to cope with my exile was psychotherapy and the birth of my daughter. Therapy allowed me to rewrite my story, to discover my own trajectory from the lineage of my family history, and to accept and embrace my contradictions. My daughter enabled me to come to peace, to be at home, in a way I never thought possible, and probably feared. Becoming a mother, though surely not as naturally as some people claim, has enabled me to connect to my womanhood albeit in a quite traditional way. It enabled me above all to reconnect with my mother but on a completely new level of consciousness and understanding. My mother became closer despite the distance, as if I could understand her better, hear her better, feel her closer.

My daughter, above all, gave me a sense of home in ways that I still cannot put in words. She enabled me to reconnect with my language, Arabic, the language I found myself speaking naturally to her. At times I wondered why that was the case, for, after all, my mother is Italian and the first language I heard was Italian. But Arabic has been the language of my soul, of my childhood and youth, of my consciousness of the world. Arabic is the only language I speak without an accent. With my daughter I found myself speaking Arabic every day, overcoming the need to search for a friend with whom to speak my language, the weekly phone calls of my father, the news on Al-Jazeera. Meanwhile, though, I sang to her in Italian, for those were the songs I was taught as a child.

My exiles are not over, but I am learning to live with them. The many faces of exile are shaping up, and they are finally bearable.

Gender and Violence

A Reflective Sociology of How Gender Ideologies and Practices Contribute to Gender Based Violence

Jacquelyn Knoblock

University of Massachusetts Boston

jacquelyn.knobloc001@umb.edu

Abstract: Gender based violence can be viewed—using the sociologist C. Wright Mills's conceptual vocabulary introduced in his book *The Sociological Imagination* (1959)—as both a personal trouble and a public issue. Gender based violence was most often seen in the past as a personal trouble, a private matter between couples. Of course for the women who endure this violence it is very personal and very troubling to their safety and damaging to their whole sense of being and self-worth. Only recently has gender based violence come to be seen as a public issue. In the 1980s domestic violence was found to be the leading cause of injuries to women, and the Surgeon General deemed domestic violence, "the most serious health risk facing women" (Disch 2006:471). It is not only domestic violence that is an issue in gender violence but the trafficking of women and girls through countries as sex slaves and prostitutes, and governments' inaction to stop this. This writing has allowed me to learn from my own person troubles with domestic violence, and to see the ways in which gender based violence is a bigger social issue. Putting this paper together helped me to see that it wasn't just my own personal "defects," or just his insecurities, that caused me to experience a violent relationship, as "the individual abuser and the victim do not operate in a vacuum; rather, they are nested within the supportive circles of social institutions and culture."

Gender based violence can be viewed—using the sociologist C. Wright Mills's conceptual vocabulary introduced in his book *The Sociological Imagination* (1959)—as both a **personal trouble** and a **public issue**. The **Sociological Imagination** is the ability to see the link between history and biography and to shift perspectives from the sociopolitical to the psychological (Mills 1959).

Gender based violence was most often seen in the past as a personal trouble, a private matter between couples. Of course for the women who endure this violence it is very personal and very troubling to their safety and damaging to their whole sense of being and self-worth. Only recently has gender based violence come to be seen as a public issue. In the 1980s domestic violence was found to be the leading cause of injuries to women, and the Surgeon General deemed domestic violence, "the most serious health risk facing women" (Disch 2006:471). It is not only domestic violence that is an issue in gender violence but the trafficking of women and girls through

Jacquelyn Knoblock is an undergraduate junior at UMass Boston, majoring in Social Psychology. She wrote this paper while enrolled in the course Soc. 381: "Sociology of Gender," instructed by Anna Beckwith (Lecturer of Sociology at UMass Boston) during the Spring 2008 semester.

countries as sex slaves and prostitutes, and governments' inaction to stop this.

I became a victim of an abusive relationship during my junior year of college when I was twenty years old. It was not my first relationship, but the first real relationship that I had had during my college years. As the abuse continued and worsened, I thought I was alone in a personal struggle. At the time I did not realize the extent of violence towards women. After getting away from this relationship I decided to take classes in psychology and sociology, with the intentions of figuring out "what was wrong with me." Asking myself why I had allowed myself to become a victim, and why I stayed even when the abuse became physical, I used my education as a way of empowering myself with knowledge.

Through taking classes I did learn things about myself that may have contributed to my own personal situation. But I also learned that this same experience was all too common for women around the world. I have started to understand that it is some of the ideologies of masculinity and femininity that contribute to domestic violence. Violence against women is a product of the oppression of women in most facets of life in most places in the world. Feminists see that there is a "continuum of violence," meaning that all violence is interconnected, coming from the same social root of gender inequality and the resulting sexualization of women's bodies (Marchbank and Letherby, 2007:270). In this paper I will explore some of these ideologies and of hegemonic masculinity and femininity, the practices people engage in to maintain these gender differences and inequalities, and some of the oppressions of women in different faculties of life. I will use the sociological imagination to apply these issues to my own life and my own experience of gender, focusing on ways in which these ideas relate to gender based violence.

Gender based violence is much more than just a personal issue between couples; it occurs because of some of the basic ideologies we hold about men and women, and the structures used to enforce these ideas. First of all we must examine the ideologies of hegemonic masculinity and femininity that create separate dichotomous spheres for men and women. If we see the ways in which men and women are viewed as different we can begin to see how women are devalued and oppressed, and how this leads to the continuation of violence against women. The ideas of what constitutes masculinity and femininity are not only different, but are essentially in opposition to one another. According to class discussions and handouts in two courses I am taking on Sociology of Gender and on Family Violence, feminine characteristics include softness, cooperation and concern for relationships, being emotional (absence of control), love and nurturing, dependency, intuition, harmony and connection. Masculine characteristics include reason, rationality, intellect, self-control, autonomy, toughness/strength, competition, aggressiveness, and separation.

It is not that these characteristics are inborn, but they are socially constructed. In the article "The Social Construction of Gender" Judith Lorber explains the ways in which everyone is "doing gender." Lorber defines gender as a process, a stratification, and structure. By a **process** she means that gender creates social differences that define what it is to be a "man" or a "woman." Creating these differences is something we continue throughout the lifespan through behaviors, norms, and expectations which are enforced or sanctioned by peer groups and authority. **Gender as a stratification system** means that men are ranked above women within the same race and class (Lorber 2003:116); this is what creates the inequalities between men and women.

As a **structure** "gender divides work in the home and in economic production, legitimates those in authority, and organizes sexuality and emotional life" (Lorber 2003:117). This structure is an important component in the stratification system. "When gender is a major component of structured inequality, the devalued genders have less power, prestige, and economic rewards than the valued genders" (Lorber 2003:117). In our society men have "more power, more prestige, and more property than the members of the disfavored group" (Lorber 2003:117)—the disfavored group in this case being women. This stratification is important in the idea of gender violence; the power men have and the lesser status of women are reasons it continues to occur. Violence within personal relationships is about power and status. I know that in my own situation, I knew that I had less power, and that he was in control of the relationship, possibly because he felt that he didn't have control in other aspects of his life.

Holly Devor's article "Gender Role Behaviors and Attitudes" is especially useful in explaining how the differences between masculinity and femininity are based on unequal status. She states that **masculinity** is a competitive thirst for power which can (but doesn't have to) lead to aggression, and that **femininity** involves a "quest for harmony and communal well-being, which can, but need not, result in passivity and dependence" (Devor 2003:485). Here again, traditional masculinity reinforces aggression, and femininity can lead to dependence and passivity, both characteristics leading to the maintenance of a violent relationship. She takes it further than this though, saying that the United States' "patriarchal gender schema" (2003:485) gives valued attributes the quality of maleness, while leaving for females the characteristics that maintain heterosexuality and nurturing (2003:486). Power in this system is a masculine attribute. The practices and **roles** of femininity "expressed through modes of dress, movement, speech, and action which communicate weakness, dependency, ineffectualness, availability for sexual and emotional service, and sensitivity to the needs of others" (Devor 2003:486). Thus, femininity, Devor states, is most fitted to satisfy "a masculine vision of heterosexual attractiveness" (2003: 486).

The way women hold their body and move, they take up small spaces and use small steps, they don't look back when looked at, they are more likely to smile, they wear clothes that show more bare skin and inhibit movement, all of these practices serve to "communicate subordinate status and vulnerability to trespass" (Devor 2003: 486). Anything that is masculine must remain untainted of any femininity, and projects the opposite, male practices serve to maintain dominant status, and physical power. In practicing femininity women are conveying vulnerability and low status. Men want to convey power and aggression.

One can see how these presentations of self can lead to an easy situation for violence. When women appear vulnerable and dependent and of a lower status they are more easily taken advantage of by men. Within my abusive relationship I was constantly trying to appease, concerned with maintaining the relationship more than my own well-being. I didn't think of myself as dependent, but I see that I was presenting myself that way by constantly trying to fix the relationship to keep it going. I see that by engaging in all these feminine roles, I, like other women, was an easy target for violence. Later in this paper I will engage with some of these individual practices that actually cause women to be more vulnerable to violence.

Growing up, I learned that there were differences between boys and girls, but also was taught to believe in the **myth of meritocracy**, that I (along with ever other citizen of the U.S.) could "do whatever I wanted as long as I set my mind to it"—that rewards were based on accomplishments. I wasn't taught to see that maleness was the norm and that femininity was "Not A" as Lorber puts it, in the same way that I was not taught to see the privilege that came with being born white. In noticing that males were often unwilling to admit that they were over privileged, Peggy McIntosh realized that whites also only chose to see racism as a disadvantage for others, but not as a privilege for themselves. In her article "White Privilege: Unpacking the Invisible Knapsack" she tackles these issues. McIntosh defines **privilege** as unearned assets that put one at an advantage (1988). She states that whites are taught to see their lives as normal and morally neutral, that they aren't taught to see themselves as oppressors. This is true. I was taught that racism was "bad," but that nowadays it stemmed from individual hatred, not from cultural dominance of a group. I was blind to the **interlocking oppressions** in their "embedded forms" (2006:75). Within this system of interlocking oppression I didn't see my racial advantage, just as I didn't see my gendered disadvantage, all to maintain the myth of meritocracy.

My early education taught me that racism and sexism were problems of the past that had since been overcome, that now everyone was equal. Not until taking college level courses did I begin to see the fallacy of this myth. Also, being white and therefore privileged, I did not think that I could take advantage of typical resources that were available for battered women, I felt that they were created for those who "had less" than I had, who needed it more, that I would be turned away. Again, I didn't see myself as the image of a stereotypical battered woman, being that I was middle class. When I finally turned to my family, they did have the means to help, which may not have been available otherwise. My parents paid for me to see a therapist, who helped me to see that it wasn't just me, that all abusers are the same, and who slowly coaxed out of me what was really going on and encouraged me to leave. She said that it would be hard, there would be a mourning period, but that I should do it as long as I could know that I would be safe.

At almost the same time my aunt and uncle offered me a place to live at their house in South Carolina. My uncle offered to enroll me in community college in their town and help me get back on my feet. The fact that my family and especially my uncle could afford to help me like this is one of the major reasons I was able to leave the relationship. I didn't have the money to enroll in school again myself. I knew that I could be safe because I would be living over a thousand miles away. This made me realize some of the unearned privileges of class. If I had had no one with the available funds to help me, if I had been dependent on my boyfriend for income, or if I had had children, I probably wouldn't have been able to leave the abusive relationship.

Although early education doesn't teach us to see the bigger picture of the systems of oppression, there are practices which teach/reinforce gender roles and segregation in elementary schools. This problem is discussed by Myra and David Sadker in their article "Missing in Interaction." They describe how elementary school teachers inadvertently give more attention, both positive and negative, to boys in the classroom, and that the girls, although appearing to be doing better than boys in terms of better grades and less punishment, are "reinforced for passivity" (2006:355), maintaining a feminine quality that in the future does not help them retaliate against violence directed towards them. **Gender segregation** occurs also at this age, boys play with boys and girls play with boys; it is almost taboo for them to play with one another. When allowed to choose where to sit, children sit with members of their own sex, and the boy who sits with girls is generally outcast by members of his own sex, or forced to do so as a punishment from the teacher. In "He Defies You Still: The Memoirs of a Sissy" Tommi Avicolli describes being taunted for being effeminate and being called a sissy, a faggot, and a girl; he recalls that these terms made a boy

> Vulnerable. Feminine. And feminine was the worst thing he could possibly be...no boy in his right mind wanted to be like them (girls). A boy was the opposite of a girl. He was not feminine. He was not feeling. He was not weak. (2006:150)

Important in this gender segregation are **pollution rituals,** in which children treat members of the other sex as "germ carriers" (Sadker & Sadker, 2006:358), that by touching one you will be polluted, and more often it is girls who are the "cootie carriers."

As one can see from Avicolli sentiments, and these rituals, the girls are made to be a lesser subspecies. The Sadkers state that "it is boys who work hardest at raising the walls of sex segregation and intensifying the difference between genders. They distance themselves, sending the message that girls are not good enough to play with them" (2006:359). Boys then take up more time in the classroom and space on the playground, and this is not questioned or fixed by the teachers to make things more equal, reinforcing the girls' inferior status. The Sadkers conclude that gender segregation "is a major contributor to female invisibility" (2006:360).

I remember clearly these types of pollution rituals in elementary school, frequently started by the boys; girls generally tried to maintain their relationships with the boys, while the boys tried to become distant to prove themselves to the other boys. For myself, I remember it was a privilege when the boys allowed me to play with them. To be allowed to do so you had to prove that you were good enough or tough enough, you had to conform to their ideas of play. If you got hit hard, you couldn't cry, or if you couldn't hold back your tears, you couldn't tell on the boys or they wouldn't let you play again. Then again, the girls were always thrilled when the boys wanted to play with us.

I also remember that the loudest troublemakers, which in my experience were always boys, took up much of the classroom teacher's time. This frustrated me. I can see how a lot of little girls got lost or forgotten about in the classroom. The most important thing to take away from this is that in elementary school years there is segregation which creates the superior status of boys and the inferior status of girls. This sets up the lower status of women for the rest of their lives, although it isn't always in plain view. But this lower status, and the "reinforced passivity" is what allows for violence to take place and why it isn't made an issue by the majority. I think another important aspect of the attention that little girls do receive, which the Sadkers mentioned, is that when little girls receive praise from teachers it is often based on their appearance, on how "cute" they look (2006:357).

Girls and women are first and foremost judged by their appearance. **The Beauty Myth** has imposed harsh standards of attaining beauty on women in what Naomi Wolf says is the last effort to "keep women under control by imprisoning them in their bodies" (515). Most women struggle the effort to reach these unattainable measures of beauty. In the film "Killing Us Softly 3" Jean Kilbourne states that "the body type that only 5% of women naturally have is the only one that is seen as beautiful" (2000). She says this is about teaching women that they shouldn't take up space, and not be too powerful, or be too full of themselves. In relation specifically to gendered violence, Elayne Saltzberg and Joan Chrisler point out that the styles and body types that we consider beautiful often leave women vulnerable to injury or attack, such as long hair or dangling jewelry that have gotten caught in machinery, and high heels and tight skirts and dresses which prevent women from running away from danger (163). The current body type that is part of attaining the beauty ideal is one that is extremely thin. Sandra Bartky in the article "Foucault, Femininity, and the Modernization of Patriarchal Power" states that this gives women no help in fending off violence:

> An aesthetic of femininity, for example, that mandates fragility and a lack of muscular strength produces females bodies that can offer little resistance to physical abuse, and the physical abuse of women by men, as we know, is widespread.... A woman may by no means develop more muscular strength than her partner. (2003:35)

In my own life I have subscribed to this "culture of thinness." I was naturally thin as a child, and I engaged in sports, cross country, track and field, which maintained and promoted and almost required the thin body type. I liked my thin body, enjoying the compliments I received. But it is telling that at the time of the most severe physical abuse in my relationship, I was at the lowest weight I had been in years. I didn't feel that I could fight back, I was much too small to defend myself against attacks. Although, it didn't happen for me, it must also be understood that not only does this prescription of thinness lead to easier victimization, but victimization can also lead to eating disorders, which then lead to thinness. Trauma has often been linked with eating disorders, sexual abuse being the most common relation to the origin of an eating disorder. Eating disorders give some women a sense of control over their own bodies and also a way to anesthetize the negative feelings that come with victimization (Thompson 2006:181). So the pattern can be that thinness and maintenance of the beauty ideals lead to victimization, and/or that victimization leads to eating disorders, and subsequently thinness.

There are other features of living in America that facilitate gender-based violence. One of these is the privatization of the family. In foraging and agricultural societies the work of all family members was important to the group's survival, but as societies evolved, especially Western capitalist societies, this also changed. In the last hundred years "there was a clear understanding about the obligations and entitlements each partner took on when they married. He was obliged to work outside the home; she would take care of life inside" (Rubin 2006:305). This life inside included not only the domestic tasks or cooking and cleaning, but also the maintenance of family's well-being. But the fact that the male was bringing in the income, and the devalue of the feminine tasks, set up an inequality within the home. This is reflected in the fact that domestic violence is the number one cause of injury to women, and that women are 9 times less safe in the home than out of it (Marusich-Smith, class notes 2/5/08). The privatization of the family means that "family business" is private, what goes on in the home is behind closed doors and not talked about with "outsiders." My boyfriend, for instance, didn't want me to discuss our relationship at all with anyone because it was private, it was ours, and others "wouldn't understand." I think really he didn't want anyone else to have any influence over me, or others to know what he was doing.

As society industrialized, men began to have less control in their work lives, and began to take it out on their families exerting control. As family life was their domain, many women tried to project the perception that they had a perfect family life and relationship, because they were socialized to think that this was the most important part of their lives (Marusich-Smith, "Family Violence" class notes, 1/29/08) and so wouldn't admit to abuse or talk about it. After my own relationship became abusive, I didn't want to admit that anything was wrong, because admitting it to others meant it was true, and I felt that I had failed, or that something was wrong with me because it had happened. I tried to give the image that everything was fine, that our relationship was perfectly happy. Also, since he didn't approve of any places where I could work, I became in charge of "domestic" life, while he brought in a paycheck. I often ended up stuck inside the apartment all day, afraid to go anywhere else and trying to at least keep it clean, and do things to try to make him happy. Our relationship lost any sense of equality. Abuse is less likely to occur when the relationship is egalitarian, or the woman earns more than the male (Marusich-Smith, "Family violence" class notes 1/29/08) As women have entered the workforce, things have started to change, but there are still barriers in employment which make it harder for the woman to be an equal earner, and which maintain the patriarchal structure of the family and of society.

Women's domestic duties play out in the fact that "women's family responsibilities [are] certainly the most important ... cause of sex differences in earnings" (Crittenden 2006:418). These differences in earnings are significant, "The average earnings of all female workers in 1999 were 59 percent of men's earnings" (Crittenden 2006:417). This stems from the fact that women with children often work part time, or have taken leaves in their careers; the differences in wages for these reasons is termed the **cost of motherhood**. It is also caused by the fact that jobs which require the feminine quality of nurturance (which also are the occupations with the highest percentage of women) are devalued by our patriarchal culture which gives value to masculine traits, and are "the most systematically underpaid, relative to their educational and skill demands" (Crittenden 2006:421). This makes it hard for women to find equality in the workplace, and therefore difficult to find equality in the home, the equality which, as I mentioned before, would lessen the likelihood that violence against the woman would occur. Another issue of inequality of power that plays out in the workplace is that women are often sexually harassed, and sometimes blamed for their own harassment (Allen 2006:501). Women who hold jobs in traditionally male fields and who are assertive and competitive are often targets of sexual harassment, because they challenge their male co-workers' masculinity and have stepped out of their feminine roles (2006 1995:506).

Another arena in which the inequalities between males and females are played out is in the way we use sex acts to define our masculinities and femininities. Don Sabo says that often, "Sexual relationships are games in which women are seen as opponents, and his scoring means her defeat" (1994:276). So sex is seen as the man wins, the 2006 loses. How unequal is that? In the article, "How Men Have (a) Sex" John Stoltenberg states that there are no discrete singular differences between what we call males and females, but that sexuality creates a gender (2006:270). That "the act of fucking makes their sexual identity feel more real than it does at other times" (2006:269) "and it's expected that if you're the man, you fuck. And if you don't fuck, you're not a man" (2006:273). This means that the male must be masculine in the sex act, and be the one "acting" while the female is passive, and is "acted upon;" this helps to maintain the "lie" that there are innate differences between the sexes. Stoltenberg describes this type of sex:

> Violence and hostility in sex help the lie a lot too. Real men are aggressive in sex. Real men get cruel in sex. Real men use their penises like weapons in sex. Real men leave bruises. Real men think it's a turn-on to threaten harm. A brutish

> push can make an erection feel really hard. That kind of sex helps the lie a lot. That kind of sex makes you feel like someone who is powerful and it turns the other person into someone powerless. That kind of sex makes you feel dangerous and in control (2006:271).

These actions are examples of the characteristics of aggression, strength, and control in masculinity. This "acting" and being "acted upon" even plays out between members of the same sex in sexual acts. Within prison culture, a smaller less powerful male may become the "woman" (submissive partner) of a dominant tough male inmate (Kubler, 500). In the novel *Stone Butch Blues* by Leslie Feinberg, the main character Jess, who identifies herself throughout much of the novel as a butch lesbian, can't handle the idea of a former fellow butch friend, Frankie, dating another butch, Johnny. Her main issue with their relationship was "Who was the femme in bed?" (Feinberg, 1993:202). Even for Jess, someone who was challenging gender assumptions, in sexual acts she felt that one person (the butch, masculine partner) had to be the aggressor, the one acting, while another person (the femme, feminine partner) was the one taking, or being acted upon. Many men that I have been with have felt this way or acted in these ways (which Stoltenberg described above) during sex, and I have heard many male friends recall incidents (and probably exaggerate) times in which they have acted like this. I know that personally, I generally don't enjoy this type of sex as it makes me feel used, disposable, and takes away a sense of power over my own body. But then again, I am sometimes uncomfortable with voicing this issue; perhaps because I am a female, and not used to asserting myself, perhaps because I am concerned with the relationship, and am possibly afraid that the male will feel demasculinized if I bring up the issue. In this way I am complying with sex that reinforces my gender role, that does not create mutual respect and enjoyment. I feel that these type of sexual relationships may have led in part to my maintaining an abusive relationship. Many of my previous college relationships had been purely of a sexual nature, and I was trying to hang on to anything that had some emotion attached to it.

Also, pornography which socialized many adolescents and boys to the ways in which sex acts are "supposed" to be played out, often objectify women and "sexualizes their inferior status" (Disch 2006:472). The film *Killing Us Softly 3* also portrays the ways in which women's bodies are sexually objectified in pornography and in advertising. Often only one part of the body is focused on. Jean Killbourne states that this creates a climate that allows for violence. She says that turning a human being into a "thing" makes them inhuman, and is the first step in violence (2000).

All of these gendered ideals and inequalities lend themselves to a culture in which gender based violence is pervasive. This violence plays out both outside the home and within the home. Outside the home women are being trafficked and sold into prostitution, 200 million people worldwide suffer this fate (Goodwin, 2006:492). Inside the home, women are ten times more likely to be abused. This is what happened to me, the abuse occurred in private, and I tried to hide it. I was ashamed. I didn't think that this was supposed to happen to someone like me, it was something that happened to someone else. I was intelligent, college educated, middle class, white, an athlete. I didn't picture myself as an "abused woman."

I remember seeing a presentation on dating violence in middle school and thinking if anyone ever hit me, I'd leave him for sure, and I'd probably hit him right back. So what happened? Why did I stay, Why didn't I hit back or defend myself? I stayed because of emotional attachment; wanting the relationship to work; feeling that I had failed especially as a woman; isolation from others; and embarrassment of admitting what happened. I did try to hit back sometimes and defend myself, but found that because I wasn't strong enough or trained in fighting/defending myself, it only made things worse. Like Melanie Kaye/Kantrowitz mentions in her article "Women, Violence, and Resistance," many women won't even think about using violence, because they think it's wrong, or they are afraid to (479). I can understand this. I know that when I used violence, it worked against me, and many women are afraid of that happening. Most women tend to see violence as masculine, and not in their domain, and therefore wrong (479). I see Kaye/Kantrowitz point, that women need to have the ability to use violence in order to protect themselves and not comply with their own oppression (**the violence of rebellion)** (479), but then again I wish that there was a world in which we did not have to use violence, one in which males did not have to use violence. However, I suppose that, being realistic, Kaye/Kantrowitz is right for the time being. To resist violence, women need to be able to use violence to protect themselves. Maybe if men see women as equal threats in violence, they will be less likely to attack them.

Feminists say that men gain from this culture of violence because it "leaves women in a constant state of fear" (Marchbank and Letherby, 2007:272). They say that this fear can come either from experiencing beatings or harassment, or from a general anxiety of incurring the former or sexual assaults. Women cannot know which male is a potential attacker (Marchbank and Letherby, 2007:273). Although men as a whole may gain from this culture, not all male individuals do; in fact this culture of violence and the construction of masculinities also hurts males. In the essay, "Just Walk On By: A Black Man Ponders His Power to Alter Public Space," Brent Staples soon learns that when walking late at night he is viewed by other pedestrians, and particularly women, as a threat. A woman ran away from him in fear once, which made him feel embarrassed and dismayed (2006). He says he felt like "an accomplice in tyranny" (2006:191). He was not a threat, he knew that, but he began to take precautions to make himself appear less threatening to others (2006). But he knows that this contradicts ideologies of masculinities, that "it is only manly to embrace the power to frighten and intimidate" (2006:193). This masculinity also restricts males in the ways they are taught to communicate.

In the article "Real Men Don't Cry And Other 'Uncool' Myths" Phil W. Petrie maintains his masculinity through remaining "cool" in the face of crisis. He states that men must be rational, able to make decisions, and that in particular, for black men, being "cool" is a "metaphor for power" (2006:224). Cool on the outside, inside he was pained, "I was frustrated and wanted to scream…cry…I needed to be soothed as well as she" (Petrie 2006:221). According to Petrie, men must give up some of their power, and lose their fixed role of masculinity in order to come to terms with and express their feelings (2006). It doesn't help that men feel that the only emotion they are allowed to or encouraged to express is anger. My abusive ex-boyfriend would hold emotions in until he would erupt in rage, and he would try to appear rational and maintain control when his demands were totally irrational and nonsensical. The model of hegemonic masculinity also encourages men to perform acts of violence on each other, to find out who is stronger and who is weaker. This ideal of dominant masculinity, along with the idea of submissive, dependent, femininity, must be changed in order to curb gender-based violence.

This writing has allowed me to learn from my own person troubles with domestic violence, and to see the ways in which gender based violence is a bigger social issue. Putting this paper together helped me to see that it wasn't just my own personal "defects," or just his insecurities, that caused me to experience a violent relationship, as "the individual abuser and the victim do not operate in a vacuum; rather, they are nested within the supportive circles of social institutions and culture" (In class handout, "The Power Wheel"). The Power Wheel diagram allows one to see how abuse plays out in the micro world of the individual relationship. The abuser gains power through use of intimidation, coercion and threats, isolation, male privilege, minimizing denying and blaming, using male privilege and emotional and economic abuse coinciding often with physical abuse. The power Wheel then displays how this individual relationship is in the center of a surrounding macroworld. That abuse exists in a society where the culture consisting of values, rituals, language, and norms lend themselves to the institutions of education, economics, work, government, media etc, which then allow for the individual acts of gendered violence (Class handout, "The Power Wheel"). Gendered violence is an issue of unequal power and inequality and this difference in power is seen in the way that society has constructed masculinities and femininities (through values, norms), and the practices that enforce these constructions. Estelle Disch sums up these institutions that are blamed for violence against women:

> They include the system of gender inequality that creates an image of women as inferior objects worthy of disrespect; pornography, which sexualizes women's inferior status and presents women as fair game for sexual abuse; privacy since it encourages a lot of violence to remain behind closed doors, [and] women's unemployment and poverty, which keeps women from leaving abusive men. (2006:472)

These institutions along with the subscription to women of the current beauty ideal, the invisibility of girls in school, and the fact that most women to not even consider using violence to protect themselves, contribute to the violence against women that occurs worldwide. Many feminists view violence "not just as a product of women's subordination but as actively adding to that subordination—in other words, women can never have equality until gendered violence has ended (Marchbank and Letherby, 2007:273).

The issue is very important, both for the individual position and psyches of women who have experienced, are experiencing and/or will experience gender based violence, and for the situation of equality for women as a whole, because until the structures of inequality, based in the power of patriarchy and subordination of women, are destroyed, the destruction of women's lives through violence will continue.

WORKS CITED

Allen, Robert L. "Stopping Sexual Harassment: A Challenge for Community Education." Pp. 501-511 in *Reconstructing Gender: A Multicultural Anthology*, edited by Estelle Disch. Boston, MA: McGraw Hill.

Avicolli, Tommi. 2006. "He Defies You Still: The Memoirs of a Sissy." Pp. 147- 152 in *Reconstructing Gender: A Multicultural Anthology*, edited by Estelle Disch. Boston, MA: McGraw Hill.

Bartky, Sandra Lee. 1988. "Foucault, Femininity, and the Modernization of Patriarchal Power." Pp. 25-45 in *The Politics of Women's Bodies: Sexuality Appearance and Behavior*, edited by Rose Weitz, 2003. New York: Oxford University Press.

Crittenden, Ann. 2006. "Sixty Cents To A Man's Dollar." Pp. 417-424 in *Reconstructing Gender: A Multicultural Anthology*, edited by Estelle Disch. Boston, MA: McGraw Hill.

Devor, Holly. 2003. "Gender Role Behavior and Attitudes." Pp. 484-489 in *Signs of Life in the USA*, edited by Sonia Massik and Jack Solomon. Bedford: St. Martins.

Disch, Estelle. 2006. *Reconstructing Gender: A Multicultural Anthology*. Boston, MA: McGraw Hill.

Feinberg, Leslie. 1993. *Stone Butch Blues*. Los Angeles: Alyson Books.

Goodwin, Jan. 2006. "The Ultimate Growth Industry: Trafficking in Women and Girls." Pp. 491-495 in *Reconstructing Gender: A Multicultural Anthology*, edited by Estelle Disch. Boston, MA: McGraw Hill.

Kaye/Kantrowitz Melanie. 2006. "Women, Violence, and Resistance." Pp. 478-490 in *Reconstructing Gender: A Multicultural Anthology*, edited by Estelle Disch. Boston, MA: McGraw Hill.

Kilbourne, Jean. 2000. *Killing Us Softly 3*. Film. Edited & Produced by Sut Jhally. Media Education Foundation.

Lorber, Judith. 2006. "The Social Construction of Gender." Pp 113-120 in *Reconstructing Gender: A Multicultural Anthology*, edited by Estelle Disch. Boston, MA: McGraw Hill.

Marchbank, Jennifer and Gayle Letherby. 2007. *Introduction To Gender: Social Science Perspectives*. Harlow, England: Pearson Longman.

Marusich-Smith, 2008. Class Notes. Sociology 316: Family Violence. University of Massachusetts Boston, Spring.

McIntosh, Peggy. 2006. "White Privilege: Unpacking the Invisible Knapsack." Pp 71-76 in *Reconstructing Gender: A Multicultural Anthology*, edited by Estelle Disch. Boston, MA: McGraw Hill.

Petrie, Phil W. 2006. "Rean Men Don't Cry… And Other "Uncool" Myths." Pp. 221-226 in *Reconstructing Gender: A Multicultural Anthology*, edited by Estelle Disch. Boston, MA: McGraw Hill.

Rubin, Lillian. 2006. "The Transformation of Family Life." Pp. 304-313 in *Reconstructing Gender: A Multicultural Anthology*, edited by Estelle Disch. Boston, MA: McGraw Hill

Sabo, Don. 2006. "The Myth of the Sexual Athlete." Pp. 274-278 in *Reconstructing Gender: A Multicultural Anthology*, edited by Estelle Disch. Boston, MA: McGraw Hill

Sadker, David & Myra. 2006. "Missing In Interaction." Pp. 354-360 in *Reconstructing Gender: A Multicultural Anthology*, edited by Estelle Disch. Boston, MA: McGraw Hill.

Saltzberg, Elayne A. and Joan C. Chrisler. 2006, "Beauty is the Beast: Psychological Effect of the Pursuit of the Perfect Female Body." Pp. 162-172 in *Reconstructing Gender: A Multicultural Anthology*, edited by Estelle Disch. Boston, MA: McGraw Hill.

Staples, Brent. 2006. "Just Walk On By: A Black Man Ponders His Power to Alter Public Space." Pp. 191-194 in *Reconstructing Gender: A Multicultural Anthology*, edited by Estelle Disch. Boston, MA: McGraw Hill.

Stoltenberg, John. 2006. "How Men Have (A) Sex." Pp. 264-274 in *Reconstructing Gender: A Multicultural Anthology*, edited by Estelle Disch. Boston, MA: McGraw Hill.

"The Power Wheel." http://womenscenter.virginia.edu/sdvs/domestic/power_wheel.pdf

Thompson, Becky W. 2006. "'A Way Outa No Way': Eating Problems among African-American, Latina, and White Women". Pp. 176-188 in *Reconstructing Gender: A Multicultural Anthology*, edited by Estelle Disch. Boston, MA: McGraw Hill

Wolf, Naomi. 2003. "The Beauty Myth." Pp 451-459 in *Signs of Life in the USA*, edited by Sonia Massik and Jack Solomon. Bedford: St. Martins.

Wright Mills, C. 1959. *The Sociological Imagination*. Oxford: Oxford University, Press, Inc.

The Snail's Pace of Racial Progress in America
Sociological Insights from a Participant Observer

Henry Mubiru

University of Massachusetts Boston

henrymubiru@yahoo.com

Abstract: This self research essay is a personal reflection on what I know of and how I know the experiences and racial relations in America generally and in my immediate surroundings particularly. This self exploration will focus on several special incidents that are pivotal in my understanding of racial relations. Based on the exploration undertaken, I would like to submit that racism is real, alive and well in our society. The prospects for improved racial relations are there, but unfortunately they do not cut across the larger United States. In much of the rural and countryside America, in the face, discriminative and prejudiced racial practices continue to shape the relations between white and black America. On the other hand, in the urban and suburban areas crude, in the face racist practices like denial of seats in public arena's is fading only to be replaced with the much crafted institutional forms of racism championed by state agencies. Personally I see the race relations improving though at a snail's pace. The election of people of color to public office, more minority students at universities and colleges and the increasing number of inter-racial relations are examples. The more people become enlightened of each others contribution to the whole, the more we are likely to accept our differences and forge ahead. I see friendships of all races at major colleges and campuses and personally I've extended my hands to reach my white brothers and sisters to openly discuss social issues and what we can do to improve them. With the minorities of yesterday becoming the majorities of today, race harmonization is a matter of priority. The optimism espoused by presidential candidate Barrack Obama of a color blind America is indeed a step in the right direction that should be followed by those charged with policy execution, for it is these policies that pertain to housing, the criminal justice industry and the job market that have a crucial influence on racial matters.

INTRODUCTION

This self research essay is a personal reflection on what I know of and how I know of the experiences and racial relations in America generally and in my immediate surroundings in particular. This self exploration will focus on several special incidents that are pivotal in my understanding of racial relations.

First and foremost, as a black child growing up in the late 1970s and early 1980s in Uganda on the African continent, my knowledge of race issues was very limited. My immediate family, the schools I at-

Henry Mubiru is an undergraduate sophomore at UMass Boston, majoring in Criminal Justice. He wrote this paper while enrolled in the course Soc. 321L: "Racial and Ethnic Relations," instructed by Anna Beckwith (Lecturer of Sociology at UMass Boston) during the Spring 2008 semester.

tended and the people I interacted with were basically of the same **race**, a homogenized society. However, there are people from different ethnicities that make up the Ugandan population. These different ethnic groups have a different language and culture quite different from mine that I became aware of as I grew up. The Bantu ethnic group that I belong to control the power and resources of the nation and as such inevitably there is constant friction between the different ethnic groups for survival.

Apart from occasionally seeing white people on television and newspapers, my parents did not dwell so much on explaining to me why there were people different from me, and whether for that matter anybody's color mattered at all. I suspect they had other crucial issues to think about; especially if there was something for me to eat, how to pay my school fees and how to deal with the ever present fear of malaria. Later in the mid 1980s, I recall the rising calls to end the powerful forces of apartheid in a segregated South Africa and how Nelson Mandela was championing the cause of black South Africans to freedom and independence. This knowledge for the first time gave me an insight to think about the racial dichotomy of black and white and to ask my teacher then, "why the white man would oppress a black man in his land?" I further gained valuable knowledge through learning about the history of **colonialism**, where country after country on the African continent was occupied by the white man, exploited of its resources and the native people left largely exposed to the horrors of generational poverty and the divisive politics of religion. The savagery and barbaric acts of slavery that stripped the African continent of most of its resource—the people only to be reduced to sub-humans working on the white man's plantation in the Americas—also gave me a snap shot on **racial relations** in society.

HOUSING

In 2001 I migrated to the United States. Since then I've been struggling with the question of who I am in a racialized society while dealing with the most critical issues that cut across the racial divide including: housing, immigration, employment, the criminal justice system and interracial romance.

For many immigrants, their first real feel about the tenets of **racism** is when they are looking for housing. I and my cousin set out to look for some nice apartments especially those that had the "for rent signs" posted. This was in the city of Waltham largely occupied by middle class white Americans. We were always received with smiling faces but once we opened our mouth, we were told "how we had an accent and asked where we were from." I remember one landlord asking me if we had tried public housing or thought about the Dorchester area where "public transportation is easily accessible." To me this was blatant racial discrimination that was intended to inevitably bundle me with my own kind. The point had been made that the accommodation I was seeking was not meant for people like me even though I could afford it. When we eventually got an apartment for rent through an agent who charged us a month's rent, we were excited and moved into an area we revered for its safety but also a growing number of African immigrants. To our disbelief when we moved in, the white neighbors in the opposite apartment who rarely talked to us moved out. This trend consistently continued as more immigrants moved in, almost all white families left as if in protest. In fact, just like in "The Making of Groveland," "There would be about seven or eight of us (blacks) and about thirty of them (whites). But, you know, you'd see a white family move out and a black family move in, and soon it was more of us"(247).

In discussing the issue of housing and how it invokes racist undertones, many of the immigrants in my community have stretched further to have a shot at the bigger American dream, and have failed albeit systematically. Many realtors sold poor, high interest recurring, sub-prime mortgages to immigrants promising them "the best deal ever." However, as reality is sinking in, many of the immigrants specifically targeted by the shear looks of their ethnicities have lost the houses and the "little" savings they had put in. According to the article "Here Today, Gone Tomorrow: The Impact of Subprime Foreclosures on African American and Latino Communities" published by the Poverty and Race Research Council Washington DC., African Americans and people of color are three times as likely as whites to receive a subprime loan.

If ownership of a house is an important indicator of one's social standing in the American society and indeed a stepping stone to the creation of wealth, then a great deal of my friends in the sprouting immigrant community and to a large extent the minority population have been left out of this important process. The resultant wealth gap is indeed seen as a racist design to promote the majority whites while exploiting the minority races. This systematic racial impoverishment and the corresponding enrichment is also best described in the documentary, *Race: The Power of an Illusion Part III*, in which both white and black soldiers returning from the World War II were encouraged to take advantage of the housing initiative, popularly dubbed the Fair Housing Act, without having to put down a huge down payment. Again as in so many other times the blacks lost out on this opportunity for they were deemed unworthy and "would inevitably lower the value of these properties if they were allowed to move in." This created the concepts of **red lining** ratings for the poor minority and **green lining** good ratings for the white majority.

Indeed this had been a life time opportunity for the federal government to level and deal with the issue of racism once and for all, but this chance was lost too. The fact that the federal government failed to take any action in condoning what the Department of Housing was doing under its very nose, was to me the formal sanctioning of segregated living in America. Today in our society I see the same trend of denying minority races equal opportunity in housing with many city and town laws aimed at curtailing black people from moving into suburban areas predominantly occupied by the whites. Some of these include high property taxes where many blacks can't afford and the city zoning laws that demarcate exclusive white neighborhoods.

IMMIGRATION

My **immigration status** has also made me realize how the immigration laws of the United States have consistently prioritized who is urgently needed in this country and who is not. On the one hand, those of us from the poor continents mainly inhabited by people of color have got a raw deal as far as the immigration process is concerned, waiting in lines for years before their cases are approved or even reunited with their families. Those from the developed continents of the world, on the other hand, especially western Europe, have had a red carpet rolled out for them. Less is required of them since their skin color ultimately gives them a free pass to most of the important life-changing resources like access to employment and education. In this clearly calculated public policy, a white man (with the same age as mine) is unfairly propelled in the system for his likeness to the majority white Americans whereas unfairly denying me (of the minority race) the same equal opportunity. This prioritization is a huge determinant of who enjoys the **privileges**, resources and the wealth of this nation. Such privileges include for example quick and easy access to bank loans, cheaper credit lines and corporate employment. While I struggle with eighty hours of work a week to survive in a capitalist economy, my fellow immigrant from Europe has a leg up enjoying the benefits of **institutional racism**.

In addition to these facts, as an immigrant I have been labeled **alien** in many of the places that I've worked. I've been told, "to go back to Africa and deal with our natural poverty." It has become a heated debate in America today about whether immigrants are any more welcome as it was at the beginning of the century. In many cities and towns across America, white Americans in particular are up in arms protesting the lack of **assimilation** to the American way of life by inferior immigrating subgroups. Just like me, many of the immigrant communities have succumbed to the stereotypes labeled at us: inferior, savage, uncivilized, diseased, and with low intelligence, in comparison to the majority whites. In the article, "Covering Immigration: Popular Images of the Nation" by Leo Chavez (213), the concept of immigration wars seems to be surging. There is a "demonstrable shift to an increasingly anti-immigrant public debate and public policy initiatives during the last quarter of the twentieth century…As Oscar Handlin once wrote, immigration is the history of the nation. But immigrants are also newcomers whose difference and 'otherness' do not go unquestioned or unremarked upon. Their very presence raises concerns about population growth, economic competition, and various linguistic and 'cultural' threats" (214).

Immigration is another area that I have clearly seen the invisible yet powerful forces of racism in the American society. Having been told earlier on that America is a nation of immigrants, I thought I was indeed at the forefront of those who would be welcomed with open arms. I can't recall the number of times I underwent **racial profiling** by the police because I looked foreign and my immigration status was highly questionable. One particular incident still memorable to me was when I was riding on a train in the sub way of New York City. The officers asked everyone to have their identification at hand. I was seated next to a white man who did not have an identity card or a passport, both of which I carried. The officer looked at my license and at me and asked where I was going, where I lived, and what in particular I was doing in the United States. To my own embarrassment I answered all these questions in the full presence of all the passengers. Within me I felt a sense of being rejected from a country I loved and respected. My white neighbor, whom I later learned was a 'visitor' from Romania, was told, "to travel with identifying papers next time" and was not profiled like I was.

The notion of being **prejudiced** against people from the point of view of their nationality or origin, as much as is perplexing in America, is not a new phenomenon in the trek of immigrants. According to Herbert Blumer, race prejudice is "the feeling on the part of the dominant group of being entitled to either exclusive or prior rights in many important areas of life. The range of such exclusive or prior claims include: choice lands and sites, right to certain jobs…the claim to certain positions of control and decision making as in government and the law and the right to exclusive membership in institutions such as schools and the church" (171). As a black man I have often felt less welcome at Museums, at the city park in my town or at the Veterans post where I predominantly see the white race.

In the film *Mississippi Masala*, we see how in the 1970s Indian Ugandans were discriminated against and forced out of a country they called home. They were seen as non-Africans and "Uganda was for black Africans." The reality of the matter is that these immigration sentiments and emotions are not going to end tomorrow just as I realized myself. In fact, in the United States, black immigrants like myself face constant prejudice and discrimination. The task before immigrants is to quickly learn and blend into a racialized society, and jump on any opportunity towards the American dream.

According to Howard Taylor,

> Race is multiply defined in this society; that is, there is no one single definition, but several definitions. All these definitions apply simultaneously, and no one definition takes precedence over another. These definitions do however have one thing in common; they are all creations of society. They have been put in place by humans and by their social interactions and their societal institutions. (47)

Earlier, I pointed out that in this essay I will try to show how for quite some time I struggled with defining my race. If I define **race as biology**, the color of my skin would qualify me to be a member of the black race which I openly embrace. According to Clara E. Rodriguez and Hector Cordero-Guzman, "Placing Race in Context," "Race is something more than phenotype and genotype and was influenced by contextual factors such as class, education, language and birthplace. Thus race may be more of a social construction than is generally admitted" (90). For example there are black people with a light skin and there are white people with full lips and curly hair.

When I interacted with different people especially on the jobs, many referred to me as a negro, while my fellow blacks called me incognito because I hanged around white folks and I lived in a place predominantly white. To them my social status was not in any way compatible with the black race and therefore I was not exactly a true "Brother." Here, **race as a social construction** depends on how different groups of people interact with others. This social process can lead to stereotypes about people just like many of my black friends who didn't see me as a member of the black race.

Race as ethnicity also defies logic. For example, I identify myself as a member of an ethnic group (African) on the basis of a shared common culture. Yet, there are other groups of black people that do not identify as African even though they share some elements of African culture. These include the Haitians and Cape Verdians, for instance. So, to ethnically identify them as racially black Africans would be un acceptable to many of them.

There are people whose **race** in society is defined by their **social class** especially invoking aspects of wealth and status. In some societies like Brazil, those who have been lucky to access these wealth creating resources have been quickly initiated into the white race and those on the bottom rung of the wealth hierarchy, even when physically white have been designated black. In the United States, however, race is primarily defined through biology.

In my own observation I find all these definitions contradictory, and just like Michael Omi and Howard Winant in "Racial Formation," "The designation of racial categories and the determination of racial identity is no simple task. For centuries this question has precipitated intense debates and conflicts in the United States; disputes over natural and legal rights, over the distribution of resources, and indeed, over who shall live and who shall die" (75).

EMPLOYMENT

Employment is another critical area where the issue of race has distinctively surfaced. Race is an important determinant of life enhancing opportunities in the labor market. Many Americans depend on their jobs for their survival and that of their immediate families. From the little savings out of the jobs they get, income which is used to create substantial wealth depends on the nature and type of jobs. It is important to note that a few of my work experiences indicate that work has to a great extent been racialized and indeed shaped by racism. The concept of **racism** is defined as, "a principal of social domination in which a group that is seen as inferior or different because of presumed biological or cultural characteristics is oppressed, controlled and exploited—socially, economically, culturally, politically, psychologically—by a dominant group" (Wilson 1973, as quoted in Higginbotham and Andersen 83). This principle of domination is alive and well in the labor markets today with the dominant white majority determining which race is employed where and how.

In my short work experience as a person of color, there are jobs that I have tried to access only to reach a dead-end. Many of these jobs are found in the affluent suburban communities, offering relatively higher wages, and are strictly reserved for the whites. While searching for a job a few years ago through an employment agency in Billerica, the staff coordinator informed me of an open position in a manufacturing firm for Air controls in Newton. She actually sent me there and the production manager told me that indeed there was an open slot on the assembly line, but he couldn't make a decision then but would call me. I waited for a call and made calls to him until I was told that they needed somebody who lived "close to the plant" and it would be difficult for me to commute from Medford where at the time I was living. It was interesting to note that one could be denied a job for not living in and around the job location. To me this was a flimsy excuse to politely say, "no you are not welcome." Also a close look at the employees of this factory whom I saw from the parking lot revealed that a majority of them were white while a minority Asian. To me such jobs that are exclusively reserved for the dominant white majority remind me of the blatant discrimination of the past in which there were signs and posters at job sites that read, "Whites only Blacks need not apply." Just like my experience with this case, discrimination on the jobs today is not as blatant or straight in your face, but it has taken on "a more elusive and less apparent stance with many companies expending millions of dollars to conceal it" (Herring 277).

Another way work has been racialized through the use of advertisements. In many of the jobs that are exclusively held by the white majority into which access for blacks or other minorities is limited, advertisement has been used as a tool to deny access to those groups. Instead of advertising in the main stream media, for example major newspapers like the *Boston Globe*, many companies are concealing the availability of openings by using area newspapers. In the cities of Waltham and Newton for instance, outlets like the *Waltham Chronicles*, the *Tribune* and the *Daily Sun* are used to advertise jobs in the many manufacturing and biotech companies in the area. Many of the minorities who live in these affluent communities, as recently told by a close friend searching for a job in the nearby factory, do not read these newspapers that are many times delivered through subscription at the door steps of the residents. In addition to this concept of concealed advertisement, my experience has led me to realize how the use of employment networks has racialized the workplace.

There are outstanding channels through which the white majority pass on information about job availabilities, such as pending interviews, who is to carry them out, the possibility of promotions, and how one can improve on their chances of promotion. These include a steady line of well maintained friends, exclusive membership to clubs and other recreation places, and the extra help extended to the white students by the teachers in terms of referrals, direct job recruitment and vacancy information. This is popularly referred to as **white privilege** as it makes it easier to access resources necessary for success including jobs. Unfortunately this kind of networking is distinctly absent for the minority races, the African Americans in particular. According to Deirdre A. Roster in "How White Networks Exclude Black Men from Blue collar Jobs,"

> ...[The] municifent flow of various forms of assistance ... served to convert institutional ties (as teachers) into personal ones (as friends) that are intended to and do endure well beyond high school. (289)

> ...[The] black males could not effectively use the institutional connection to establish successful trade entry...As a result, black men sought employment using a truncate, resource-impoverished network consisting of strong ties to other blacks. (290)

My experience has also led me to realize that even the types of work minorities of color do and the corresponding status pegged on them is by far **racialized**. Different races have been largely assigned to particular jobs. Low status and less paying jobs without benefits mainly go to minorities of color, while high status jobs with opportunity at promotions and better pay go to whites. I tried on several occasions to get a white collar office job that would provide me with a decent salary and benefits, for I thought with my modest education credentials (an associates degree) I would easily join the entry level rungs. Similar to the first instance where I was rejected for the color of my skin, this time my foreign accent and the lack of knowledge of the base ball game, played an additional role.

I was subsequently told to understand by a close friend that white collar office jobs were not for us and that he had a place with a ready job for me. This place turned out to be a nursing home catering primarily to elderly white residents. What surprised me was majority of the wage workers were exclusively black and all the senior nursing positions went to the whites. This scenario of events seems to be the trend in the larger American labor market—with risky, no benefit, manual oriented, and less opportunity for promotions based jobs going to the blacks and other minorities, while the high technology, skilled, high wages with benefits and the possibility of upward mobility are reserved for the dominant white majority. According to Cedrick Herring in "Is Job Discrimination Dead?"

There are both primary and secondary jobs, the later reserved for the Whites and the former to the blacks by employers but sometimes by Unions who fear blacks will drive down their wages.

> As a consequence, African Americans are locked out of jobs in the primary sector labor market, where they would receive a better pay and treatment, and they tend to be crowded into the secondary sector. As these disparities compound overtime, as primary sector workers enhance their skills and advance, secondary sector workers stay mired in dead -end jobs. (282)

According to Lawrence D. Bobo, the labor market has further been racialized through **laissez-faire racism by** assuming that the government has no hand in racism (either in its creation or its destruction), rather it is the invisible hands of supply and demand in the market which fuel racism.

The idea that everything is to be determined by market forces of demand and supply and that government has limited or no input is quite self-defeating in itself. Yeah, I can see some progress in achieving a leveled ground for all races to access opportunities. However, I also see institutional racism pro-actively twisting the achievements and systematically threshing out people of color from the opportunity process. Examples of segregated housing and poor inner city public transportation are just points to highlight.

The labor market in America has to a great extent been racialized by the invisible market hands of racism, exclusive networkings for the white majority, and the apportionment of secondary jobs to minorities and primary jobs to the white majority. In essence the hard, less rewarding, bottom of the rung, dirty jobs tend to be predominantly done by the minorities. I've taken on labor intensive jobs like clearing construction sites, snow shoveling and dish washing for very long hours yet earning peanuts out of it. Duty and responsibility bound, full-benefit jobs go to the white majority.

CRIMINAL JUSTICE

Racism is a powerful force at play in the labor market today; though not very visible, it is structurally rooted. This spirit of group domination has further been used in the criminal industry with large percentages of blacks and other minority races jetted into prisons and jails, not given any opportunity at redemption as a systematic way of ridding them from the social fabric. Many of my friends and people I know in various communities have succumbed to this fatal scenario, where they are either jailed or deported for clearly non-threatening crimes. According to Manning Marable,

> The informal, vigilante-inspired techniques to suppress Blacks were no longer practical. Therefore, beginning with the great depression, and especially after 1945, white racists began to rely almost exclusively on the state apparatus to carry out the battle for white supremacy. Blacks charged with crimes would receive longer sentences than whites convicted of similar crimes. (Marable, as quoted in Eduardo Bonilla-Silva 370)

It can be precisely said that race relations in the United States continue to portray the tensions and mistrust between the white and black races, pushing the wedge in the racial relations further. The numbers **incarcerated** in the prisons for minorities particularly African Americans drive this notion of a systematic group position, and Angela Davis points it out in a most blunt way in her article, "Race and Criminalization: Black Americans and the Punishment Industry," The Democratic and Republican officials have successfully put in place an out of control punishment industry with a huge racial imbalance assuming criminality of Black people and thus evidence of structural racism.

INTERRACIAL ROMANCE

Let me move next to the issue of interracial romance and how it is pivotal in understanding racial relations in our society. Personally I have not been involved in any bi-racial relationships, but who knows what tomorrow holds for man! However, many of my close friends have and are actively involved in these mixed relationships. Their experiences have been quite diverse as the relationships themselves; many have been welcome in these families with open arms especially the women married to or dating white men. On the other hand, some of the black guys dating white women have faced un-believable resistance from general rejection as inferior and uncivilized to the more racist descriptive undertones of being called the N word. A lot of times the reasoning is white women are for white men and exclusively so. At other times where a bi-racial family has been formed, as is the case of my friend Charles, the offsprings have had a rough time to socialize as they are frequently asked about their parentage and why they for example have black hair or skin. This scenario underscores the mentality widely held of a purified white race, dominant with a group position on many of the social, political and economic undertakings of American society. In the novel *Indian Killer* by Sherman Alexie, we see John a native American Indian adopted and raised by white parents. When he starts to date white girls, "The girls' fathers were always uncomfortable, and grew more irritated as he continued to date Mary, Margaret, or Stephanie…

> 'Hon,' a father would say to his daughter. "What's that boy's name?"
>
> "Which boy, daddy"?…
>
> "…Well, adopted kids have so many problems adjusting to things, you know. I've read about it. They have self-esteem problems. I just think, I mean, don't you think you should find somebody more appropriate?" (17, 18)

This is indeed a veiled racist remark made by the white parent to the unsuspecting daughter which eventually continues to strain the race relations in society.

In the film, *Mississippi Masala*, Mina, an Indian-Ugandan immigrant-refugee, is dating a black man unknown to her parents. When they find out, they ferociously fight this relationship and order her to end it. To the parents, "Indian women are for Indian men and not for the inferior poor Blacks." Through my personal experience, it can therefore be said that, when it comes to bi-racial romance many times in our society we have pegged onto them a negative indicator further stratifying us along racial lines and inevitably reducing the opportunities towards assimilation.

IDENTITY

In "Finding the Universal: Reflections on a Multi-Prismed Identity," author Mark Hagland explores the notion of identity construction by an Asian American. "I often find myself with multiple identities; 'I'm a man, an American, Asian American, an adult Korean adoptee, gay person, …journalist and spiritual seeker.'" (39). Through these differences there are "multiple assumptions that people make regarding racial, ethnic and sexual identity that impact how they view and interact with me personally" (39).

I've on several occasions been described as a male, African American, student, single, non-alcoholic and spiritual. This, I've found out, determines my interaction with a cross section of society. Living in the suburbs of the Boston, I've realized as a minority I'm always underlooked, like "not much is expected of me." Kids I meet at the community field are always inquisitive to know whether there are any houses and cars where I come from! The general lack of acceptance of personal diversity to the extent that instead of celebrating it, we just trash it, is in itself another indicator of the domination and exploitation of dominant majority race to subjugate and eliminate the minority races"—another form of racism. The differences in perception that I receive depending on the nature of the identity involved, brings with it further confusion and isolation unknown to others with the same differences in the majority race.

These issues of identity based on skin color, national origin, gender, sexual orientation, and faith all lead to wars and conflicts, exacerbating the racial tensions devastating to personal and group interaction.

In fact, whether it's the majorities or minorities in society today, we all need each other and are indeed co-dependents. The minorities of yesterday will be the majorities of tomorrow, and the majorities of yesterday may be the minorities of tomorrow! Racism targeted towards minority races still rampant in our society, is likely to proportionately decline especially as the number of minority populations swell significantly. I see a comparably high number of ethnicities and races at job sites, in apartment buildings, in the hospitals and much more pronounced in our school system. It is important to note that the gains anticipated from an increasingly racially diverse society could easily be erased by both the structural and institutional forms of racism prevalent in our society. According to Farai Chideya, in "A Nation of Minorities: America in 2050,"

> America is facing the largest cultural shift in its history. Around the year 2050, whites will become a "minority." This is uncharted territory for this country, and this demographic change will affect everything. Alliances between races are bound to shift. Political and social power will be re-apportioned. (11)

The teens and twenty-somethings of the millennium generation are the true experts on the future race, for they are re-creating America's racial identity every single day and are more likely to interact with people of other races and backgrounds than other generations. Critically important is, a third of this generations is non-white, not just black, but Asian, Latino, Native Americans and multi-racial.

CONCLUSION

In conclusion, I would like to submit that racism is real, alive and well in our society. The prospects for improved racial relations are there but unfortunately, they do not cut across the larger United States. In much of the rural and countryside America, in the face, discriminative and prejudiced racial practices continue to shape the relations between white and black America. On the other hand, in the urban and suburban areas crude, in the face racist practices like denial of seats in public arena's is fading only to be replaced with the much crafted institutional forms of racism championed by state agencies.

Personally I see the race relations improving though at a snail's pace. The election of people of color to public office, more minority students at universities and colleges and the increasing number of inter-racial relations are examples. The more people become enlightened of each others contribution to the whole, the more we are likely to accept our differences and forge ahead. I see friendships of all races at major colleges and campuses and personally I've extended my hands to reach my white brothers and sisters to openly discuss social issues and what we can do to improve them. With the minorities of yesterday becoming the majorities of today, race harmonization is a matter of priority.

The optimism espoused by presidential candidate Barrack Obama of a color blind America is indeed a step in the right direction that should be followed by those charged with policy execution, for it is these policies that pertain to housing, the criminal justice industry and the job market that have a crucial influence on racial matters.

WORKS CITED

Blumer, Herbert, "Race Prejudice as a Sense of Group Position" in Charles A. Gallagher (ed.), *Rethinking the Color Line, Third Edition* (McGraw Hill, 2007), Pg. 169-175.

Bonilla-Silva, Eduardo, "Keeping Them in Their Place: The Social Control of Blacks Since the 1960s" in Elizabeth Higginbotham and Margaret L. Andersen (eds.), *Race and Ethnicity in Society, Third Edition* (McGraw Hill, 2006), Pg. 370-375.

Chavez, Leo R., "Covering Immigration: Popular Images and the Politics of the Nation" in Elizabeth Higginbotham and Margaret L. Andersen (eds.), *Race and Ethnicity in Society, Third Edition* (McGraw Hill, 2006), Pg. 213-217.

Chideya, Farai, "A Nation of Minorities: American in 2050" in Elizabeth Higginbotham and Margaret L. Andersen (eds.), *Race and Ethnicity in Society, Third Edition* (McGraw Hill, 2006), Pg. 11-16.

Davis, Angela Y., "Race and Criminalization: Black Americans and the Punishment Industry" in Charles A. Gallagher (ed.), *Rethinking the Color Line, Third Edition* (McGraw Hill, 2007), Pg.314-323.

Davis, Delvin, "Here Today, Gone Tomorrow: The Impact of Subprime Foreclosures on African-American and Latino Communities," in *Poverty & Race Research Action Council, Vol. 16, No. 3, May/June 200 7*, Pg. 1-2 and 12-13.

Espiritu, Yen Le, "Asian American Panethnicity: Bridging Institutions and Identities" in Charles A. Gallagher (ed.), *Rethinking the Color Line, Third Edition* (McGraw Hill, 2007), Pg. 113-121.

Hagland, Mark, "Finding the Universal, Reflections on a Multi-Prismed Identity" in Jane Jeong Trenka et al (eds.), *Outsiders Within* (South End Press, 2006), Pg. 39-42.

Herring, Cedric, "Is Job Discrimination Dead?" in Elizabeth Higginbotham and Margaret L. Andersen (eds.), *Race and Ethnicity in Society, Third Edition* (McGraw Hill, 2006), Pg. 277-284.

Higginbotham, Elizabeth and Margaret L. Andersen (eds.), *Race and Ethnicity in Society, Third Edition* (McGraw Hill, 2006).

Kwame, Hassan, et al, "Toward a Transformative View of Race: The Crisis and Opportunity of Katrina," in Chester Hartman and Gregory D. Squires (eds.), *There is No Such Thing as a Natural Disaster* (Routledge, 2006).

Omi, Michael and Howard Winant, "Racial Formation" in Elizabeth Higginbotham and Margaret L. Andersen (eds.), *Race and Ethnicity in Society, Third Edition* (McGraw Hill, 2006), Pg. 74-78.

Pattillo-McCoy, Mary, "The Making of Groveland" in Elizabeth Higginbotham and Margaret L. Andersen (eds.), *Race and Ethnicity in Society, Third Edition* (McGraw Hill, 2006),

Rodriguez, Clara E. and Hector Cordero-Guzman, "Placing Race in Context" in Charles A. Gallagher (ed.), *Rethinking the Color Line, Third Edition* (McGraw Hill, 2007), Pg. 85-91.

Royster, Deidre A., "Race and the Invisible Hand: How White Networks Exclude Black Men from Blue Collar Jobs" in Elizabeth Higginbotham and Margaret L. Andersen (eds.), *Race and Ethnicity in Society, Third Edition* (McGraw Hill, 2006), Pg. 285-294

Sherman, Alexie, *Indian Killer* (Warner Books, 1998).

Taylor, Howard F., "Defining Race" in Elizabeth Higginbotham and Margaret L. Andersen (eds.), *Race and Ethnicity in Society, Third Edition* (McGraw Hill, 2006), Pg. 47-58.

Wilson, William Julius. *Power, Racism, and Privilege: Race Relations in Theoretical and Sociological Perspective.* New York: Macmillan, 1973).

Films:

Race: *The Power of An Illusion, Part III, The House We Live In.*, California Newsreel, Executive Producer Larry Adelman, 2003.

Mississippi Masala by Mira Nair, 1991.

No Longer Adding to the Problem

Changing Society's Racialized Structure from Within

David Couras

University of Massachusetts Boston

davidcouras@gmail.com

Abstract: In this paper I critically explore the role race has played in my personal life. Whether the events portrayed paint me in a light of perfection or not is not the issue at hand, but how the misperceptions of society in regard to race can easily sway the uninformed mind is essential to understand. The goal of this paper is to show how "race" is socially constructed and is not some inherent characteristic that separates us from one another. In becoming aware of this reality we can free ourselves from being prejudicial and racist, and begin to see people for who they truly are and not what we have been told to think they are.

INTRODUCTION

This self research paper is an exploration of the personal events in my life that have brought me to the current position I hold in regards to race in the North American society. In the earlier part of my life I viewed race "as an essence, as something fixed, concrete, and objective" (Omi and Winant, 75); However, as I matured in my thinking and further immersed myself into more than the miniscule world of Somerville, Massachusetts, my view of society and race became different, and more complex. This process not only transformed my way of thinking, but I believe has also allowed me to evolve as an individual.

I was able to accomplish this because the more exposure I had to the non-white world the more evident it became to me that race is not a creation of any genetic traits such as skin color, brain size, nose shape, etc., but is something more subjective and abstract and is in fact a social construction in which racial formation has occurred in order to decide who has what and who goes without.

In order to explain the evolution of my attitude toward racism in the past to the present, I would argue that it has taken place in five distinct phases of my life. These phases are very similar to the ones developed by William Cross, author of *Shades of Black: Diversity in African American Identity,* in which he argues that the five stages of racial identity development are pre-encounter, encounter, immersion/em-

David Couras is an undergraduate senior at UMass Boston, double-majoring in Political Science and Criminal Justice. He wrote this paper while enrolled in the course Soc. 321L: "Racial and Ethnic Relations," instructed by Anna Beckwith (Lecturer of Sociology at UMass Boston) during the Spring 2008 semester.

ersion, internalization, and internalization-commitment (see Tatum, pp. 54-55). This model was referred to as the psychology of nigrescence, or the psychology of becoming black, because this model was developed to explain how blacks have come to view themselves in the racialized world of American society. I think, however, this model can be applied in another way to the development of my own psychology as a white American. I can apply the model to understand how my attitude toward race and racism developed ever since my adolescent years and into my present adulthood.

I am presently undergoing the internalized-commitment stage in my life in which I see the misdeeds of many of my past actions in which I at the time had not perceived as wrong, and am making the commitment as an individual to help change the injustice of society while at the same time being comfortable with my identity as a white male.

PRE-ENCOUNTER

In the first stage of racial development, we have the pre-encounter phase, which takes place normally around the ages of 8-10, but certainly will vary depending on a person's specific experiences. I remember one of the last instances in which I was still present in this initial phase of racial development. I was about 7 or 8 years old and my friend Brian's mom took him and me to the local playground, probably to stop bothering her. When we got there a couple of kids in our age group asked us if we wanted to play freeze tag with them, and of course that being the game of the time we were more than happy to participate. Before we started, Brian's mother said we should offer them some of the leftover donuts that we had bought earlier, because she rationalized that her son and I would have an easier time catching these kids if they were weighed down by the donuts. That itself seemed logical to me even for the age of seven, but then she said something to us that at that point in time did not make perfect sense to me. "Make sure to give one to that black kid, you will need whatever you can to slow him down." I instantly thought to myself "hmm, I wonder what she meant by that as both of them are tall and skinny and look like they would be much faster than me and my short pudgy friend." It did not occur to me that she was basing this statement solely on the fact that this kid was black, and not for the obvious physical reasons that would give him and his white friend a certain edge on my friend and me.

It is in this stage that the issue of race is of no concern to an individual for one of two reasons. The first being that they are too young to grasp the understanding of the social connotations that come along with being a member of a specific racial group or scenario; and second, their exposure to other groups has been very limited, because they come from a racially homogenous neighborhood in which diversity is non-existent. The former of the two reasons would be more aligned with how I grew up, because in my immediate neighborhood there was a decent number of families of color around, and certainly I would encounter black kids at the parks and playgrounds, but in these early stages my view of race was not yet developed. In this pre-encounter stage, the individual truly lives in a color-blind society because while the physical differences are visible, the realities of many racial injustices are not yet apparent (see Gallagher, pp. 96-99). I saw race at this time in a very simplistic way because I simply saw everyone the same. I was too young to comprehend the social significance of any racial actions or comments, such as the one made by Brian's mom. In these early stages of my life, I was aware of the visible differences such as skin color that existed between me and this kid at the playground, but I would not be aware of the social stigma that came along with his black skin until I was much older.

ENCOUNTER

The innocence of the pre-encounter stage gave way to the harsh realities of the encounter stage. It is at this stage that individuals become aware of the different societal roles that come along with being a member of certain racial groups. This realization for me took place shortly after the playground incident, and it took place in two separate incidents. In the first incident, I was still about 8 years old or so. This encounter stage began with a very routine day-to-day encounter in which my younger sister's friend Lisa came to our steps as she had numerous times before to ask us "is Valerie around to play?" Sitting on the steps was, my older brother, older sister, and I just sitting there talking, and we told Lisa that "Valerie went to the grocery store with my dad and that she would be back later." As soon as she walked away I remember hearing coming out of my older brother's mouth a word that I would come to hear many times throughout my life "I can't stand that fucking *N----r*." Being the innocent yet inquisitive kid that I was I quickly responded to my older brother's statement by asking what the word meant. I was expecting to hear this long drawn out elaborate definition of what this word meant, but my brother simply responded "A black person."

The second incident of the encounter stage could not have happened without the existence of the first, and this second incident was perhaps the most intense racial encounter of my life. Fast forward roughly two months later from the first incident and we have my older brother and three of his friends playing street hockey, and one of the three happened to be black. Therefore, I thought perhaps naively that, well, if a black person is a "N----r" then well he must be one as well, so for some reason I thought like most young kids who hear a new word I would like to try it out. Lets just say that my calling him using that word did not go over so well and perhaps the only reason I left that scenario unharmed was because my brother and his two other friends were able to calm him down and convince him that I was a child who did not know any better.

At the time I time felt conflicted by this whole scenario, after all how could my brother tell this kid I didn't know any better when he was in fact the one who taught me this obviously painful and dangerous word. In retrospect, I realize that what my brother was trying to do was convince this kid that I did not know any better at my young and naïve age to comprehend that the word is such an ignorant and racially harmful expression to use. The fact of the matter is that my brother honestly meant that I didn't know any better to censor myself and refrain from using such words in the presence of black people. At the time, I could not have possibly perceived what all this meant, but now that I am older, I realize this incident truly embodies American society. I say this because America no longer practices Dejure segregation as used in the Jim Crow South, but we now practice what Lawrence D. Bobo (2006) would argue Laissez-faire racism.

This type of racism is not as open and blatant as the Jim Crow form in which it was legally acceptable to have whites and blacks segregated from one another; nevertheless this type of racism still promotes the idea of inequality. The only real differences of these two types of racist tools is suggested in the article by Eduardo Bonilla-Silva titled "Keeping them in their Place" in which he suggested the widespread lynchings that occurred during the Jim Crow South are now being legitimized by the state's legal system in which over half of all death-row inmates are black in spite of the fact that in society blacks make up only 12 percent of the population. This suggests that while we as a society don't blatantly enforce unequal policies we still as a society have constructed a system that punishes minorities, especially blacks unequally. It is hard to say what is worse and what is better, because both forms lead to such unfair treatment, but surely today's racism is not quite as bad; after all we refrain from using the N word until blacks are absent from the conversation.

White attitudes such as the one my brother had developed and I was developing truly were a product of this type of racism of which Bobo speaks. This type of racism not only has created an unjust system but has tricked white people into believing their actions have not been unjust. This accomplishment is evident in the sense that as of a 2001 Gallup Poll nearly 70 percent of whites said that in regards to the workplace blacks are treated "the same as whites," while at the same time 60 percent of blacks reported experiencing racial barriers at the work place (Herring, pp. 278-9). The reason for this is as my brother would have explained back then—that it was not wrong for him to use the N word because he didn't mean anything by it and it was just a word to use to define black people. Nevertheless these actions and words have consequences and have come to reflect society's behaviors. We may say that we care to change things and that we don't believe in inequality, yet our actions don't support the changes that are necessary for this to happen such as government interference in overturning the wrongs of the past and present. In fact many white Americans especially in the North agree with many issues that would further the cause of equality for minorities, such as in 1986 when 93 percent of all whites supported the idea of integrating white and black schools; yet this was highly problematic in the reality in which only 26 percent of all whites supported government interference to pursue these ends (Bobo 2006).

Over the next couple of years, my thoughts, and views on the differences between me as a white person and blacks, further developed through the use of media, my society, school, etc. My view of race by the time I was in high school progressed to the point as discussed by Herbert Blumer in his article "Race Prejudice as a Sense of Group Position"[1] in which he makes the argument that racial prejudice is a tool largely to maintain and preserve the status quo that currently exists among different groups. In our current racial society there are two roles that are being played off each other, on one side of the coin you have the dominant group played by the whites and on the other side of the situation you have the subordinate race, largely a cast consisting of blacks and other people of color in the starring role. The dominant group seeks to preserve this relationship for numerous reasons according to Blumer, such as a feeling of superiority in comparison to their minority counterpart, a feeling of intrinsic difference between the subordinate race, a belief that they are entitled to certain rights and privileges that do not extend to all spheres of life, and lastly a fear and suspicion that the subordinate race has plans to reign as the dominate group in American Society.

IMMERSION/EMERSION

These thoughts of white superiority and the need to preserve this dominance traveled with me and developed further into my high school years. By now I was full swing into the third stage of racial Development, that of Immersion/Emersion. It is in this stage in which one becomes aware of their societal position in terms of race and starts embracing this position. For me I

[1] Herbert Blumer in his article "Race Prejudice as a Sense of Group Position" discusses positions of the superior and the dominate group.

was embracing the position of white superiority, and my actions this time were aligned with those of furthering the cause of inequality whether I was aware or not. I remember sitting at the lunchroom in high school, my friends and I were stupid teenagers and were well aware of the presence of the segregated white and black tables. At my table during the freshman and sophomore years there were eight of us sitting there and except for our table all the tables in the cafeteria were completely segregated. But at our table one of the eight kids sitting was black. One day we were having a debate about this and the question was posed "Why are all the black kids sitting together at the cafeteria?" My friend Ryan who was the one black kid at the table immediately interjected and said well "I am black and I am sitting with you guys." Immediately after Ryan said this we all looked at each other and thought the same thing, but I was the one to convey it aloud. "Ryan you might as well be white!" By making this comment, I assumed that those of us at the table and the black kids at the school all agreed that he was an "atypical" black student due to his being an Honors student, speaking an extensive vocabulary, using minimal slang terms, and not wearing baggy clothes.

Ryan did not embody any of the mannerisms or slang that has been traditionally associated with being "black." For us sitting at the table a great deal of our perceptions in regards to what is black was shaped by our media intake. We were all familiar with movies such as "Boyz N the Hood," "Blue Hill Avenue," etc., as well as many Rap videos on MTV and grew accustomed to the notion that blacks by and large were typically aggressive, violent, inarticulate, and for the most part uneducated. Ryan did not fit any of these stereotypes in which the media has portrayed as being black, and as far as we were concerned at the table he was not black. I did not think of the significance of the remarks I had made at the time. It just came out as a natural response to his observation. Ironically I meant my comment of his being "white" as a positive statement, and not as a negative one. This is very similar to the situation in the novel *The Indian Killer* in which the main character John (who was Native American adopted at birth by a white family) was referred to as "Chief" by one of his construction co-workers. It is not quite clear which way his co-worker meant this word, but I believe based on the prior actions of his white co-workers to invite him out for drinks this man was actually making an extension to John to be a part of the group at work and no longer feel isolated from everyone else. John, however, took this comment especially from a white person to be demeaning and an affront to his Native American heritage (p. 65). When I said this I was in fact granting Ryan acceptance into the dominant white group, as Blumer wrote about, and was allowing him to feel comfortable as just another one of us sitting at the table. But what Ryan saw was me stereotyping black people by saying essentially they are the opposite of what he is, and therefore he is not black. "No stereotype is a good stereotype."[2] This instance shows the harms of stereotyping because it forces people a lot of times to act in a certain manner, because they believe that by not embracing these stereotypes they are not a part of this group. I am not surprised that by junior year Ryan was no longer sitting at our table, and was now wearing baggier clothes, and lost his spot as one of the top kids in our class; in turn he had become increasingly more popular among the black kids who the year before rarely associated with him.

This third stage of development may have been the most emotionally painful, because looking back I realized my comments had threatened Ryan's security as a

[2] This is a statement from professor Beckwith, made in one of her lectures.

black male when I referred to him as white. The irony of this whole situation is that I had a similar encounter in which my identity as a white male and a member of the dominant group was put into question by my peers during my high school years. I was at this point a junior in high school and I went to a Catholic High School in Somerville/Medford Massachusetts in which many kids that attended were from Somerville, Medford, and various Boston neighborhoods. The most represented neighborhood in the school was the heavily Irish ethnic enclave of Charlestown in which this neighborhood was bombarded with houses decorated by Irish flags and shamrocks painted on the front of numerous buildings as well as plenty of Irish pubs scattered throughout the tiny square mile area that Charlestown was comprised of. Needless to say, the people from this area are very proud of their Irish heritage and were not afraid to show it, and were not afraid to let you know especially if you were Italian that you were nothing more than a Guinea, a Wop, etc.

When I encountered this situation myself I was in shock. I felt as if I as an Italian was below them. Sure I was Italian, but most importantly I was white, and throughout the entirety of my life always felt compelled to circle or check the white/ Caucasian box on any form I had ever filled out but I was not their type of white. I was shackled to the *Hyphenated identity* of being Italian-white or Italian-American, if you will, in which these two identities conflicted with one another;[3] this, as far as they were concerned, was just not good enough for them as members of the dominate Irish group, or at the very least they represented the most dominant group in my neighborhood. In recalling this memory, I cannot help but compare it to the PBS Documentary, *Race: The Power of An Illusion*, in which in the 1900s the Massachusetts legal system ruled in favor of Armenians being white. What this ruling essentially did was give Armenians access to all the benefits that were associated with being white and allowed them membership into the dominant group of American society, the white group. I reflect on this, because I feel as if in this instance these two Irish kids were acting as the judges in this case and by their choices of words referring to me as a Guinea were making the ruling that I was not white, or at the very least I was not part of the superior group that was entitled to the best treatment within our society.

I exited out of my high school years with a racial identity that was very confused and bitter; I was unsure of myself regarding where I belonged on the racial and ethnic hierarchy. Sure I saw the color of my white skin as being the same as those kids who called me a Guinea, but was my white as good as their white or was I experiencing the same phenomena that many Southern and Eastern European immigrants had experienced years prior?—an experience in which many ethnic groups such as the aforementioned Armenians, the Irish themselves and the Jews were not considered white. These groups—such as the Jews depicted in Karen Brodkin's article "How Did Jews Become White Folks?"—went through years and even generations before they assimilated into American society, that is, before they were accepted by the Anglo-whites as being a part of their racial order. After this experience in my life, I felt as if I was not apart of this order and that I, as my ancestors before me, had to prove myself as white. I exited the phase of immersion confused between my dual identities of being white, and being Italian, as well as pondering the puzzling question of whether any of it really mattered.

[3] In "A Hyphenated Identity" Harry Kitano discusses the trials and tribulations he experienced as being both Japanese and American.

INTERNALIZATION

In the early part of my college years I entered the internalization phase of my life or as James Marcia would define as the moratorium phase (Tatum 1997), which is a state of active exploration of roles and beliefs in which no commitment has been made to any specific role. It is at this stage that I became acutely aware of a lot of my past prejudices and the fact that I was blinded by my own whiteness. I even remember my first day at the University of Massachusetts, when I looked around the campus observantly, and immediately thought to myself "wow so college is where all the Asian people in Massachusetts have been hiding." I was unaware of the damage of these thoughts and my past actions.[4] Until this stage, I never saw my actions as wrong, nor did I even view them as racist per se, even when I would casually refer to black people using the N word; after all that is what I was taught they were. I never could get past my own whiteness and that's all that seemed to matter until this phase of my life, because by this point I wasn't even sure how to view myself, but I was certainly searching.

As I matured in my thinking and in my behaviors certain things in my life started to click. At work, I received two promotions in the same year, the first to a shift manager, and the second to an Assistant Manager. I initially saw these promotions as a sign of my hard work, my intelligence, etc. Even though my good fortune was blossoming, the sociologist in me could not help but pay attention to the injustices that were occurring in the workplace. In my first year and a half at my place of employment I received two promotions during this period, yet two other people I had been working with had yet to be promoted once during this same time. Initially I perceived the justice in this to be the mere fact that they were not as hard working as I, but as I got to know them better and work with them more, and especially once I became one of their bosses, I realized my hypothesis was far from the truth.

I may often talk of the hardships of growing up in a family of seven in which my parents combined income was less than $65,000 per year (based on today's numbers), but nevertheless in spite of my working class upbringing I was afforded certain access to society that many black kids were not allotted. I went to a high school that taught college prep classes, and was able to prepare me for the challenges I would see in the future. I also received special attention from teachers who recognized my intellectual abilities and talents such as the argument made by Deirdre A. Royster in the article "Race and the Invisible Hand." Essentially, she argues that whites are more likely to be assisted by white teachers who pay attention and are willing to assist their white students in bettering themselves (287-291). For my co-workers this was very much so the case: both of them came from rough neighborhoods in Boston, and were by their own admission neglected by many of their teachers and as a consequence didn't put forth as much effort in their schoolwork as I had—as someone who was constantly given positive reinforcement by my high school teachers as well as my employers. I am not necessarily making the claim that solely for the reason that these two co-workers of mine being black was the basis for being passed over per se, but it is significant to realize that because of their status and role in society as blacks they were not given the structural, organizational and educational background that many white people in higher classes are given or even for that many whites in similar class backgrounds such as myself.

Through these experiences in this phase, I realized that maybe in regards to those two kids from Charlestown I was not

[4] See "Blinded by Whiteness" by Mark A Chesler, Melissa Peet and Todd Sevig.

white, but as far as society was concerned, I was apart of the advantaged group. My role in society became evident to me as I progressed in my life while the lives of many of the minorities around me remained static because I had the advantages of being white in a white dominated society. The upbringing I had—even though it was a working class one—was still one in which the advantages of being white are present, as it is pointed out by Royster in "Race and the Invisible Hand." This is because I had access in my white neighborhood to the larger white society that many black people don't have access to such as my co-workers, teachers an education, and white people with good societal standing that could vouch for me if I needed them to do on any job application. As far as society was concerned regardless of my ethnicity or my class upbringing that I was born into, I was indeed white.

INTERNALIZATION-COMMITMENT

Now that I was aware and certain as to which role I adhered to in society, I was able to make the transition to the fifth and final stage of my racial development—that of internalization-commitment. It is at this stage that I am becoming aware of the daily injustices that exist in the racialized world of American society, and am trying to eliminate my part in constructing these injustices, and trying to broaden my horizon as well as expand my views.

As a youth growing up in Somerville, Massachusetts, I stayed away from the projects and ghettos of nearby neighborhoods such as Rindge Ave in Cambridge, or Mystic Ave in Somerville, because they were so gross in appearance due to their run-down look and frankly these neighborhoods were downright scary. It was not until I started working in the Rindge Ave area rather recently that I have moved past my hang-ups of this place being a horror movie within a white reality. In many ways this neighborhood is friendlier than others, people know each other, give high fives and "show some love" towards one another. However, I still hear people say, "wow you actually walk through there late at night to get home?" We hear this from many people because there is a negative stigma about such places as Rindge Ave because these neighborhoods are often associated with drug dealing, violence, etc., as well as being predominantly black. As a consequence of this reality it has been racialized in the minds of white society that geographic locations such as ghettos and low-income housing is an African American phenomenon. The reality, however, is much more complex, because such areas are a construction of white society's racist policies of isolationism and segregation of African Americans from the larger society.

In the article titled "The Making of Groveland" by Mary Pattillo-McCoy we see an example of how white society has created a disadvantaged black culture. The Chicago neighborhood of Groveland is a prime of example of society at work. Groveland in the 1960s had a total population of 12,710 people in which only six inhabitants were African-American. In just 10 short years the neighborhood transformed into being over 80 percent white, and today is a staggering 98 percent black (p. 248). This rapid transformation is largely a biproduct of real-estate agents utilizing selling tactics such as "blockbusting"[5] in which they prey on the fears and concerns of white homeowners that as the presence of blacks in the area will increase, property values fall. While this tactic for the homeowner may not be racially based, certainly is of an economic concern, given that one's house is a great source of wealth and probably a significant reason as to why white families have an average net worth of

[5] A term used in the documentary *Race: The Power of an Illusion.*

$140,700 whereas blacks have an average of $20,400.[6]

Tactics such as blockbusting and redlining are also used to discriminate based on race by refusing to grant loans, mortgages, or insurance to blacks to buy houses in white advantaged neighborhoods. These tactics have created separate societies within the larger societal structure in which blacks are isolated from the white society where much of the economic power exists. For the Burnett family who wished to purchase a home in Levittown, Long Island in the 1950s this was impossible because this area was constructed to be the perfect ideal white society and blacks were not allowed to "penetrate through."[7] This is very much the opposite the situation of Groveland. It turned a middle-class neighborhood that was largely white into a low-income area with a lot of social strain and a greater likelihood of crime. It did this because it removed the economic middle class of the area and replaced it with a poor class of people. Ironically the first set of African-American families in this neighborhood were as affluent and as well off economically as their white counterparts, but the scare tactics of real estate agents drove off the white families who were bombarded with images of young black mothers pushing strollers in their neighborhoods, and other such negative visuals they could use to drive off the white families and make the sell.

Understanding these injustices is all apart of the process that goes along with this last step of commitment. The last step I believe is a lifelong battle because as long racism exists we will be constantly dealing with race. Racism is not something that is going to stop easily in our society; it is a day-to-day phenomenon. Even earlier on this very day of Sunday May 18, 2008, I encountered a situation that was troublesome to me. Three young black kids came into my store today causing a disturbance. One of the three kids started yelling the N word, and I asked him politely to stop yelling in the store as well as to refrain from using such an inappropriate word. When I said this he argued back at me by saying the word "aint a swear, it's a word for black people." Once again I was encountered with a very simplistic definition of perhaps one of the most troublesome and complex words ever to be constructed in America. The reality of all of it is that words such as the N word, Guinea, etc., do not really have any true meaning; they are all words that have been created over time to which no fixed meaning is attached. These words have no meaning until we as a society give them meaning. For the eight year old Dave, the N word was just a word I never heard it until others told me that it was a word for black people. This very idea embodies Howard Taylor's essay entitled "Defining Race" in which he argues that race is a social construction that has been created through human interactions with one another throughout the course of history. Until recently I was not aware of the transient nature of race, but now I realize it is all subject to historical time as well as individual experience and interpretation. I as an Italian was not considered "white" to the Irish kids I went to school with.

CONCLUSION

These issues in regard to race will eventually need to be answered as America becomes "A Nation of Minorities" (Chideya 2006) in which the non-Hispanic white population that currently represents 75 percent of the U.S population is expected to decline. The decline is a result of exponential rise, during the last 30 years, in the migration from non-white areas of the world

[6] Information on wealth comes from a reading handed out in class titled "Race, Wealth and the Commons."

[7] A term used in the documentary *Race: The Power of an Illusion.*

such as Asia, and Latin America, to the U.S. This massive wave of immigration is similar to the influx of foreign born peoples that transpired at Ellis Island in the late 1800s to the early 1900s in which masses of southern and eastern Europeans such as Jews, Italians, Poles, etc., flooded the Island of New York (Foner 2006). These people were initially not perceived as "white" and therefore denied participation in many essential sects of society such as politics, certain jobs, etc., but the more next generation assimilated into America and became successful, the more they were accepted as "white." If groups such as Armenians, Italians, and Jews can become viewed as "white" and be allowed entrance into the privileged classes of society who is to say that eventually Hispanics, Asians, blacks, etc., won't be eventually allowed similar participation in our society.

Before entering this class just four months ago my whiteness would have blinded me to these realities, but in this final stage of development my eyes are opened and I am well aware that we do not live in a color blind utopian society in which stereotypes don't exist. In being aware of these injustices, I must make a conscious effort as a member of the dominate group in society to at the very least change my own prejudices if I cannot help transform those of society. My actions are significant, since as an employer who has power to hire people I have to be willing to look at more than what is on their resume, as well as to see beyond their skin color, their hair style, etc., and view them on equal footing.

In the past, I was unable to do this. I often would go into an interview and already have an answer in the back of my mind just by seeing names such as Rashaad, Ntianu, etc., on the applications. Now that I am aware of my mistakes, as well as of the reality that my whiteness is not an inherent trait but a construction of society, I must be willing to work on the role I play in the racialized world—a role that is significant because of its potentially damaging consequences. However, I am making a commitment to work on my behaviors, and at the very least if I can't change society's racial structure I certainly don't want to be adding to the problem.

WORKS CITED

Alexie, Sherman. *Indian Killer.* (Time Warner Company, 1996).

Blumer, Herbert. "Race Prejudice as a Position," in Charles A. Gallagher, *Rethinking the Color Lines*, Third Edition (McGraw Hill 2007), pp. 169-175.

Bobo, Lawrence D. "Race Relations in America at the Close of the Twentieth Century" in Elizabeth Higginbotham & Margaret L. Andersen (eds.), *Race and Ethnicity in Society: The Changing Landscape* (Thomson Wadsworth, 2006), pp. 87-95

Bonilla Silva, Eduardo. "Keeping Them in Their Place" in Elizabeth Higginbotham & Margaret L. Andersen (eds.), *Race and Ethnicity in Society: The Changing Landscape* (Thomson Wadsworth, 2006), pp.370-375

Brodkin, Karen. "How did the Jews become White Folks" in Elizabeth Higginbotham & Margaret L. Andersen (eds.), *Race and Ethnicity in Society: The Changing Landscape* (Thomson Wadsworth, 2006), pp. 59-66.

Chesler, Mark A, Melissa Peet and Todd Sevig. "Blinded by Whiteness" in Elizabeth Higginbotham & Margaret L. Andersen (eds.), *Race and Ethnicity in Society: The Changing Landscape* (Thomson Wadsworth, 2006), pp. 156-165.

Chideya, Farai "A Nation of Minorities" in Elizabeth Higginbotham & Margaret L. Andersen (eds.), *Race and Ethnicity in Society: The Changing Landscape* (Thomson Wadsworth, 2006), pp. 11-16

Muhammad, Dedrick and Chuck Collins., "Race, Wealth and the Commons," in *Poverty & Race Research Action Council, Vol. 16, No. 3, May/June 200 7*, Pg. 1-2 and 3-7.

Foner, Nancy. "From Ellis Island to JFK" in Elizabeth Higginbotham & Margaret L. Andersen (eds.), *Race and Ethnicity in Society: The Changing Landscape* (Thomson Wadsworth, 2006), pp. 219-225

Gallagher, Charles A. "Color-Blind Privilege" in

Elizabeth Higginbotham & Margaret L. Andersen (eds.), *Race and Ethnicity in Society: The Changing Landscape* (Thomson Wadsworth, 2006) pp. 96-100.

Herring, Cedric. "Is Job Discrimination Dead?" in Elizabeth Higginbotham & Margaret L. Andersen (eds.), *Race and Ethnicity in Society: The Changing Landscape* (Thomson Wadsworth, 2006), pp. 277-284.

Kitano, Harry. "A Hyphenated Identity," in Elizabeth Higginbotham & Margaret L. Andersen (eds.), *Race and Ethnicity in Society: The Changing Landscape* (Thomson Wadsworth, 2006), pp. 136-145.

Muhammad, Dedrick, and Chuck Collins. "Race, Wealth and the Commons." *Poverty and Race* 16 (2007).

Omi, Michael and Howard Winant. "Racial Formation" in Elizabeth Higginbotham & Margaret L. Andersen (eds.), *Race and Ethnicity in Society: The Changing Landscape* (Thomson Wadsworth, 2006), pp. 74-79

Pattillo-McCoy, Mary. "The Making of Groveland," in Elizabeth Higginbotham & Margaret L. Andersen (eds.), *Race and Ethnicity in Society: The Changing Landscape* (Thomson Wadsworth, 2006), pp. 247-251.

Royster, Deirdre A. "Race and the Invisible Hand" in Elizabeth Higginbotham & Margaret L. Andersen (eds.), *Race and Ethnicity in Society: The Changing Landscape* (Thomson Wadsworth, 2006), pp. 285-294

Tatum, Beverley Daniel. *Why Are All the Black Kids Sitting Together in the Cafeteria?* (Basic Books, 1997)

Taylor, Howard F. "Defining Race" in Elizabeth Higginbotham & Margaret L. Andersen (eds.), *Race and Ethnicity in Society: The Changing Landscape* (Thomson Wadsworth, 2006), pp. 47-54

Films:

Race: The Power of an Illusion: Episode 2: The Story We Tell PBS.

Race, the Power of an Illusion: Episode 3: The House we Live in. PBS.

Money Whitens But Doesn't Erase

A Reflective Sociology of Racism from the Middle of the American "Melting Pot"

Dima Kurin

University of Massachusetts Boston

blinelaxd27@yahoo.com

Abstract: Engaging various sociological perspectives, concepts, and literature, this paper deals with my experiences with racism and racial prejudice ever since I moved to U.S. at the age of nine. The points that are addressed deal with my experience living in Russia as a member of the white majority without any exposure to other races or ethnicities, and my experiences with race and racialization at the schools and various social circles I entered in the U.S. Another point that is discussed in depth is my family's struggle with welfare and the struggles that that system presents. The essay concludes by exploring my experience with racism in the workplace as well as in the real estate market. The essay finishes by reflecting back on the journey that my family took and where we have ended up because of it.

From an early age until recently it was always hard for me as an immigrant to this country to feel sympathetic to the minorities in this country and their struggles because my family was busy fighting its own fights and stereotypes. In this paper I will explore my views on race and racialization in the U.S. from the time when I was categorized with my parents as an immigrant to now when I am no longer classified as an immigrant.

I can see how my parents who have not **assimilated** nearly as well as I have into the white American culture are still targeted as immigrants and foreigners in different facets of their lives. Having assimilated into the American culture thoroughly, I discovered that even when I went back to Russia only four years after coming to the U.S. I already felt like I did not belong in the country where I grew up. My realization was that while America preaches tolerance and unity, in reality it imposes penalties, both physical and emotional, on those who do not conform and fit the roles that have been designed for them during the last several hundred years. It is sad to see and grasp the idea that no matter what my parents accomplish in their lives, their language barrier and demeanor—so entrenched in their behaviors from spending the majority of their lives in another country—will follow them for the rest of their lives and mark them as **forever foreigners** no matter how accepting they become of American culture.

Dima Kurin is an undergraduate senior at UMass Boston, double-majoring in Management and Criminal Justice. He wrote this paper while enrolled in the course Soc. 321L: "Racial and Ethnic Relations," instructed by Anna Beckwith (Lecturer of Sociology at UMass Boston) during the Spring 2008 semester.

My first encounter with race was when I came to the United States in 1995. In Russia I was completely oblivious to the idea of **race** as a social construction as defined by Taylor or **racism** of any form because my community appeared to be completely homogenous, where everyone was middle class white, living in very similar apartments and earning very similar wages. No one had any reason to differentiate themselves between one another in a negative way which is what the idea of race and racism fosters. In fact, I do not remember seeing a person of a different race than me until arriving in America.

The one thing that separated me from the majority of my friends in Russia was the fact that I was Jewish. However, my parents made sure to hide that fact by giving my brother and I my mother's last name of "Kurin" instead of my father's last name which is "Movshits." This change plus my relatively young age kept me from having any early exposure to racism.

One of the first early things that I had observed when moving to America was that being white was a privilege just as being Christian was in Russia. Deviance from Christian white standing was not beneficial to an individual, which is why our Jewish background was hidden for my brother and me by our parents. My brother and I always thought of our **race as self-defined** and thought of ourselves as white because we were socially conditioned to do so (Taylor). However my grandparents always told us that we were Jewish first and white second even though we should act and assimilate into the white culture, but never forget that we are Jewish. My grandparents would tell us that we are lucky that **anti-Semitism** is not as prevalent in our generation as it was in theirs and that we would never have to live through an event such as the Holocaust.

It is as if by going through a catastrophic event such as the Holocaust the Jewish people were given **honorary white** status similar to the status attained by Asian Americans as discussed by Min Zhou in "Are Asian Americans Becoming 'White'?", where Asian Americans gained their "white" status by becoming educated and more "American"—"American" being synonymous with being white. For all their suffering and perseverance as productive members of society, the Jewish people have gained similar status through education and entering the middle class. When people find out now that I am Jewish they tell me to be proud of that and to identify myself as such; but I have been instilled both a reason to identify as white and an internal fear of identifying myself as Jewish. However, years of adamantly calling myself white have left their mark and it feels uncomfortable now to call myself anything besides that. America was not the **melting pot** of society that I had expected it to be. Like Seattle in the novel *The Indian Killer* by Sherman Alexie, America—which would appear on the surface to be a melting pot of Native Americans and Whites—is only advertised as a melting pot of all cultures seamlessly "melting" together to form one mass. In reality, like Seattle in *The Indian Killer*, America is a country with many different groups sharing the space and waiting for the one spark to ignite a conflict between two or more racial groups.

After moving to America my family was placed on welfare and in addition to having to deal with the fact that we were immigrants, we now had a stigma attached to being welfare recipients. On a daily basis I was subjected to teasing and badgering by fellow classmates about the fact that I did not speak English, and my clothing and my daily habits were different. The school system separated me from the general student body further by putting me in ESL classes with other students who did not speak English, and therefore grouped us all together as separate from the rest of the student body. The further separation from the main student body occurred with the help of such things as reduced lunch, which alienated all of the foreign born kids further and grouped us all together as the "foreigners" even though we were all from different countries. Eventually we embraced the role of foreigner and all sat together and did not engage with the rest of the students. We were alienated and spent our lunch breaks and recess playing together, sharing stories of our home countries and different acts of racism and **prejudice** we had experienced. We discussed how we had been prejudged and labeled without doing anything to earn those labels, and the different put-downs that we had been subjected to in America. Through these daily conversations I realized that although I was an immigrant, I had **white privilege** because I had been treated a lot better than other foreign born kids who were not white, but were Indian, Black or Latino. While they were blatantly excluded from certain activities such as kickball and gym activities, I was not excluded but was picked last or teased during the activities.

It was during high school that I experienced the greatest understanding of racism and race, even though by that time I had for the most part assimilated into the American culture. It was during my high school years that I saw racial and ethnic segregation of students in the lunch room; something that I had never experienced before. Seeing the lunch room divided into African Americans, Asians, Whites, Russians and Latinos was a surreal experience and I was automatically sucked into the segregation by sitting down at the table with all the other Russian kids without ever thinking twice about it. Suddenly my whiteness became an issue when faced with a blatantly segregated environment and I assumed a racial role that everyone else has as well. Because Brookline High School was not a racially diverse school in itself, many of the African American students were part of the METCO program and were bused to the school from intercity communities. METCO was the primary desegregation plan implemented at BHS to diversify the school. However, this did not do anything besides isolate the black and white kids further from each other because being part of the METCO program and being put into an affluent white community only made the majority of the black students look inward and create their own social circles. As described by John O'Neil in Beverly Daniel Tatum's book *"Why Are All the Black Kids Sitting Together?"* since black students had a much more similar background both economically and geographically, they tended to stay together because they had formed a relationship between each other based on their economic status and they were often from the same area. The black students also found their shared experiences regarding racism as something to unite them by sharing their stories of being discriminated with each other. The white students, however, saw this as "self-segregation" based on race and never took into account the extreme disparity in economic standing, and thereby immediately racialized the reason behind the separation. The white students in the meantime separated themselves from the black students because they also saw them as "invaders" on their turf, as if they did not belong at the school in the first place. It was quite clear that the integration attempts by the school were not successful, expecting students who had grown up in different environments without much interaction with people of other races to all of a sudden intermingle with people of other races. This was foolish and further promoted apprehension on both sides.

A racially charged issue that came up in my high school that I never understood until recently was the issue of our using an American Indian as our mascot because our teams were known as the Warriors. My school had a very small minority of American Indian students and they had protested on numerous occasions that our mascot offended them and their families. Because the

Native American students were a very small minority fighting an issue that had been prevalent for a very long time, the rest of the student body did not take them seriously and thought that they were exaggerating an issue to get some attention. In fact, just as Charles Fruehling in "Playing Indian" mentions, when those who oppose the mascots make their opposition public they get threatened by those who have a lot of vested interest in sports and their school. At my school the minority of American Indian students began to get harassed by the rest of the students, especially the athletes of the various sports. During that time I did not understand why they had been so adamant about the mascots offending them, but in retrospect I had never been subject to something resembling their situation and could not understand it; on the contrary I blindly followed the masses against the minority.

One of the biggest sacrifices that my father made when we moved to America was abandoning his job as an accountant to become a pizza delivery driver to support our family while my brother and I went to school and my mom tried to pass the board exams to become a doctor. When we originally moved to America my mom was very hopeful that it would only be a matter of months before she could get a job as a doctor since she had been one in Russia for over twenty years. However, the reality of the situation was that my mom could only get minimum wage jobs because of her immigrant status and language barrier. Linda Burnham in "Welfare Reform, Family Hardship, and Women of Color" states exactly the problems and stigmas that came with being on welfare. The fact that we were on welfare and needed assistance with housing and food stamps affected my parents on a daily basis and they felt very ashamed that they could not provide for their family without government assistance. That embarrassment spread to me as I began to feel more awkward getting my free lunches at school, and soon either brought food from home or paid full price for my lunch to avoid having those feelings. My older brother was also often forced to call and decipher different documents that my parents would receive from the welfare program; without him they would have had no idea what to sign and send back in to continue their welfare coverage.

Eventually my mom gave up on finding a job and began a four year long process of studying and preparing for the medical board exams in hopes of becoming a doctor and not having to work minimum wage jobs. To continue our welfare coverage my dad had to work long hours and my brother had to pitch in as well so that we had enough working hours to continue to receive aide. My mom would always hesitate to tell people outside her circle of friends that she did not work and studied because she faced scrutiny from Americans as well as other immigrants for not holding a job while she prepared for her exams. I would find her upset numerous times over comments made to her about her being a slacker and just leeching off the system and not contributing back. This racist attitude towards my mom based on her being an immigrant and not seeking immediate minimum-wage work took an emotional toll on both of my parents and built a resentment in them towards the average American individual that is still present today. This is a very real **"cost" of racism** that is very often overlooked because it is very hard to measure.

Racism is more often hidden behind economic terms or by how much property or work mobility someone has. Emotional tolls that people experience are something that has to be seen first hand and not something that can be quantified by a survey or a study. Furthermore, looking back at the welfare system and its effect on my family it is hard not to recognize that the system is designed to hold people down while still allowing them to engage in secondary sector jobs that support the economy, yet provide no upward mobility. While the **welfare reform** on the surface appears to support the immigrants and the needy of America, it has many safeguards that people are not aware of, that aim to prevent people from reaching the middle class and beyond. Some of these mechanisms are the inability of those on welfare to pursue educational opportunities as I had discussed previously; some others are that those on welfare are not allowed to have over a certain amount of money in the bank, which in turn prevents them from ever buying their own house, which is a major stepping stone in securing financial wellbeing in the future. Had my brother and I been younger and unable to help the family and had my mom not been educated enough to eventually become a doctor, our family would have been further mired in the welfare program without ever having the ability to escape into the middle class.

In addition to facing discrimination because of our dependency on welfare and associated programs, my parents also faced scrutiny because of their dual citizenship status. When my parents got their citizenship from America in 2002 they decide to keep their citizenship with Russia because they had pension benefits that they wanted to collect while still feeling some loyalty to their country. Peggy Levitt describes in "Salsa and Ketchup" the issues that my family faced because they embraced being **transnational**. For a long time my parents tried to hide the fact that we participated in elections back in Russia as well as sending money back to help out my grandparents who were living in the Ukraine. When our neighbors found out about my parents' involvement with politics as well as the economy back in Russia, they started making derogatory comments to my parents about our being communists and whether we had any loyalty to America, or whether we were just here to exploit the system so that we could pump the money back into the Russian economy. In reality that could not have been further from the truth because we had supported my grandparents while we lived in Russia, and we continued to have that duty when we moved to America. In fact, in the grand scheme of things it was very hypocritical for people to treat us poorly because we were sending money back to Russia to help our family, when they went every day to the store to buy products that were foreign made by foreign companies, essentially pumping money into he Japanese, German or another countries economy.

One of the most misinformed issues that I have had to deal with was the difference between my **nationality** and my **ethnicity**. It is always extremely frustrating to tell someone that I am both Russian which is my nationality and Jewish which is my ethnicity because many people do not understand that I can be both. Another problem that I faced is illustrated well by Michael Omi in "Racial Formation," which is that even by self-defining myself I was open to interpretation by other people, such as when I told people that I was a U.S. citizen they were further confused because I had also said that I was a Russian immigrant. I believe that the biggest reason behind this confusion is the intertwining of race, ethnicity and nationality. Whenever I have had to fill out any important forms I am still obligated to put down my place of birth which is Moscow. This falls right in line with the ideas of Michael Omi which is that no matter that the government decided that I am a U.S. citizen I will always be classified as a past immigrant—something that I will never be able to relinquish.

Where high school opened my eyes to the world of racism, my college years and work experiences kept my eyes open to the many faces of racism in education, work, and family. One of the first impressions made on me in the corporate world was the idea of accelerated or **forced assimilation**.

While this was not as brutal or intrusive as the forced assimilation of the Native American children—as described in Christine Lesiak's PBS documentary *In the White Mans Image* by where they were specifically shipped to boarding schools and isolated while being forced to live as a white person—my forced assimilation was equally intrusive because it emphasized that to succeed I needed to assimilate at a faster face. While in high school I felt that I could assimilate into the American culture at my own pace, once I entered the corporate world as an auditor and night manager of a hotel I realized that I had no choice but to completely immerse myself in the white culture and leave all remnants of my Russian heritage behind. What I realized when I entered my management position was that all the departments at my company were highly racialized. All the management personnel were not only white, but euro-white so immediately I was forced to accelerate my assimilation so that I could fit into my new role. It is important to note here that Russia has for centuries been seen by the West as an outsider or on its margins.

The only departments in the company which were completely white were the management and the security department. At the beginning I was constantly asked about my name in a rather straight forward manner. This was also the first time that I had become aware of the behind the scenes job discrimination that was present in my workplace as illustrated by Cedric Herring in "Is Job Discrimination Dead?"—which was that black applicants to the security department were being called back at a much slower rate, usually when all white applicant choices were exhausted. While I was hired with no prior security experience and at a much younger age than is general present in the high end security field, even qualified African American applicants who worked on a part time basis were not offered a full time position. Their part time status also left the African American employees open to termination when downsizing during the slow times was implemented.

My immigrant status excluded me for the most part from the dating scene in high school. The only limited dating exposure that I had in high school was with girls who were Russian like me. It was sad to see that despite the fact that I wanted to date girls of all backgrounds I was stuck, and more so expected exclusively to date girls of a Russian background. There was a certain amount of **border patrolling** done by my parents as well as my friends to ensure that we were all dating the same people from the same circles without allowing any outsiders in. Another reason for the insignificant amount of interracial dating in my high school was that everyone always thought the same thing when there was a party. This was put best by Mark Chesler in "Blinded by Whiteness"

> My black friend invited me to a party with her. And the first thing I could think of was how many white people are usually there. I remembered thinking, this is probably going to be uncomfortable, and I would rather go out with my white friends.

This mentality was rampant in every racially divided circle in my high school and because of it everyone continued to stay separate, and without engaging with someone of another race outside of school in a social setting it was impossible to get to know them on a more personal level. It was through my dating that I finally saw my parents for the **fair weather liberals** that they were. While never openly speaking out against people of a different race, nationality or ethnicity than us, they were not shy about showing their dissatisfaction when my first serious relationship was with a girl who was part German. Even though they had always preached tolerance to my brother and me and always said that all that they cared about was our happiness, once it came to light that my girlfriend had German heritage there would not be a day that I would not hear some sort of comment about it. The constant barrage of comments ranged from the need to keep our Jewish heritage pure, to the idea that my girlfriend's grandparents were the ones who killed my relatives in World War II.

My second full time job exposed me to a larger world of racism and race prejudice. A year ago I got a job as a loss prevention specialist at Star Market. My primary responsibilities were to apprehend shoplifters and keep an eye on employees to make sure there was no employee theft. In essence I was expected to stereotype people based on their appearance and demeanor. This basis of my job opened up a whole new world of racism and stereotyping that is very prevalent in the loss prevention field. Herbert Blumer describes in "Race Prejudice as a Sense of Group Position" the feelings **of race prejudice** as:

> (1) A feeling of superiority, (2) a feeling that the subordinate race is intrinsically different and alien, (3) a feeling of proprietary claim to certain areas of privilege and advantage, and (4) a fear and suspicion that the subordinate race harbors designs on the prerogatives of the dominant race. (170)

Each of those feelings and ideas was prevalent with everyone in my department as well as all management associated with my department. The majority of the employees in the loss prevention department were not intrinsically racist on an individual level, but were prejudiced on a basis of **group position.** Many of my coworkers felt that all of those who used EBT cards to get their welfare benefits were automatically leeching off the government and that we were the "suckers" and that those who worked for a living should be entitled to more benefits than those who do not work. My coworkers also made a habit of pointing out that those who used welfare money tended to buy expensive products and drove nice cars, when in reality it was a very small percentage of those who used food stamps. These comments and the disdain that they projected made for a very uncomfortable work environment where I felt that I could not talk about my past for fear of being scrutinized by my coworkers.

Systematic racism was very prevalent in my department as we were often instructed to follow or observe black customers and watch black employees even though statistics showed time and time again that white customers and employees were stealing at a higher rate. The figures were kept close, however, by the simple fact that through **racial profiling** we were simply following and observing a disproportionate number of black individuals even when we had no reason to do so. Just as Ellen M. Barry describes in "Parents in Prison, Children in Crisis," that the system is designed to single out minorities such as African Americans, our system in my department was designed to do the same thing. We were instructed to follow those with bubble jackets and loose clothing as well as those who wore baggy jeans, and it appeared that this was the kind of apparel worn by African Americans. A very popular observation that my coworkers always brought up was that on shows such as "cops" there were mostly black people committing crimes and that it was a good example of our society where people of color and minorities were committing all the crimes; so they also used this as a justification for our targeting of minority groups. One of the observations by Farai Chideya in "A Nation of Minorities" is:

> One study that tracked a year's worth of network news coverage found that sixty percent of images

> of blacks were negative, portraying victims, welfare dependents and criminals. This is a far cry from the reality of the black community.

It is clearly obvious that these images are playing a major role in continuing the prejudiced attitudes that transfer over into the workplace and when paired with people in position to target African Americans, it continues to perpetuate the problem by feeding the system.

The last facet of my life that has contributed to my understanding of racism and prejudice around me is my experience with my parents buying a condominium as well as my mom becoming a doctor in about the same time frame. The building complex where we chose our condominium was a rather exclusive space and there were and still aren't many minorities living in that area. My parents noticed a very prominent way of **gate keeping** and **steering** as referred to in the documentary *Race: The Power of an Illusion, Part III*; my parents were shown apartments facing the ocean which were not only bigger but much more desirable because of that. My parents also observed that African Americans and other minority potential buyers were steered toward the apartments facing the parking lot which were smaller and had a much less desirable view. My mom also commented on the fact that she had overheard the realtor agents telling minority buyers about other opportunities in my town where the living spaces were much more affordable and were in a "better" area where they would like them more. It was very sad to understand that the realtor was concerned with keeping these people out of the building even though he could make more money doing business with them. This goes to show that racism and prejudice goes beyond even the economic forces that drive our economy.

My whitening was not complete, however, until my mom became a doctor and in essence we had achieved the American dream of a good job, a good house and both kids in college. When word had gotten around that my mom was now a doctor and that we owned a condominium in a prestigious community it was suddenly as if the floodgates of communication had opened from American whites from management as well as all white collar positions. While my mother, my brother and my having good jobs and owning property put us in better position to interact and secure some of the things that affluent white Americans had around us, we still understood an important trend that is illustrated by Melvin L. Oliver in "Wealth and Racial Stratification" which is that because we were the first generation in America we had not had the time to amass the wealth that those around us had. Even though on the outside we seemed equal to those around us, in reality it was not so since we did not have the stocks or multiple paid off properties to our name. After all the struggles through hard economic times and enduring racism, we had realized that although we had reached close to a pinnacle of what my parents could accomplish in this country, we still trailed by a large margin those who lived side by side with us but had hundreds of years of family wealth passed down to them.

In conclusion, I think that it is important to reflect on one's past armed with the knowledge one gets throughout all the experiences that are gathered over the course of a lifetime. By reflecting on where I came from and the path that I took to get to where my family and I are now, I came to understand that we had navigated many hurdles the system poses for immigrants. We escaped the welfare system and succeeded in putting ourselves in a position to pass our income and wealth first to me and from my parents and from me to my children in order to gain equal footing with those who have lived in this country for generations.

The American dream is not something that is attained when someone comes here from another country looking for a better life; it is a battlefield of structural constaints. This "melting pot" that is America sees immigrants not for the benefit they bring to the nation, but as parasites that must be kept down and under control. In reality money truly does whiten but it does not erase, and it shouldn't because we should always remember where we came from in a positive light no matter how many people want us to be ashamed or forget.

BIBLIOGRAPHY

Alexie, Sherman. *Indian Killer* (Warner Books, 1996).

Barry, Ellen. "Parents in Prison, Children in Cris," in Jane Jeong Trenka, Julia Chinyere Oparah, and Sun Yung Shin, *Outsiders Within* (South End Press, 2006), pp. 59-74.

Blumer, Herbert, "Race Prejudice as a Sense of Group Position" in Charles A. Gallagher (ed.), *Rethinking the Color Line, Third Edition* (McGraw Hill, 2007), Pg. 169-175.

Brodkin, Karen. "How Did Jews Become White Folks?" in Elizabeth Higginbotham & Margaret L. Andersen (eds.), Race and Ethnicity in Society: The Changing Landscape (Thomson Wadsworth, 2006), pp. 59-66.

Burnham, Linda. "Welfare Reform, Family Hardship, and Women of Color" in Elizabeth Higginbotham & Margaret L. Andersen (eds.), Race and Ethnicity in Society: The Changing Landscape (Thomson Wadsworth, 2006), pp. 314-319.

Chesler, Mark A. "Blinded by Whiteness" in Elizabeth Higginbotham & Margaret L. Andersen (eds.), Race and Ethnicity in Society: The Changing Landscape (Thomson Wadsworth, 2006), pp. 156-165.

Chideya, Farai. "A Nation of Minorities" in Elizabeth Higginbotham & Margaret L. Andersen (eds.), Race and Ethnicity in Society: The Changing Landscape (Thomson Wadsworth, 2006), pp. 11-16.

Frankenberg, Erica. "Race in American Public Schools" in Elizabeth Higginbotham & Margaret L. Andersen (eds.), Race and Ethnicity in Society: The Changing Landscape (Thomson Wadsworth, 2006), pp. 347-353.

Fruehling, Charles. "Playing Indian" in Elizabeth Higginbotham & Margaret L. Andersen (eds.), Race and Ethnicity in Society: The Changing Landscape (Thomson Wadsworth, 2006), pp. 115-119.

Herring, Cedric. "Is Job Discrimination Dead?" in Elizabeth Higginbotham & Margaret L. Andersen (eds.), Race and Ethnicity in Society: The Changing Landscape (Thomson Wadsworth, 2006), pp. 277-284.

Levitt, Peggy. "Salsa and Ketchup" in Elizabeth Higginbotham & Margaret L. Andersen (eds.), Race and Ethnicity in Society: The Changing Landscape (Thomson Wadsworth, 2006), pp. 334-341.

Oliver, Melvin L. "Wealth and Racial Stratification" in Elizabeth Higginbotham & Margaret L. Andersen (eds.), Race and Ethnicity in Society: The Changing Landscape (Thomson Wadsworth, 2006), pp. 240-246.

Omi, Michael. "Racial Formation" in Elizabeth Higginbotham & Margaret L. Andersen (eds.), Race and Ethnicity in Society: The Changing Landscape (Thomson Wadsworth, 2006), pp. 74-80.

Tatum, Beverley Daniel. *Why Are All the Black Kids Sitting Together in the Cafeteria?* (Basic Books, 1997)

Taylor, Howard F. "Defining Race" in Elizabeth Higginbotham & Margaret L. Andersen (eds.), Race and Ethnicity in Society: The Changing Landscape (Thomson Wadsworth, 2006), pp. 47-54.

Zhou, Min. "Are Asian Americans Becoming "White"?" in Elizabeth Higginbotham & Margaret L. Andersen (eds.), Race and Ethnicity in Society: The Changing Landscape (Thomson Wadsworth, 2006), pp. 233-239.

Films:

In the White Man's Image. PBS American Experience (1992).

Race: The Power of an Illusion. California Newsreel (2003)

Beyond "Simply Understanding"

Sociologically Reimagining and Reconstructing the Meaning of My Education

Kathleen R. O'Brien

University of Massachusetts Boston

kat4rose27@yahoo.com

Abstract: In our society, where people tend to have a very linear and individualistic way of thinking, I should have graduated college years ago, be married or about to get married, live on my own, or with my boyfriend, and have a full-time career. But none of that is happening! And I sense that people often feel the need to bring that up to me. Usually it's people I haven't seen for a while, like distant family or old friends from high school. The reality is that I live at home with my mom and my brother, work part time while taking a full course load at the University of Massachusetts (UMass) Boston and am surrounded by people who love and care for me, without many of whom I don't think I could survive. But I can't help feeling that a lot of people look at me like I am just doing what I am doing because I don't want to deal with the "real world"—like a "spoiled brat" who doesn't want to grow up. However, I have learned too much in my years in school to be satisfied with "simply understanding" (Gordon, 9). I believe that the social sciences need to be taken more seriously in academia not just as interesting subject matter, but also as a means of understanding the world in a political, economic and historical context. I agree with the *critical theorists* of the Frankfurt School that "people's ideas are a product of the society in which they live" and that what "intellectuals...should adopt is a critical attitude to the society they are examining" (Wallace and Wolf, 120). It is in the study and application of C. Wright Mills's sociological imagination that we can realize the importance of this.

In the beginning of life, when we are infants, we need others to survive, right? And at the end of life, when you get like me, you need others to survive, right? But here's the secret: in between, we need others as well.

—*Morrie Schwartz, quoted in Albom (1997: 157)*

The word "spoiled" has a lot of connotations, most of them negative. The Oxford Dictionary defines "spoil" as both "to harm the character of (a child) by being too lenient" and "to treat with great or excessive kindness, consideration, or generosity." And most times, at least in our society, it is followed by the word "brat."

It strikes me as rather humorous that our society can believe someone to be treated with "*excessive* kindness, consideration, or generosity." But of course, a society that stresses hard work, consumerism

Kathleen R. O'Brien graduated from UMass Boston in Spring 2008, double majoring in Anthropology and Social Psychology with a Certificate in Environmental Studies. She wrote this paper while enrolled in the course Soc. 341-1: "Elements of Sociological Theory," instructed by Mohammad H. Tamdgidi (Assistant Professor of Sociology at UMass Boston) during the Fall 2007 semester.

and wealth cannot really value kindness, consideration or generosity. It is due to this dichotomy that I sometimes wonder if I am spoiled or just very lucky to have lived the life I have lived. My life has by no means been perfect, but I seem to have a way of bouncing back from the bad things that have happened in my life and make things turn out in a good way.

In our society, where people tend to have a very linear and individualistic way of thinking, I should have graduated college years ago, be married or about to get married, live on my own, or with my boyfriend, and have a full-time career. But none of that is happening! And I sense that people often feel the need to bring that up to me. Usually it's people I haven't seen for a while, like distant family or old friends from high school. The reality is that I live at home with my mom and my brother, work part time while taking a full course load at the University of Massachusetts (UMass) Boston and am surrounded by people who love and care for me, without many of whom I don't think I could survive. But I can't help feeling that a lot of people look at me like I am just doing what I am doing because I don't want to deal with the "real world"—like a "spoiled brat" who doesn't want to grow up.

When I began writing this paper, I focused on my packrat "problem," but now I see it as not only a problem, but also another insight into why I am the way I am. I am the biggest packrat I have ever known. I don't mean I keep things for a little while and then eventually throw them away, or keep really meaningful things in a few shoe-boxes to memorialize the "good old days." I mean that I am 99% sure that somewhere in storage I still have the wrapper from the gum I was chewing the first time I kissed a boy—in 1994. And that is probably wedged between an old bank statement and a pair of socks. Part of the reason I have so much stuff is that I have moved from my childhood home and back again a couple times. So many of the stuff that I have are just leftovers from my childhood that I haven't gotten the chance to go through yet.

This may not appear that big of a problem at first glance, but as I prepare to graduate, move onto graduate school and enter the "real world" this problem looms in the back of my mind as I think about how much *stuff* I have, and that this *stuff* may represent the very core of my subconscious "rejection" of growing up and moving into the next phase of my life. I have also realized, while planning to write this paper, that this seemingly funny "problem" has connections to another habit in my life: failing to prepare for the future. Both of these habits seem to be related to holding onto the past or some sort of nostalgic attempt to keep things the way they used to be. Using the sociological imagination as a guide, in what follows I hope to analyze the interrelation of these in the framework of a focus on my educational/career goals and plans.

I graduated from high school in 1999. At that time I planned to go to the University of Connecticut (UConn) for four years, graduate, get a job, get married and have kids by the time I was 30. Well...that didn't quite happen as planned! I went to UConn for two years, took semesters off here and there, and technically (according to our **social norms** of post-high school education) should have been done with my undergraduate education by now. However, I am now 26 years old, four years into a romantic relationship, a "super senior" at UMB, living at home and scared to death of the next four years! The biggest question that keeps running through my mind is, "Am I scared because I'm not where 'I' want to be with my life, or because society tells 'me' I should be somewhere else?" Wallace and Wolf (2006) explain how George Herbert Mead's view on this effort in **self-interaction** is important: "The internal conversations one has with oneself are an essential part of the Meadian perspective because they are the means by which human beings take things into account and organize themselves for action" (208). This question that I continue to ask myself does not only "organize myself for action" but also allows me to dig deeper into who **I** and **me** really are.

Mead explains that "the attitudes of the others constitute the organized 'me,' and then one reacts toward that as an 'I'" (Wallace and Wolf, 206). Because the outside world may perceive "me" in a certain way, "I" find myself working harder to defend my choices and prove that I am where I want to be. "The others" may see "me" as a person who is disorganized, messy, or emotionally attached, but I like to argue that "I" am just doing my own thing and who cares what others think? In researching my packrat tendencies, I have found that, yes, I may be holding onto the past in some way, but I am also allowing my true self to show through the seemingly unconscious ways I decide what to keep and what to throw away.

In the article "To Retain or Relinquish: Exploring the Disposition Practices of Packrats and Purgers," Coulter, et al. (2003) explain that "packrats and purgers have very different 'core beings,' value systems, perspectives about product meanings, and attitudes toward being future versus past focused, wasteful versus industrious" (42). Their research has hinted at a correlation between possessions and empathy. They have concluded that when "packrats" do eventually get rid of their stuff, it is either by donation or handed down to a friend or family member, to "ensure that they are remembered by the new 'caretaker'" (42). Packrats may be "past-focused," but it does not always mean we *dwell* on the past. I have learned to deal with my past, and I enjoy going through my old possessions and reminiscing while thinking about how my sense of self has changed over the years. In *Tuesdays with Morrie*, Morrie and Mitch discuss something very similar to this. While discussing the past and forgiving others, Morrie says that "Love is the only rational act," and I agree. What I choose to keep is an act of **rational choice**, where I see the benefit of keeping certain things that remind me of those I love as much higher than the cost of throwing them away and possibly forgetting them.

Erving Goffman's analysis of the **self** focuses on resistance and creativity. He states that "our sense of being a person can come from being drawn into a wider social unit; our sense of selfhood can rise through the little ways we resist the pull. Our **status** is backed by the solid buildings of the world, while our sense of personal identity often resides in the cracks" (Wallace and Wolf, 236). I agree that I convey my uniqueness, my self, in the ways that I "resist the pull." I do not allow myself to become caught up in social norms that praise consumption and penalize frugality, which may be why I keep so many things! I like to keep things because I believe they will have some utilitarian value in the future, either by me or by someone else. I see no sense in buying something new when I have so many things that could be reused. Many people in our society find this strange because we put so much emphasis on detaching ourselves from emotion, adopting what Georg Simmel calls the **blasé attitude**, following a schedule of what we should be doing in some certain time in our lives, which often manifests itself in our consumption patterns. And as we "grow up," we're "supposed to" forget the things that brought us joy in our younger years because we're "supposed to" be focusing on the future.

I never realized it until coming to UMass Boston and learning about **relations of power** in many of my classes that my status as a woman in a patriarchal society such as ours has a tremendous impact on why I continue to ask myself the question of whether or not I'm "supposed to" be where I am in my life. I have experienced what Dorothy Smith calls the **line of fault** "between what [I] know and experience as [my] everyday life and what is official knowledge, as expressed in the symbols, images, vocabularies, and concepts of the patriarchal culture" (Wallace and Wolf, 294). I know many 26-year old male students at UMB and none of them seem to ask this question of themselves. It is in our society's collective unconscious that women around my age are supposed to be planning for marriage and children, along with having a successful career. While I am *thinking* about these things, right now I need to focus on finishing my undergraduate degree, and I am lucky enough to have a boyfriend who agrees! Smith explains this phenomenon as **bifurcation of consciousness**; she notes that when women do become involved in public life they still have to continue to pay attention to what is going on at home; we are expected to perform "traditional" roles as caregivers along with our "modern" roles as career women.

This issue becomes especially complex when people ask me what I go to school for and I tell them one or all of my interests, then find myself defending the social sciences I have come to know and love. I am double majoring in anthropology and social psychology, and enrolled in the environmental studies certificate program. I have come to expect to hear "What are you going to do with that?", along with a confused or disapproving look. This look has come to be a stimulus I have to constantly interpret in order to respond to in every situation. According to Wallace and Wolf, Herbert Blumer's "**symbolic interactionism** inserts a middle term into the stimulus-response couplet so that it becomes stimulus-interpretation-response" (Wallace and Wolf, 214). This theory stresses that many situations are interpreted differently by different people, and during these interactions I have a list of retorts lined up in my head, then I choose one or more depending on who asks the question or the situation we are in. This **face-to-face interaction**, stressed by Berger and Luckmann's **phenomenological approach**, "is where the real action is" (Wallace and Wolf, 278).

Similarly, in "My Choice of a Lifetime: Finding True Love in a Sociological Imagination," UMass Boston student Katherine Heller uses this theory of symbolic interactionism to describe her "problem" of finding true love and states "love and a successful relationship are both terms that can be defined differently by different people. Perhaps I am too much into looking at my relationship in terms of how other people say a perfect relationship should be" (21).

In their explanation of **social constructions of reality,** Berger and Luckmann examine "the processes by which any body of knowledge comes to be accepted as 'reality'" (Wallace and Wolf, 277). We have been socialized to believe the "reality" that certain things need to be done in a predetermined span of time, a very **modernist** perspective, and it is during these conversations that I find myself the most uncomfortable with my position in life. Most times I just say, "Whatever I want," and leave it at that. I feel as though I can see and hear their thoughts: "Shouldn't she be working in an office somewhere? When is she going to grow up, get a "real" job and raise a family?" In our society it is somewhat acceptable for a woman to forgo having a family *if* she has a successful career already in place. It is during these uncomfortable moments, I am ashamed to say, that I succumb to what Charles Horton Cooley describes as the **looking glass self.**

Cooley defines this concept as being constituted of the following three elements: "the imagination of our appearance to the other person; the imagination of his judgement of that appearance; and some sort of self-feeling, such as pride or mortification" (Wallace and Wolf, 195). When people react to my educational / career goals (or perhaps my lack thereof in their view) I feel a mixture of anger, fear and defiance. I realize that I may just be imagining that they are seeing me in a negative light, but I have become so used to this response that I cannot help but internalize it. I am angry because I truly enjoy what I am doing; I am fearful because I then begin to doubt the happiness I feel and see it as a cop-out or a "spoiled" emotion; and I am defiant because they don't know me and have no idea what I go through on a day-to-day basis. I feel as if these different emotions are fighting across the **front and back regions** of my everyday life and personality—using concepts drawn from Erving Goffman's **dramaturgical perspective**, pointing to the ways in which we present and conduct ourselves as if on and off stage. In the back region I am hiding my fear and anger, while in the front region I try to only display defiance or indifference. In "Why I Smoke: Sociology of a deadly habit," SUNY-Oneonta student Emily Margulies explains "the individual acts according to how he wants the **generalized other** to react to him. In order to call upon the desired attitudes of others we must present ourselves in a way that the symbols and attitudes we project will arouse the desired reaction in them" (3). This **impression management** is really tough for me because I would like to appear confident to those who may doubt my choices, but I can't help but be hurt by their lack of understanding.

Abraham Maslow's "holistic-dynamic theory" assumes that "people are constantly being motivated by one need or another and that they must at least partly satisfy lower level needs such as hunger and safety before they become motivated by love and esteem needs...Moreover, some people are driven by neurotic needs to perpetuate the status quo rather than to move in the direction of psychological health or self-actualization" (Feist and Feist, 493). I think that what some people may call "spoiled" relationships in my life have allowed me to satisfy my safety, love and esteem needs so I could become motivated by "cognitive needs," which are described as "a desire to know, to solve mysteries, to understand, and to be curious...to find out how something works just for the satisfaction of knowing" (Feist and Feist, 502).

Many people in our society are driven by the "neurotic needs to perpetuate the status quo," as shown in the film *Affluenza*. Affluenza is a disease of consumerism that has been caused by the never-ending struggle to attain the "American Dream." It has symptoms such as exhaustion, anxiety, constant permanent dissatisfaction and fractured family lives. I do not feel these symptoms in my life, which is why I continue doing what I am doing. Perhaps by staying in school and "rejecting" the "American Dream" I am working to ensure that I am not motivated by consumerism's "neurotic needs."

When I am at school I really feel like I am in my element. My friends and family lovingly poke fun at me and call me a professional student, but the truth is, if I could be, I would! Learning new ideas and being validated on my own ideas is like a high for me. To me there are few things more rewarding than working really hard in a course and getting an A, or laboring over a paper for days and getting it back with affirming comments from a respected professor. These feelings are stressed by George Homans, a **social exchange** theorist whose first of five propositions, the **success proposition,** states that "for all actions taken by persons, the more often a particular action of a person is rewarded, the more likely the person is to perform that action" (Wallace and Wolf, 306). I am rewarded for my hard work with good grades, which in turn are rewarded with a scholarship, so I continue to do what I have been doing.

However, I didn't always feel this way about school. My first two years of college, at UConn, were an academic mess! I failed courses and received papers back with Ds or Fs on them. But the funny thing is, it was during that time that I also learned many of the things that make me the person I am today. I didn't grow up in a house full of college graduates. In fact, neither of my parents went to college, but both were fairly successful in their careers and provided for my brother and me well enough. But I did grow up in a house full of encouragement and love. I made pretty good grades in school and was rewarded with high-fives and "I-knew-you-could-do-it," while a lot of my friends would get money or gifts when they did well. It was just a given that I would go on to college and continue to "live up to my potential." My experience at UConn didn't quite happen the way it was planned, and I look to the theory of **cultural capital** as an insight into my experience there and beyond.

Pierre Bourdieu argues that parents provide children with cultural capital. He believes "those from privileged homes have the attitudes and knowledge, especially cultural knowledge, that make the education system a comfortable, familiar sort of place in which they can succeed easily" (Wallace and Wolf, 113). Though my family was and continues to be very supportive of my educational career, I didn't fully understand the goal of my educational career until I was in college for a while around like-minded people. I was lucky enough to have had the money to go to a good college out of high school, but the environment was filled with kids who *had* grown up in a houseful of college graduates, and I could feel that in every class I attended.

I would like to quote an old journal entry I recently found—which I wrote during the summer after my last semester at UConn, in 2001—in order to highlight the change I went through at this time:

> *...Maybe the whole time I knew that I was there for reasons other than academia...I learned so much from the people I surrounded myself with. I've learned how to appreciate myself for who I am and not to focus on the things I can't change; I've learned that money is not all that matters in the world...it's nice to have a decent supply to have a good time with, but the goal of college is not to study something you don't like just to make a lot of money down the road. Money does not equal success. To be successful one must accomplish goals she sets for herself, not goals someone else has set for her, or goals she thinks she has to set to make others happy. I've learned that learning is one of the most stimulating activities one can pursue. I've learned to appreciate nature for more than a stomping ground for humans...I've found the therapeutic sensations of walking through the woods and finding different kinds of trees and staring in utter amazement at how they grow so strong and tall...sitting at a dam in the river and watching how the water stops flowing from a man-made device, or a stream with massive rocks causing rapids to gush toward a steep drop...*

In a way, I was taking a **phenomenological** approach and decided to "suspend or 'bracket' my learned cultural notions" (Wallace and Wolf, 262) to define my social situation. It was then that I took an oath to reject the disease of consumerism highlighted in the film *Affluenza* and concentrate on the non-material aspects of happiness. I refused to allow my life to be "taken up by caring for things instead of people" (*Affluenza*).

"**Phenomenology** asks us not to take the notions we have learned for granted, but to question them instead, to question our way of looking at and our way of being in the world...[it] asks us to assume the role of the stranger" (Wallace and Wolf, 262). I was assuming the **role of the stranger** at UConn because everything I was learning was new to me but "normal" to the people from whom I learned them. When I first got to Storrs I was very unhappy and homesick; I was a city girl stuck on a farm. In fact, there were so many girls with my name there that people just started calling me Boston. However, by my second year I had made such amazing friends that I did not want to leave! It was as if I had woken to a whole new world: from music to nature to animals to literature to science; the learning was never-ending with my friends. No matter where I went I was learning something. Walking or driving through the woods I learned about trees and soil, around the farm I learned about animals, at the clubs and in the dorms I learned about music. In retrospect it was by far one of the most intellectually stimulating environments I have ever been in. My **stock of knowledge** completely changed and enabled me to think of the world in a different way.

I never thought I would ever experience anything quite like UConn again. However, as strange as it may sound, due to the differing physical environments, I have felt almost the same way about my experiences at UMass Boston. Sure there are no dorms, or farms, or even much grass, but the people I have met here are equally as intelligent, interesting, and fun to be with. What is even better is that my courses and professors have also been a huge influence on the way I see and think about the world. I would have to say that the overall intellectual experience I have had at UMB has been a continuation of my first taste of such feelings at UConn. My personal enlightenment at UConn paved the way for my academic enlightenment at UMass Boston.

I think the main reason for the above is the overall academic tone at UMass Boston. I don't feel like a "nerd" or a "geek" when I spend hours on end debating scholarly issues with friends. SUNY-Oneonta student Emily Margulies explains this feeling in terms of **exchange theory** in her paper "Why I smoke: Sociology of a deadly habit." "People's interests find expression in social groups. People are attracted to a certain group because of the perceived reward they may receive from the association. Sometimes just the association itself is the reward. Often times, it is not what people do together, but the fact that they do it with others that makes an experience enjoyable. Most human pleasures have their roots in social life" (5). I also feel very comfortable at UMB because many of the people I have met have had similar experiences: going to college after high school, taking time off, and then coming to UMB with a serious purpose.

However, it is those times that I am not at school that the trepidation sets in. In her paper, UMass Boston student Katharine Heller states "I can see now that some of my dilemma has come about due to the fact that society has been feeding me information about love and relationships that may be totally unrealistic but that I have come to internalize nonetheless" (25). Although Heller knows that many of the aspects of "love" she has been seeking have been socially constructed, she cannot help but be disappointed that she is not able to find it. In my situation, I know that society's focus on "hard science," business, or medical majors are also socially constructed; I cannot help but **internalize** the negative feelings people project on me.

In *Keeping Good Time*, Avery Gordon explains C. Wright Mills's idea of how "'The Social Role of the Intellectual' was to refuse 'the job' the institution offered and the position of the **scared employee**. The scared employee, ruled by a 'general fear…sometimes politely known as 'discretion,' 'good taste,' or 'balanced judgement,' remains mired in 'the job'—'the pervasive political sanction and censorship of most middle class intellectuals'" (9). The workers in Michael Moore's film *The Big One* are caught up in a similar situation. They had become so loyal to their companies that when they got fired they felt as though they were being betrayed by an old friend. Emile

Durkheim (Wallace and Wolf 2006) calls this a ***corporate voice***, where people end up pursing the interests of their company instead of their own. When that tie is broken they feel a sense of ***anomie***, or normlessness.

I have learned too much during my extended time at school to allow myself to work in a dead end job and become a "scared employee." The reason I have stayed in school so long is because I have become fixated with the prospect of "changing the world," or at least doing some good in it. Another one of Morrie Schwartz's conversations with Mitch in *Tuesdays with Morrie* hints at this idea as well. When the young Morrie went to his father's factory and saw the way people were being **exploited** by the employer controlling the **means of production** intended to create **capital** he vowed he "would never do work that exploited someone else, and he would never allow himself to make money off the sweat of others" (Albom, 78).

I have learned too much in my years in school to be satisfied with "**simply understanding**" (Gordon, 9). I believe that the social sciences need to be taken more seriously in academia not just as interesting subject matter, but also as a means of understanding the world in a political, economic and historical context. I agree with the **critical theorists** of the Frankfurt School that "people's ideas are a product of the society in which they live" and that what "intellectuals...should adopt is a critical attitude to the society they are examining" (Wallace and Wolf, 120). It is in the study and application of the **sociological imagination** that we can realize the importance of this.

C. Wright Mills writes that the "first fruit of this imagination—and the first lesson of the social science that embodies it—is the idea that the individual can understand his own experience and gauge his own fate only by locating himself within his period, that he can know his own chances in life only by becoming aware of those of all individuals in his circumstances" (Mills 1959: 349). I now know that the period in which I am currently living—often described as "post 9/11" or in terms of global climate change—is situated in a particular historical, economic and political context that shapes the way I feel and act on a daily basis. Mills describes how for a long time "hard sciences" were the "common denominator of serious reflections" and that "other terms and other styles of reflection seem mere vehicles of escape and obscurity" (351). I think that the problem I am facing (feeling a certain way but being told by society that I should feel another way) is enmeshed in these old ways of thinking, and that the sociological imagination of the social sciences will soon become the norm. Then people will no longer ask me "what are you going to do with that?"

In "The Roots of Procrastination: A Sociological Inquiry into Why I Wait Until Tomorrow," UMass Boston student Jennifer Kosmas states "if we can become aware of our own social constructions, then we can also break out of the self-destructive habits by knowing ourselves" (1). In the film *Twelve Angry Men*, Henry Fonda's character appears to also reject the idea of "simply understanding" and instead employs many of the concepts we are learning in our theory course, such as being aware of social constructions and looking at a situation through the eyes of the other. Because of Fonda's character's strong **sense of self**, a man's life was spared and eleven other men became aware of their own socially constructed views and prejudices. This film highlights **symbolic interactionists'** theory that different people can interpret the same situation differently, and that getting "just the facts" is rarely as easy as it sounds.

Since I have started becoming serious about school (within the last 5 years or so) I have learned much about the world (and this country's place in it) that I never knew before. And many things make me so mad that I just can't sit around and wait for them to change! Learning history in grade school and high school was nothing compared to what I have learned in college. I have this feeling as though I was lied to in those schools, and now that I am beginning to learn a little bit of the "truth" I am addicted to finding out more. The 'truth" I have learned is that there is no one "truth."

Knowing and understanding that there are other knowledges that exist and function in the world has been a complete eye-opener for me. I have realized how hard it is to see a way of life as just an **ideology** that was conjured by some political force to legitimate a certain agenda. We take for granted the notion that a certain society is "natural" when it is really a sociopolitically constructed perspective serving particular interests (class notes 11/9/07). We have been **socialized** to believe that our individualistic, capitalist, and "democratic" society is *the* way and all other societies should function as ours. However, some may say that ours is a heavily commercialized, media-driven society constructed to serve the interests of the "dominant group," the white upper-class males. The movie *The Matrix* is an allegory for this, where machines have constructed an illusory world to keep the human population docile. I used to feel like Neo in *The Matrix*, like someone who "accepts what [s]he sees because [s]he's expecting to wake up" (*The Matrix*). It was as if I had been stuck in "the world that had been pulled over [my] eyes to blind [me] from the truth" (*The Matrix*). Now that I have learned about ideologies other than those occupied by a "developed" capitalist society, I find that I am extremely empathetic to the plights of others and want to do all I can to fight social injustices. I would never have realized all this if I had not stayed in school for so long.

I think this empathy is the greatest connection between my self as a packrat and my self as a "professional student." I'm still not exactly sure where it came from, but I do know that the emotional attachment I put onto "things" is connected to the emotional attachment I feel toward many issues I have learned about in school. In fact, I never even sell back my school books because I feel like I will lose that information forever; even if I don't have the answer right away I know I can go to my office and look it up in one of my many books! The very act of selling back textbooks only strengthens the feedback loop of capitalism. Publishers of textbooks must realize that many students cannot afford these expensive books, and that students count on that extra cash at the end of every semester. So they make deals with college bookstores to buy them back at a fraction of the price, and then in turn resell them for more than that the next semester! This is not to serve the best interest of the student; it is to serve the best interest of the publishers and booksellers—to make more profit.

Our capitalist society blocks emotion, and when people internalize this blockage they become alienated from themselves and others. According to Wallace and Wolf, Marx believes **alienation** happens when "people view each other in terms of the narrow standards of the workplace rather than as full 'species beings'" (Wallace and Wolf, 88). Many people in the U.S. are so fixated on their work that they have no time or energy left to do the things they really like to do or to face the things they really need to face. In *Tuesdays with Morrie*, Morrie and Mitch have a discussion about forgiveness, and Morrie tells Mitch to "forgive everyone now; don't wait!" and that "the culture we have does not make people feel good about themselves. We're teaching the wrong things. And you have to be strong enough to say if the culture doesn't work, don't buy it. Create your own" (Albom, 36). Morrie impresses upon Mitch that emotion and the ability to convey it to one another is such an important part of being human.

Sadly, many of the "wrong things" we are teaching reinforce the idea that love and emotion, or "*excessive* kindness, consideration, or generosity" (Oxford Dictionary) spoil people. The fact that I allow my emotions to guide my life and letting others know how much I love and appreciate them is how I have created my own culture, and I am going to try my hardest to continue making it that way and pass it on to others.

REFERENCES

Albom, M. (1997). Tuesdays with Morrie. New York: Broadway books.

Coulter, R.A. (2003). To retain or relinquish: Exploring the disposition practices of packrats and purgers. *Advocacy in consumer research*. 30: 30-43.

Feist, J. and Feist, J. (2006). Theories of Personality. Boston: McGraw Hill

Gordon, A. (2004). *Keeping good time: Reflections on knowledge, power and people*. Boulder: Paradigm Publishers

Heller, K. (2004/2005). My choice of a lifetime: 'Finding true love' in a sociological

imagination. *Human Architecture: Journal of the Sociology of Self-Knowledge*. 3(1&2): 21-31.

Kosmas, J.M. (2003/2004). The roots of procrastination: A sociological inquiry into why I wait until tomorrow. *Human Architecture: Journal of the Sociology of Self-Knowledge* 2(2): 74-81.

Margulies, E. (2003). Why I smoke: Sociology of a deadly habit. *Human Architecture: Journal of the Sociology of Self-Knowledge*. 2(1): 1-11.

Wallace, R. and Wolf, A. (2006). *Contemporary sociological theory: Expanding the classical tradition*. Sixth edition. New Jersey: Prentice Hall.

Films:

Affluenza. 1997. Bullfrog Films.

The Matrix. (2000). Universal Pictures.

Twelve Angry Men. (1957). MGM.

The Big One. (1999). Miramax Home Entertainment.

Tuesdays with Morrie. (1999). Touchstone.

4.0: *Self-Doubt, the Fear of Failure, and the Power of Symbols*

Nicole Jones

University of Massachusetts Boston

nicole.jones001@umb.edu

Abstract: I am hoping that by writing this research paper and applying various sociological theories, perspectives and concepts to my personal trouble, I will be able to step back and gain a deeper understanding of an issue that has left me constantly trying to prove to myself and others that I can be better, do better and accomplish more. My continuing struggle with the belief that I am completely incapable of success, despite being a graduating student with a 4.0 grade point average, has resulted in an enormous amount of self-doubt that leaves me questioning my abilities each time I have an assignment to complete. As a result of this exploration, I find that while I do not feel as though I am chasing the wrong things, I do feel as though I am not putting as much effort into those things that mean the most to me. I have realized that I can be successful with both my grades and in my relationships; it is possible to balance the two. Even more, being successful and being perfect are two different things. In the past I have viewed the two as being interchangeable, which is not the case. My definition of being successful has differed as I have realized that it takes more than having perfect grades, but rather living up to my full potential, growing, becoming a better person, maintaining strong relationships, making time for the little pleasures in life, and truly being happy, not only with my accomplishments in school, but outside of it as well.

My name is Nicole and I am not perfect. I eat way too much junk food, I never return telephone calls and I say all the wrong things at all the wrong times. I am okay with that; I can deal with the fact that I am carrying around some extra chunk or that my friends think that I am flaky, but what I cannot deal with is failure.

Although failure is considered to be unacceptable and of great disappointment in our society, my definition of failure, and even that of success, differs greatly from those who surround me. This, according to Herbert Blumer and his symbolic interactionist perspective, is the result of **interpretation**, the idea that "human beings act toward things on the basis of the meanings that the things have for them" (Wallace and Wolf, 217). The term "failure" can be seen as being highly ambiguous and each individual possesses a different meaning for the word, which will ultimately have an impact on the way he or she views a situation involving failure.

For example, in the article, "Exiting the Self-Destructive Highway: A Sociological Path Back to a Future Career," UMass Boston student Paul Conner states that he

Nicole Jones graduated from UMass Boston in Spring 2008, double-majoring in Sociology and Criminal Justice. She wrote this paper while enrolled in the course Soc. 341-2: "Elements of Sociological Theory," instructed by Mohammad H. Tamdgidi (Assistant Professor of Sociology at UMass Boston) during the Spring 2008 semester.

"would deliberately do things to derail any progress in his life" (Conner, 137), showing that to him, failure was an inability to get ahead. However, for me, the meaning of failure is much more complex and has become a major issue in my life, one that has become extremely overwhelming, causing immense anxiety and, in turn, is almost unbearable to handle.

The movie *Twelve Angry Men* also provides us with an understanding of how people can view situations differently based on their preconceptions, backgrounds and interpretations. This film tells the story of twelve jurors who make sense of everyday life each from their own standpoint and eventually attempt to put themselves in the shoes of an eighteen year old boy who is accused of committing the murder of his father. It is not until the jurors take the steps of putting themselves in each other's shoes and really understanding one another that the votes begin to change and the jury is able to make a verdict regarding the boy's guilt.

I am hoping that by writing this research paper, I will be able to step back and gain a deeper understanding of an issue that has left me constantly trying to prove to myself and others that I can be better, do better and accomplish more. My continuing struggle with the belief that I am completely incapable of success has resulted in an enormous amount of self-doubt that leaves me questioning my abilities each time I have an assignment to complete.

For a moment, or even as I write this research paper, I would like to take on a **phenomenological** approach and set aside the thoughts and ideas that I have constructed regarding failure, and instead question the very things I have taken for granted. Following this perspective, I will "assume the role of a stranger, like a visitor from a foreign country" (Wallace and Wolf, 262). In doing so, I will attempt to disregard the patterns and routines that I have become accustomed to in order to understand why it is that I have such an extreme fear of failure, where it came from and how I can find a resolution. In doing this research, I plan to use concepts associated with various sociological perspectives seeking to explore the relationship between my individual experiences (**personal troubles**) and the larger society (**public issues**). The sociologist C. Wright Mills called the ability to recognize such a link between micro and macro social processes, the **sociological imagination**.

I think it would be appropriate to rewind about eight years and re-visit the end of my senior year in high school and into my time spent in community college in order to make sense of my life before my fear of failure erupted.

I left high school and entered community college with the purpose of proving that I was different—different than the rest of my family. I was the daughter of low-income, undereducated parents who continued to live their lives through poor decisions. The support, direction and guidance that parents are expected to provide were non-existent and my grades reflected that. In her article, "Confessions of a Maineiac: The Family, Academia, and Modernity," Macalester College student Jessica Sawyer expresses that the family is "intended as a place to enrich children, so they are then able to leave and reach their full potentials" (Sawyer, 195). However, my family unit differed greatly from this. Instead, the neglect on my parents' behalf left me with no one to look up to, no inspiration and no path toward a future goal. A study done by an assistant professor at Brooklyn College found that students who come from destructive homes, especially girls, "do not have the strength to survive in an academic world" (Watson, 103). The way I saw it was that I was headed nowhere, destined to fail from the very beginning.

I decided to attend a local community college, something I saw as a **functional alternative**, a notion developed by Merton. While expressing that "societies must have certain characteristics in order to survive" (Wallace and Wolf, 53), Merton stresses that at the same time "particular institutions are not the only ones able to fulfill these functions" (Wallace and Wolf, 53); functional alternatives may serve as substitutes. I knew I was not capable of attending a university, for I felt I did not have the skills or knowledge needed to survive in that atmosphere; however, I knew the importance of higher education and saw community college as a way of making progress and working toward becoming something.

I never really struggled, only because I never really tried. I knew I was not capable of understanding the material, applying myself or maintaining decent grades. The lack of support that I had experienced growing up had formed severe insecurities in my personality, ones that continue to feed my self-doubt. I was two years into community college when I realized that somehow I needed to get my act together. I needed to get over the fact that I had a troubled childhood, move forward and stop making excuses for myself. My GPA then was a 2.6 and I had no idea what I wanted for myself. I had always lacked confidence, competence and motivation and did not know where to begin. However, I wasted no time. I knew that I wanted to lead a different life than that of my parents and the only way to do so was to take advantage of the opportunities that were in front of me.

For the first time in my life I struggled. It was not because I was being challenged, but rather because I was challenging myself. For the first time, I took my education seriously and went into it with the idea that I was capable of accomplishing whatever it was I put my mind to. I had made a decision to put aside the insecurities that I had developed as a child and move forward in hope of bettering myself. Similarly, on a more extreme level, in the HBO documentary *Multiple Personalities* individuals who suffered from traumatic experiences as children were prevented from living their lives in a way in which they could grow and become contributing members to society. It was not until they were able to face the issues that haunted them throughout their childhood that they could begin to heal and move onward.

For example, Gretchen, a woman who was molested as a child, continued to go to therapy to help her through the process of fusion. As she worked through her traumatic experience, she was able to continue her studies in art and eventually graduated at the top of her class. In each situation, both mine and that of Gretchen, overcoming setbacks caused by childhood experiences were necessary in order to become the people that we wanted to be. Allowing previous incidents in our lives to control our actions would only continue to hold us back.

After two more years in community college, I transferred to UMass Boston with a 3.5 GPA, the biggest accomplishment of my entire life. My acceptance was a complete shock. Although I had been applying myself to the best of my ability, I still did not believe in myself. I did not believe that I had what it took to go away to college. After the initial shock set in, I was flooded with excitement and then a familiar feeling rushed through my body—the feeling of self-doubt set in and I knew I was setting myself up for failure.

Since my move to Boston almost two years ago, I have struggled with not being good enough, trying to be better and ultimately letting myself down each time. The truth is that I realize it is absolutely absurd for a student with a 4.0 GPA to be writing a paper about failing, but it truly is my biggest fear in life and it is something that affects the decisions that I choose to make each day.

According to the **rational choice theory**, individuals tend to do things that benefit them more than they "cost" them. That is, "individuals are rational and base their actions on what they perceive to be the most effective means to their goals" (Wallace and Wolf, 303). In my case, I am constantly weighing the pros and cons of every decision that I make based on the idea of whether or not my actions will result in even the slightest degree of failure. My habit of basing my decisions on whether they will bring me rewards (grades) or punishments (failure) has affected my life in many ways. I have neglected relationships with friends and family as I have decided that my education is the most important thing in my life, leaving little to no time for others. This is something that is common among honors students who "tend to be compulsive, driven people, with few or no satisfactory interpersonal relationships" (Watson, 104). I think one of the major things that I have given up in the past two years is the chance to experience college and a city that is entirely new to me. I spend my afternoons, evenings and weekends focusing on my studies and have not made it a priority to make friends, explore my new home or even relax and enjoy how far I have come in the past few years. There is so much work to be done and no time to focus on anything else.

In realizing that I am ruining important relationships and have been missing out on what should be the best years of my life, I have tried to manage my time and give precedence to other aspects of my life, but I cannot. George Homan's concept of **value proposition** explains my behavior by stating that "the more valuable to a person is the result of his action, the more likely he is to perform that action" (Wallace and Wolf, 315), meaning that the importance of my grades is so valuable to me that I will do anything in order to maintain them, even if that means giving up things that most people could not live without.

These very decisions that I am faced with each day are made through the process known as **self-interaction**, "the internal conversations one has with oneself" with the purpose of "taking things into account and organizing themselves for action" (Wallace and Wolf, 208). When the pressure of my school work sets in and I feel as though I cannot deal with the failure that is bound to come, I unconsciously begin a conversation with myself in order to not only come to a rational choice but also to calm myself down so that making a rational choice is possible. Often, these internal conversations are not calming or soothing but rather they consist of telling myself that I am not cut out for this, that I should not be here and that I am fooling myself and everyone else around me. These conversations, whether talking myself up or down, have become a normal pattern in my everyday life of dealing with failure.

It seems as though each day I am attempting to rationally evaluate situations that could destroy all that I have accomplished. I find myself facing the same dilemma each time things get tough and having to make a choice—a choice that is fogged by my anxieties, self-doubt and fear of failing. Each time I struggle with a decision because I am scared, in fact, I am terrified of making the wrong one, the one that will lead to failure—anything less than a 4.0. To me, my 4.0 is somewhat of a **symbol**, something Mead suggests as having "the same meaning for all individual members of a given society or social group" (Wallace and Wolf, 211). In our society there is a huge emphasis on achievement and success; how much someone has accomplished in a sense establishes their social status or worth. Those who have been educated and have found well paying jobs are considered successful, while hard working Americans receive less recognition. The importance of being the best and getting ahead is apparent in our culture and many individuals will go to great lengths to do so.

According to Robert Merton, this leads to **deviance,** because those who do not have access to monetary success will find other means of achieving their goals (Wallace and Wolf, 56). The concept of deviance is a major reason I made it a priority to continue with my education even if it had to begin with community college. Growing up, my parents fell under the category Merton referred to as **retreatism**, that is they "reject both monetary success (approved goals) and the means to it (approved means)" (Wallace and Wolf, 56). I, on the other hand felt it was important to **conform** to societies goals of achieving success through accepted means such as going to college, working hard and saving my money. It is interesting that in Merton's classification (Ibid.) conformity itself may be regarded as a form of "deviance," since from an alternative point of view, conformity may impede a "normal" need society has (or should have) for change.

Our society is so materialistic that the need to be successful and have an income that will support this way of life has become extremely attractive. The film, *Affluenza,* demonstrates the idea that Americans care too much about what they have and how much of it they have rather than about things that they should view as more important, namely family and friends. "Shopping fever" has become problematic as individuals spend an average of six hours per week shopping and only forty minutes during the week with their children. In order to fuel their shopping habits, Americans have begun saving less and working more, leaving less time for their children. Many spend great amounts of money on items that are stored in the back of their closets and are never actually used. However, they continue to spend money believing that the increase in consumption is the key to happiness.

What most people do not realize is that through such behavior they increase their **emotional work**, work performed for use value rather than for exchange value (which is differently called **emotional labor**). With constant consumption individuals are overwhelmed with an overload of stress and debt that require much emotional work on their part to deal with the situation. This idea can be related to what Arlie Hochschild refers to as the first, second and third shifts. According to Hochschild working women not only work to earn a living (**first shift**) and at home to raise a family (**second shift**), they also have to emotionally deal (as a **third shift** of emotional work) with the stresses arising from the conflicts between the first and second shifts. Using Hochschild's distinctions, one can see how affluenza causes additional strains requiring more emotional work in the third shift, not only in terms of an overload of debt and stress from trying to work more in order to obtain more, but also in terms of the effects of having a fractured family.

Society's pressures to succeed and the idea that an individual's worth is based on her or his achievements is most certainly a factor that contributes to my emotional work and constant worries of being anything less than perfect when it comes to my studies. According to one study, individuals who hold these views are considered to be subjective overachievers, "those who want to succeed, with social approval being their primary concern" and who are "especially interested in attaining success that will serve as a quick indication of their worth" (Arkin et al., 496). It can be assumed then that if an individual's worth or value is determined by her or his success then failing to perform well could cause the subjective overachiever great anxiety.

My GPA is a symbol of my worth, indicating that I am in fact intelligent and in some way successful. Growing up in such a dysfunctional family affected my self-esteem and in a sense depleted any sense of worth I had. With numerous step-fathers and continuous transformations within the household, I never really felt like I had a place. I never felt noticed. With such an abnormal and unstable family situation, I was forced to withdraw from friendships as well, as most of my friends, or so it seemed, had close knit families. As a child and even through my teenage years, I kept to myself and although I had many friends, I never felt like I fit in, and never truly felt accepted.

As I began to put forth effort, others began to notice and I instantly fed off their acknowledgements. For once in my life I was noticed and even praised for my abilities, abilities no one knew I had. Much like the subjective overachiever, I began to push myself, spending extra time trying to prove that I was worth something. As others accepted me, I slowly began to accept myself. This can be seen by applying Cooley's concept, **the looking-glass self**: "the imagination of our appearance to the other person; the imagination of his judgment of that appearance; and some sort of self-feeling, such as pride or mortification" (Wallace and Wolf, 203). In my case, I felt as though once I began to apply myself, people viewed me as successful and in turn assigned me as valuable or worthy. As those whom I interacted with began seeing me as someone, rather than something, I felt as though I was accepted and began feeling as though I really was worth something.

As pressure set in, the feeling of pride and accomplishment began to fade away as I began to feel as though I had somehow created a false image of myself. The person I had been pretending to be was not me, and is not me. I do not deserve to be here; clearly my success is due to some sort of error. I have people believing I am someone I am not. I feel as though I am putting on a fake front for everyone, in what Goffman terms as the front region. Goffman studied "the ways individuals in their everyday lives present themselves and their activities to others," (Wallace and Wolf, 238) and believed that **dramaturgy**, "the art of dramatic composition and theatrical representation" (Wallace and Wolf, 238), explains social interaction. Goffman uses dramaturgy as a metaphor for seeing social interaction as a theater where interactions are considered to be a play and individuals are actors performing the play.

The **front region**, he describes, is "that part of the individuals' performance which regularly functions in a general and fixed fashion to define the situation for those who observe the situation" (Wallace and Wolf, 238) and includes the part of me that comes across to my peers as studious, talented and successful. It is the part of me that I believe is an act—the side of me that is not "real." Conversely, actors also perform in a **back region**, "a place closed to and hidden from the audience" (Wallace and Wolf, 239). The concept of back region relates to the way I feel about this "scam" that I have somehow managed to pull off. It is my back region where my anxiety sets in, where I feel it is safe to break down and cry. It is my back region where I am allowed to struggle and question my abilities without anyone noticing. It is here that I have to mentally prepare myself to take on the next task, to somehow figure out a way to impress my professors, peers and family while continuing to make people believe I am someone I am not.

The idea of my front region where I feel as though I am fooling my friends, family and peers into thinking I am something I am not can be related to the imposter phenomenon, the "feeling of intellectual and professional phoniness in high-achieving individuals" (Gee et al., 2008). Those who experience these feelings do not feel successful no matter what grades they have earned, degrees they have obtained or any other recognition they have received. Imposters almost always believe that at some point or another, their true identity will be discovered and their peers will realize that they have been pretending to be someone they are not. In addition, imposters evaluate their performance constantly and as a result tend to be very harsh on themselves (Gee et al., 2008).

It is interesting to see that as I wrote this paper, my reasons as to why I am afraid to fail and where exactly it stemmed from are more complex than the simple answer I usually give people. For as long as I can remember, I have always told everyone that I am afraid to fail because, "I have a 4.0. I have never had a 4.0, and I am not ready to lose it now." Through writing this paper, I have come to realize that it is not quite that simple. The truth is that my 4.0 is who I am. It validates me; it gives me a sense of worth and is truly my identity. Without it, I feel as though I am nothing and I have failed to find a solution to this.

In her article, "Longing to be Thin: Why I Wait until Tomorrow to Change my Habits," UMass Boston student Caitlyn Boyle states that "failure is productive" and allowed herself to look back at her life and see where she failed and why. She goes on to explain that until you fail, you cannot look back at the decisions you made and learn from them (Boyle, 114). This is something I have struggled with many times as I have hoped for failure only for the purpose of being able to look back and realize that it is not the end of the world if I make a mistake. Unfortunately, I live each day under the impression that mistakes are not acceptable; the chance to look back and learn from them serve no purpose. Putting in my best effort, knowing I did everything I could and giving it all I have is simply not enough for me and I have yet to find a way to control it.

As I continue searching for answers as to what cultivates my fear of failure and ways in which to overcome something that has completely taken over my life, I have realized that I validate myself through my grades as a result of the way in which I was raised. According to Talcott Parsons, **socialization** which occurs when "societal values are internalized by society's members; that is, they make society's values their own" (Wallace and Wolf, 26), however; this is where my childhood differed greatly from most others. While most parents teach their children the value of hard work, higher education and reaching their full potential, mine did not. Instead, my early years lacked any sort of guidance or direction, something that stems from "normlessness, or a situation where rules or norms are absent" (Wallace and Wolf, 22), a term Durkheim coined as **anomie**. Growing up, my parents did not did stress the importance of education and my grades reflected that. Never once do I remember my parents making sure my homework was done before I went out and played and often I would complete my homework over breakfast the morning before it was due. When report cards came, my friends worried over the chance of getting grounded and sometimes tried to intercept the mail. I, on the other hand, had nothing to worry about and I took advantage of that.

I think that the lack of support, supervision and regulation on the behalf of my parents was largely caused by their working class position, they were what Karl Marx would label as the **proletariat**, "those who neither control capital nor direct labor" (Wallace and Wolf, 90). My parents were part of an **oppressed class** and began to see themselves through the eyes of the dominant group and internalized the messages that they received from society. To them, there were few opportunities available and the chances of advancement were little to none, and they had accepted that. Having accepted this, I feel they felt it was unnecessary to push me to my full potential. Instead, they figured I would continue in their footsteps and lead the life they had for so many years.

As I got older, I began to realize that my family was highly **dysfunctional** and was not a "cohesive unit working together for the goodness of the whole" (Wallace and Wolf, 51). My parents were unreliable, unsupportive, and lacked the responsibility they needed in order to properly raise children. As a result of my upbringing, I began to feel disconnected and **alienated**, from not only my family but also my peers. I did not feel as though I fit in with my family, who did not work to better themselves and turn their lives around, nor did I fit in with my friends who came from middle class families who instilled values in their children that mine had not. I wanted more in life and I felt as though my parents deprived me of those opportunities.

According to Pierre Bourdieu, parents provide their children with **cultural capital**, that is "the tastes, and perceptions of what is beautiful or valuable" (Wallace and Wolf, 113), which differs greatly between classes. The knowledge and experience one gains through the course of their life that allows them to succeed more so than someone from a less experienced background can be seen as cultural capital, something my parents failed to provide me with. This has contributed to my inability to participate actively in class discussions and has completely crippled my confidence when it comes to public speaking. While I am not an underachiever, I do possess the traits of one, who generally withdraws, avoids sharing in class and finds safety in hiding in such circumstances (Watson, 103). Bourdieu states that children who come from advantaged homes gain the skills and behaviors needed to comfortably handle these situations.

As I joined community college and began to straighten out my life, I became close with two very influential people in my life. Over the following few years, these individuals became my life support and brought on an **equilibrium** that stabilized my fractured family life by giving me the encouragement and guidance that my own parents could not provide. I came to view both of these individuals as my parents and have developed a relationship with them that continues to grow today. In fact, they will be coming from California to attend my graduation, whereas my own parents will not. It was their constant support that allowed for me to believe in myself and believe that there was more for me outside of the small town in which I grew up.

I think one of my biggest fears and the thing that held me to believe that I was not capable of accomplishing greater things was the idea of **class reproduction**, a concept that Pierre Bourdieu describes as "how one generation of an (economic) class ensures that it reproduces itself and passes on its privileges to the next generation" (Wallace and Wolf, 114). In my case, it was the lack of privileges that concerned me. Although my mother had come from parents who had established good careers, lived comfortably and had made good lives for themselves, my mother had somehow managed to take a wrong turn somewhere. I feared more than anything, and often still do, that I will inherit the same life and will make nothing more of myself than they have. I fear failing, I fear living paycheck to paycheck and I fear not being able to make a life for my family and myself.

As an attempt to avoid class reproduction, I participated in what Merton refers to as **role exit**, "the disengagement from a role that is central to one's self-identity and the reestablishment of an identity in a new role" (Wallace and Wolf, 50). As I made the decision to make something of myself, I left behind everything that I knew, everything that was familiar and comfortable to me and moved across the country in hopes of starting over, establishing a new life, becoming a better person, and putting everything that attempted to hold me back behind me. These included things such as the small town that held few opportunities, a stepfather who continued to reveal my flaws, and my mother who failed to provide support, and also the good things such as my friends who gave me hope, my siblings who gave me a reason to go on, and those two very important individuals who made me believe anything was possible. While giving up both the good and the bad, I was in a sense exiting the role I had played for so long in order to take on a new role and a new identity. I was leaving the person I was with the hopes of becoming the person I wanted to be.

Although I could not be happier with the success and accomplishments that have accompanied this new role, it is one that needs amending, as it has gained control of my life and has caused to me to forget many things that were once an important part of my life. In the film, *Tuesdays With Morrie*, Mitch, a sports commentator, was too busy focusing on his career and made little time to do the things that most humans value—his friends and family. This is something that I have also struggled with as I have neglected those who have always been close to me. Not only have I neglected them when I have had an overload of reading or when a test nears, I have completely withdrawn from keeping contact. I have made myself believe that if the work is not done, I should not be enjoying myself, even if I do not plan to work on the assignment. Much like Mitch, my friends have reached the point where they find it pointless to put effort into a relationship where both sides are not trying to make it work.

This was a perfect film to end the semester with, as Morrie's discussions with Mitch has made me realize that a meaningful life consists of more than getting good grades and being successful. As Morrie said in the film, "So many people walk around with a meaningless life. They seem half asleep, even when they are busy doing things they think are important. This is because they're chasing the wrong things. The way you get meaning into your life is to devote yourself to loving others, devote yourself to your community around you, and devote yourself to creating something that gives you purpose and meaning."

While I do not feel as though I am chasing the wrong things, I do feel as though I am not putting as much effort into those things that mean the most to me. I have realized that I can be successful with both my grades and in my relationships; it is possible to balance the two. Even more, being successful and being perfect are two different things. In the past I have viewed the two as being interchangeable, which is not the case. My definition of being successful has differed as I have realized that it takes more than having perfect grades, but rather living up to my full potential, growing, becoming a better person, maintaining strong relationships, making time for the little pleasures in life, and truly being happy, not only with my accomplishments in school, but outside of it as well.

WORKS CITED

Arkin, R., Lynch, M., Oleson, K., Poehlmann, K., & Yost, J. (2000). "Subjective Overachievement: Individual Differences in Self-Doubt and Concern With Performance." *Journal of Personality*, 68 (3).

Boyle, Caitlyn. (2007). "Longing to be Thin: Why I Wait until Tomorrow to Change My Habits." *Human Architecture: Journal of the Sociology of Self-Knowledge*, 5 (2).

Conner, Paul. (2007). "Exiting the Self-Destructive Highway: A Sociological Oath Back to a Future Career." *Human Architecture: Journal of the Sociology of Self-Knowledge*, 5 (2).

Gee, D., McGregor, L., Posey, E. (2008). "I feel like a fraud and it depresses me: The relation between the imposter phenomenon and depression." *Journal of Social Behavior and Personality*, 36 (1).

Sawyer, Jessica. (2004/2005). "Confessions of a Maineiac: The Family, Academia, and Modernity." *Human Architecture: Journal of the Sociology of Self-Knowledge*, 3 (1&2).

Wallace, Ruth and Alison Wolf. 2006. Contemporary Sociological Theory: Expanding the Classical Tradition. 6th Edition. New Jersey: Prentice Hall.

Watson, G. (1960). "Emotional problems of gifted students." *Journal of Personnel and Guidance*, 68 (3).

Films:

Affluenza. (1997). Bullfrog Films.

Multiple Personalities. (1992). HBO Home Box Office.

Tuesdays With Morrie. (1999) American Broadcasting Company.

Twelve Angry Men. (1957). MGM.

The Body/Mind Split in Pursuit of Beauty

Understanding Eating Disorders Through Sociological Writing

Nicole

University of Massachusetts Boston

Abstract: Why is it that we always seem to be in a constant state of war between our bodies and our minds? It is as if there is some invisible disconnect that we feel a desperate need to bridge. So we reach our hands out blindly and grasp the first thing that comes along, such as magazines featuring paper thin models portraying supposed standards of beauty. And never mind the fact that we will never get there. Sociology can give us incredible insights on problems that occur in our own lives and in the world around us. The most effective way, perhaps, of getting to the root of my own problem regarding the body-mind split is to dive directly into the heart of the storm and steadily spin outwards. Using C. Wright Mills's sociological imagination as a guide in terms of the study of personal troubles in relation to public issues, in this paper I inductively begin by analyzing the particular case of my own experience and struggle, gradually working toward relating it to the broader social world. In conclusion, I realized that there is not one solid definition of beauty and that it cannot be defined in simple, tangible terms. I think the truth is that the spectrum extends infinitely, rooted in the essential connection between the body and the mind. By writing essays like this one, it is possible to reach new understandings that can slowly and gradually help fill the mind/body gap that has been created by an eating disorder.

Why is it that we always seem to be in a constant state of war between our bodies and our minds? It is as if there is some invisible disconnect that we feel a desperate need to bridge. So we reach our hands out blindly and grasp the first thing that comes along.

Reading a magazine featuring a paper thin model, for instance, we look in awe, admiring the defined collarbones, the cheekbones, the ribs, the slenderness of her body, thinking, "this must be beauty." Thus the obsession begins. And the thrill of it all is so enticing that we begin to forget what it is to *feel*. Instead of feeling emotions that are burdensome and difficult, we can lose ourselves in preoccupation, devoting every second of every day to mold and shape our bodies to look like that model on the magazine, doing whatever it takes to feel the power, the control, the purpose to it all.

And never mind the fact that we will never get there. Other than perhaps for a split second of satisfaction at the sight of another pound lost, our body and mind does not seem to stay connected for long in

Nicole is an undergraduate junior at UMass Boston, double-majoring in Sociology and Psychology. She wrote this paper while enrolled in the course Soc. 341-3: "Elements of Sociological Theory," instructed by Mohammad H. Tamdgidi (Assistant Professor of Sociology at UMass Boston) during the Spring 2008 semester.

this way. We will never look like the model on that magazine, but we will torture ourselves trying to do so.

THE SLOW DESCENT: INTROSPECTION

Do you know what it's like to fight in a battle of mind & body in vain? You wait & you wait & you anticipate. But the memory's always there, especially when it rains.

Sociology can give us incredible insights on problems that occur in our own lives and in the world around us. The most effective way, perhaps, of getting to the root of my own problem regarding the body-mind split is to dive directly into the heart of the storm and steadily spin outwards. Using C. Wright Mills's **sociological imagination** as a guide in terms of the study of personal troubles in relation to public issues, I will **inductively** begin by analyzing the particular case of my own experience and struggle, gradually working toward relating it to the broader social world.

My own war with food, control and body image began during my first year of high school. It started slowly and innocently enough but over the course of the next six years, it bloomed into a terrible nightmare of addiction: cycles of restricting food, bingeing, purging, exercising and diet pills. Fluctuations occurred over and over between strict control and out-of-control, eating nothing and eating everything, a body that was heavy and tied down and a body that was frail and barely there.

In high school, I gained about thirty pounds; in about a year of college, I lost forty pounds. Anyone who knew me then would never have guessed, though, because I was extraordinarily good at hiding my battles. The reasons behind this war were buried deep, but not so deep that they could not be unleashed and understood.

I, like all of my peers in high school, was trying to figure out who I was. I wanted to fit in and grow up and be beautiful and independent all at once. But beauty can be a strange and extraordinarily subjective thing, especially when we are trying to determine our own beauty through the opinions that we believe others have of us. The **looking glass self**, a concept introduced by the sociologist Charles Horton Cooley, is a work of an individual's imagination: how I imagine I appear to others, how I imagine others are judging me, and the feelings that arise in me as a result. I found myself stuck and confused, so I, like many, tried looking out to others for some kind of symbolic or implicit answer.

Messages everywhere told me that the body mattered. Phrases like, "You look so thin!" or "Wow! Have you lost weight?" were never uncommon to hear. So I decided to make this my new goal, my new standard of beauty, even though in actuality, this interpretation was extremely skewed. In those early days of questioning my sense of self, I began to assume that if I could mold and shape and construct my body to look a certain way, I would be in control, confident and beautiful with no problems or worries.

In our society, the body often acts as a **cultural capital**, meaning that we use the body as a mode of judgement or differentiation that are perceived as assets to gain success in life. According to Bryan S. Turner, we have developed the "modern belief that one's body can be constructed" (Wallace and Wolf, 388) and therefore often use body shape or size in comparison to that of others to determine our own self worth. This is exactly the belief I took to heart.

I began poring over fashion magazines, trying to form an image in my mind of an ideal body type by the models I saw in the magazines. It seemed to make sense, although at the time I didn't realize how far from the truth such ideal types are. **Phenomenological sociology** encourages us to question the everyday life in order to better understand human behavior "from the point of view of the acting subject...not from the perspective of the scientific observer" (Farganis 281). Looking back on the behaviors that I began to develop at such a young age, I am hit with a startling realization that I was unconsciously creating a **subjective reality** for myself which was entirely based in my mind. The most important thing to me for so long was an increasing preoccupation with food, body image and control, a seemingly endless cycle in pursuit of an impossible ideal.

Over the years, I developed my own rules and rituals that I had down to a science. In her novel, *Appetites*, Caroline Knapp identifies these rules and rituals as "the mathematics of desire, a system of self-limitation and monitoring based on the fundamental premise that appetites are at best risky, at worst impermissible, that indulgence must be bought and paid for" (26). It became a teetering balancing act in which I was never, ever in the middle: "...eat two pieces of cake and a carton of ice cream tonight; eat nothing tomorrow or exercise for two hours today; have second helpings at dinner." If I wrote them out now, the list of rules could go on for a heartbreakingly long time. They were tiny, detailed behaviors that piled up in my head like pages of a big book of rules that became **internalized** over the years: I took them to be normal, natural, and saw this as the only way to live my life.

WAR OF THE SELVES: CONSCIOUSNESS VS. CONSEQUENCES

... & the rain falls down like fists from the sky, beating & pounding on the ground so I wait in my ark like a warrior—waiting for the grey to go away, but it's always around.

Sociologist Erving Goffman compared the interaction of everyday life to a stage. He called this point of view **dramaturgical sociology**, according to which one can identify a **front stage** and **back stage** regions in everyday social interaction; I think one can likewise identify front stage and back stage personalities and selves, expressed outwardly and maintained inwardly, respectively. My front stage persona was what everyone around me saw: friendly, smart, happy, together. Inside, or back stage, my eating disorder was what no one saw, but for a time it helped me manage my exterior façade. I could channel all of my anger, frustration and self-hatred into my eating disorder behaviors and emerge as if nothing happened. It was an addiction that became harder and harder to shake as I became more and more dependent on it.

In his article, "The Drinking Matrix: A Symbolic Self Interaction," the SUNY-Oneonta student Neo Morpheus (pseudonym) recognizes that for him, drinking was a way of functioning at first. It helped him fit in and make friends and have a good time, even though eventually it became destructive. Often, the first stages of addiction can be hidden through **impression management** in that a strategy of avoiding shame and embarrassment by appearing "cool" allows for a way to fulfill an individual's desire to look good without being branded an alcoholic. Eating disorders are often compared to alcoholism in the same way; the outward behaviors I developed helped me appear to have my life together perfectly. I could push down difficult emotions by falling into a hidden cycle of restricting, bingeing and purging in the backstage. Emotionally, it was also as if I let myself get as empty as possible, then filling up the emptiness to the point that it was overly full bringing about the need to then release all of the tension. I was left feeling numb and in control.

Although the idea of being in control was in reality the act of falling out of control, it was a need rooted in desire. Desire for control was really a desire for balance, acceptance and fulfillment, a desire not to feel pain or loneliness or longing. Louise DeSalvo, in her book *Writing as a Way of Healing* (1999), urges the importance of using desire in a constructive rather than destructive way. "If we have always wanted to write—something, anything—and if we haven't, our desire won't disappear. Our need won't dissipate. Our stories won't go away. Our traumas won't heal themselves" (33).

For me, other than the desire which fueled my eating disorder, I always have had the desire to play music. Since I was thirteen, writing songs has been a way for me to sort out my thoughts and emotions and worries. The strange thing was that I would write music that preached against falling for the false standard of beauty in the media, or for taking extreme measures to achieve a goal, or for not attempting self acceptance. I knew that what I was doing was wrong, but I was fighting what the sociologist Dorothy E. Smith calls a **line of fault**, or a split consciousness. It is a clash that many women experience "between what they know and experience in their everyday/everynight lives and what is official knowledge, as expressed in the symbols, images, vocabularies, and concepts of the patriarchal culture" (Wallace and Wolf 294).

In her article "Body Image: A Clouded Reality," UMass Boston student M.D. relates this idea to her own experience with an eating disorder. "You're pulled at both ends. You know in your mind what's more important, but society is pulling you the other way, telling you something completely different" (3). For me, it was the same: cultural symbols and internalized behaviors were telling me to keep up my eating disorder while a core instinctual self with a deeper knowledge of right and wrong told me that what I was doing was harmful and that it needed to stop. Because of this, it was very difficult to find my true self underneath the pile of internal conflict and denial.

The idea that the self can be fragmented and split can easily be linked to the film, *Multiple Personalities*, which depicts individuals living with a terrifying disorder. The subjects in the film are adults who had been severely abused as children. They developed many different characters for themselves, each of which would emerge in a particular situation. Although I clearly did not have this disorder, it did, in many ways, feel like I could be many different people in the course of one day because my moods were all over the place. I had an eating disorder "voice" that often compelled me to fuel the fire and another voice telling me to stop and change my ways. In the film, the people with multiple personality disorder had been able to "invent ways of bearing the unbearable." This was exactly what an eating disorder had brought on for me. It was a way of coping with stress, anxiety, depression, and confusion; a way to create a softer, fuzzier dreamworld for myself.

By the time I left for my second year of college, my eating disorder had begun to take complete control of my life, my thoughts, and my behaviors. I began drawing further and further away from my social network: my family, friends, and close supports. It was as if I was living in a dream where I would walk around for hours without having eaten a thing; my head was like a ball of cotton and my reality became increasingly skewed. My boyfriend and my close friends kept telling me that I was getting too thin, that I looked unhealthy and unhappy, but I chose to ignore it. I was living life like Neo in the film *The Matrix;* I knew there was something wrong with society (like Neo knows there is something wrong with the world), but it was difficult to open my eyes to my own struggle.

In the film, the Matrix is a world of computer generated control, a system of machines and artificial intelligence created by humans which nearly destroys them.

My eating disorder was something that I suppose I initially created, but it spun out of control and completely took me over until I truly had no control over my life anymore. When Neo is offered the red and blue pill and is asked the question "what is real?", he decides to take the red pill, which allows him to learn the truth. When I finally realized that I needed some kind of help, it was as if I had chosen to take the same path as Neo did. It was the harder road in the short term, but I knew I couldn't keep living like I had been for so long.

Eating disorders, like so many other disorders or addictions, can grow completely out of hand if left untreated. In "Treading Water: Self-Reflections on Generalized Anxiety Disorder," SUNY-Oneonta student Megan Murray explains that, "The day I admitted I had an anxiety disorder was the first day of a new life for me" (1). Her disorder, like my eating disorder, had gotten in the way of living her life and she knew she needed a change. After almost six years of living with my eating disorder, I finally decided that I needed help.

RESOLUTION

This beauty fight, it's just a child, a homeless night out in the wild; my eyes are searching for a sign that I'll take control now of my mind…

I withdrew from school halfway through the semester and moved into a residential treatment center for two months. It was what Goffman would have called a **total institution**: a place where I was basically closed off from the rest of society along with a group of other women who were experiencing the same problems.

Although it was a supportive and nurturing environment, it was extremely difficult to get used to at first because I was forced to follow more normalized behaviors that had become almost foreign to me. As a result of not being able to use any of my old coping methods, I experienced a flood of emotions that were more intense than anything I had ever felt before in my life. Lamenting was a healthy thing, even though it was extraordinarily painful. Louise DeSalvo wrote that, "By engaging in lament, we care for ourselves, for not to express grief is to put ourselves at risk for isolation, illness" (54). This, I believe, is absolutely true.

I was mourning the loss of a six year relationship with my eating disorder, as strange as it may sound. Since I had been putting my obsession before anything or anyone else in my life, I had pushed so many others away and had kept so many difficult emotions in a box at the back of my mind. I felt like my eating disorder was all I had left and yet I had to let it go. Treatment was a mode of **reality construction** for me in which I had to create a new system of experiences that could be "objectively factual and subjectively meaningful" (Wallace and Wolf 285). The most difficult but ultimately rewarding part of it all was that none of these new experiences could involve my eating disorder. In recreating my own reality, I was also able to search for my sense of **self**

According to George Herbert Mead, the self is comprised of the "I" and the "Me." The "I" is seen as a subject having or displaying spontaneity and initiative, and the "me" as an object which is a result of internalization of social norms and values. I was able to gradually paint a picture of who I was and who I wanted to be. Residential treatment also gave me a very strong **social exchange network** as a support system during recovery. I was living with other women who were fighting the same battle and who truly understood; we all became very close and, through sharing a struggle, we were able to help each other through the difficult time. When I was there, it was almost like being sheltered from the influences of the outside world.

As my head began to clear, I began to

realize what a materialistic society we live in, making it very hard for anyone to really feel fulfilled. The film *Affluenza* showed how much people feel like they need to buy things to be happy and satisfied. Just as the pursuit of a perfect body in my mind determined my self worth, many people portrayed in the film felt as though if they bought enough products, they would find a piece of mind. A business specialist in the film made the comment that "antisocial behavior in pursuit of a product is a good thing" and it is ridiculous that so many of us actually fall for the idea that one's sense of self can be bought and sold. In reality, when we buy in excess or become extraordinarily preoccupied with being thin, it is nearly impossible to maintain a connection between mind and body. It is almost as if these distractions force us to lose our connection with the world.

Evolution of the Mind & Body: The Price of Modernity

My body's been battered by billions of billboards which promise a product to better my life ... & lucky for me, I am bruised on the inside like millions of people with strife in their smile.

Although many recent findings have shown that the cause of eating disorders includes genetic and biological components, they are essentially a human made phenomena. From an evolutionary perspective, we were not meant to obsess over our bodies or manipulate food; in the beginning, eating, preparing and sharing food was a way to survive. We were hunters and gatherers all depending on one another. Now, however, humans have developed an increasingly complex and advanced way of life which in turn has given new meaning to food, the body and the self.

Capitalist Modernity revolves around the idea that we are living in a society in which we experience rapidly changing fashions and shifting norms. Considering the fact that the average American sees about 3,000 ads a day (Knapp 15), it is not surprising that we spend so much time wanting more. We are given thousands upon thousands of choices and are constantly seeing products flashing in lights which promise satisfaction and fulfillment. These daily obtrusions make it very difficult to locate one's true sense of self, and this phenomenon that we have created puts an entirely new perspective on what is natural or normal for humans. It may be human nature to seek belonging and acceptance, but the more we are taught to want, the harder it is to be satisfied with ourselves and with each other. Our expectations rise higher and higher until we expect a perfection that can never be attained.

Anthony Giddens described this as ***reflexivity of the self***, or the "view of oneself and one's identity as something that involves choices, decisions, and creation" (Wallace and Wolf 191). People are forced to decide how they want to live and who they want to be. This can be seen as a positive aspect of modernity, but it can also mean that, paired with this sense of constant change and increased options, we are told what is right until we feel as though we don't actually have a choice at all.

Along with modernity also comes the concept of ***disembedding***, in which tradition is replaced with expertise. This can become confusing to the human consciousness, because along with infinite choices, we also are "looking to disembedded expertise as a source of legitimate authority and to provide guidance on what one should do" (Wallace and Wolf 191). In reality, what is right for some is not necessarily right for all, but as we continue to move forward, there is an increased pressure to conform. It is almost as if we are reaching a modern day imperative of natural selection in which conformity also means survival. We are supposed to live our lives a certain way, want certain things, buy certain things until wants *become* needs. This in turn makes our sense of self entirely dependent on what we have rather than who we are.

It all boils down to ***power***. In essence, it is the "control of sanctions, which enables those who possess power to give orders and obtain what they want from the powerless" (Wallace and Wolf 122). Power is what fuels us but also what destroys us. Power creates a hierarchy of status, which in turn creates the pursuit of control and ultimately creates conflict for the people who are unable to attain it. Michel Foucault expands on this definition of power by explaining the advent of the ***power-knowledge relationship.*** The two go hand in hand in that "knowledge...brings power in its wake as it produces new types of human beings who are deemed better because they are normal" (Farganis 415). Perfectionism has become the new norm, and in particular, we seem to be obsessed with beauty as its main source.

It is not surprising, therefore, to relate this idea back to the sociology of eating disorders because control, perfectionism and the pursuit of beauty lie at the root of this disorder.This, then, is the price that we pay for living in a modern world. Max Weber called this new "highly rational and bureaucratically organized social order" the ***iron cage***, explaining that we have become trapped and are forced to conform to "a new character type, a technical as opposed to a cultured individual, a passionless, coldly calculating, and instrumentally rational actor" (Farganis 81). This describes the big picture of our society as a whole: the haves and have nots, a mix of people holding onto their high status along with people striving to climb the social latter.

The **functionalist** Talcott Parsons described the units which make up the whole as the ***personality system***, where the "basic unit...is the individual actor, the human person...[The] focus at this level is on individual needs, motives, and attitudes" and is related to the assumption that "people are self interested or profit maximizers" (Wallace and Wolf 27). Of course Parsons here was presuming the function of human personality in terms of the functioning of a "modern" capitalist society. Each one of us are acting as competitors for very limited resources. In Michael Moore's film, *The Big One*, he illustrates this problem clearly, showing examples of how we are currently experiencing an economic "survival of the fittest" due to the powerful companies which are all continuously competing to be on top. Their strategy is to maximize profit through cheap labor and company layoffs. Proctor and Gamble, for example, laid off 13,000 people since 1993 at a time of record $6 billion profits. Nike manufactures their products in Indonesia, where workers are paid less than forty cents per hour. When Moore finally gets an interview with Phil Knight, the CEO of Nike, he asks, "What's the difference between being a billionaire and a half billionaire?" Knight simply laughs, insisting that "Americans don't want to make shoes" so there would be no point in trying to build a factory in America to create more jobs. This scene in the film emphasizes the American way of life: never settle for anything less than being on top.

The accumulation of these troubling components of modernity—money, time, power, control—is what fuels our current system to the point that these human constructs have become things that we literally cannot live without. The ***dual-inheritance theory*** views cultural and biological evolution as one and the same, concluding that we "possess traits that not only help in our socialization but also make us want to be socialized" (Wallace and Wolf 412). ***Socialization*** is the process whereby we become who we are today: we cannot exist alone in society, and despite our love of individualism, a large part of our identity is shaped by the people and the environment around us.

One would like to believe that since we have evolved into socially fit beings, we would naturally find a way to cope and adapt to the stresses of modern living. The problem is that many of our developed coping mechanisms reflect the result of socialization getting out of hand; they are the negative, harmful ways in which we try to fill up feelings of emptiness and dissatisfaction with artificial means.

THE BATTLE OF NEVER ENOUGH: THE UNSPOKEN RULE OF DESIRE

Our bodies just bloom into giant balloons,
begging for release from their tethered string.
What do we want? What do we need? Why do
we feel so selfish? Why do we feel such greed?

I went to the gym today, my head filled with thoughts of bodies and the culturally defined meaning that seems to always be attached to their size, shape and structure.

To pass the time, I picked up *Vogue* magazine, which interestingly enough was featuring different articles on exactly the same topic that had been running through my mind. One article was a woman's struggle with her weight and with eating, another was about the perfect body and featured models posed with athletes. There was an article about two fashion designers who underwent a body makeover and lost a great deal of weight followed by an article telling readers that we are *not* actually what we eat, that body shape is mostly determined by a person's genes.

I skimmed through each article, occasionally looking up at the other people furiously exercising around me and suddenly I realized how contradictory all of those messages were. This is what we see every single day on TV, on billboards, in magazines; messages which over the years have slowly crept into so many people's minds and still remain as a constant, a fact, a struggle. In some sense, we are making progress by promoting ways to live a healthier lifestyle through exercising and eating a balanced diet, but at the same time, those 3,000 advertisements are continuously pulling at our insecurities. When we are consistently exposed to models who are flawless and perfectly proportioned, we begin to view ourselves as unsatisfactory and develop an strong desire to be something or someone we are not.

Popular magazines feature healthy recipes and ways to "love your figure at any size!" right next to headlines that read, "how to look ten pounds thinner or how to lose ten pounds in ten days." Which messages are we supposed to believe in? We compare ourselves to others in an attempt to determine whether or not we are acceptable and lovable. We pour over fashion magazines and sigh over the photographs of bodies that most of us will never have. And where does it get us?

Robert Merton defined ***dysfunction*** as something which has consequences that are generally maladaptive to the society. He also noted that what may be functional for some may be dysfunctional for others, and vice versa, depending on "people's interests and the degree to which these are served" (Wallace and Wolf 48-49). The body wars that have been raging now for decades are a perfect example of this. We live in a society where dieting is a norm, weight loss is a common goal, and obesity is a national epidemic. Our preoccupation with the body has become a distraction, a health risk and a threat to our sense of self.

Merton also described two different structural functions of a system which are not always negative, but can be extremely dysfunctional, especially in relation to the nation's obsession with body image. ***Manifest functions*** are consequences or outcomes which are intended. For example, a majority of companies put up ads featuring beautiful men and women with the perfect features and body types in order to sell their product. Here, the apparent purpose is to advertise for and promote healthy, beautiful bodies. However, these ads promote, in a latent way, values and products that can in turn bring about obsessive consumption of certain goods, and obsessive preoccupation with certain unattainable male (or female) ideals of beauty. Any resulting dysfunction may be unintended, but is nevertheless real, such as the fact that some of the people who see all of these ads will develop obsessive desires to buy excessive amounts of these products. ***Latent functions***, therefore, are consequences or outcomes that are neither recognized nor intended. For example, over time, beauty and the perfect body has become a symbol for acceptance, status and power. Dysfunctional consequences thus result when unrecognized disruptions of order occur, such as the rising number of women (and also men) who now suffer from serious eating disorders. This attention that we have given to body shape and size undoubtedly takes away from meanings that could enrich who we truly are.

We are daily faced with the problem of what Parsons defined as ***ascription vs. achievement***. "The dilemma here is whether to orient oneself toward others on the basis of what they are (that is, on the basis of ascribed qualities, such as gender, age, race, ethnicity) or on the basis of what they can do or have done (that is, on the basis of performance)" (Wallace and Wolf 31). This question of either/or circles around my head all the time, but I know I am not alone in this type of thinking; it is an extraordinarily confusing and often paralyzing thought for almost every woman I've ever spoken to. Do we define ourselves in terms of the accomplishments we've made or on how we look? If we're thinner or more beautiful or more fashionable, does that make us better? Can we ever have the best of both worlds? *Why does it matter so much?*

Advertisements and the media have proven time and time again that sex sells, and women's bodies are usually right in the center of this target. We see parts instead of the whole; a woman's stomach in the shape of a beer bottle or a deodorant ad with a million women in bikinis flocking to one man. Many advertisements are digitally enhanced combinations of many different women's faces and body parts to make up the perfect person. Barbara Fredrickson and Tomi-Ann Roberts called this the ***objectification theory***, or the idea that "girls and women are typically acculturated to internalize an observer's perspective as a primary view of their physical selves" (Fredrickson and Roberts 1). They describe this as "the experience of being treated *as a body* (or a collection of body parts) viewed predominantly for its use to (or consumption by) others" (174). This makes it very difficult for many women to find a self within a body.

Eating disorders take this problem to the extreme and are a way of manipulating food and one's body to represent a sort of combination between strength, selflessness, perfection and control. The ultimate goal is fueled by desire's dialogue: "If I could be the perfect weight, I would be acceptable, loved, needed, beautiful." This splitting of the ideas of good and bad, desirable and undesirable, in our minds is a result of contradictory and inconsistent values present in our culture. It is very difficult to shake the assumption that there is one standard form of beauty, and if we could only attain that form of perfection, our lives would be better. People deal with this differently, though, and desire can take many different forms: binge shopping, promiscuity, self harm, and eating disorders. Caroline Knapp sums this up brilliantly: "The methods may differ, but boil any of these behaviors down to their essential ingredients and you are likely to find a female blend of anxiety, guilt, shame, and sorrow, the psychic roux of profound—and often profoundly understood—hungers" (12).

The idea that we can manipulate our happiness through exterior means has become what Emile Durkheim would have called a ***social fact***. These are certain constraints within a society that goes beyond an individual's behavior or actions, a custom that becomes general to a given society as a whole but has an "existence of its own" (Wallace and Wolf 20). According to a new survey published online in 2008 by the *Self* magazine in partnership with the University of North Carolina at Chapel Hill, 75 percent of women between the ages of 25 and 45 have a disordered relationship with food and their bodies. 67 percent of women are trying to lose weight. 39 percent of women admit that concerns about their weight or what they eat interferes with their general happiness. And 27 percent of women say they would be "extremely upset" if they gained just five pounds. The pursuit of beauty has become a norm, but how long can we really continue to live like this?

HERE, NOW & BEYOND

The only proof we have of our existence is the here & now; that which is shifting & infinitely flying by. All of our senses, all sight & sound & taste & smell are never-ending as long as the world keeps turning. But to stop & question that which we have always assumed to be true is how we can bend this existence and make art. All we have to do is dream & in the deepest, purest depths of despair, we can create something real.

So where do we go from here?

A ***postmodernist*** would argue that it is time to reject "the idea that there can be a single, coherent rationality or that reality has a unitary nature that can be definitively observed or understood" (Wallace and Wolf 421). In other words, we need to broaden our perspectives and question our definitions of truth and reality. In order to refocus our view, Randall Collins, in contrast, urged that we need "to bring into the open the 'unexpected reality others are afraid to see'" (Wallace and Wolf 426). He called this the **sociological eye**, or the expansion of social awareness. In a world that is as complex as what ours has become, the only way that we can successfully make sense of it is to communicate with each other and to share our experiences. Louise DeSalvo stresses this point by explaining that "engaging in creative work, perhaps more than any other human endeavor, allows us to be autonomous while also providing us with opportunities for establishing a sense of our interconnectedness with others" (109).

In the film *Tuesdays With Morrie*, many of these issues and solutions are brought to light. Morrie Schwartz, the late sociologist from Brandeis, described life in terms of "tensions of opposites" in which people are constantly being pulled back and forth. He tries to teach Mitch Albom—his student and well-known sports journalist and author of the book inspiring the film—that we have a painful addiction to time which creates a web of internal conflict: conflict of roles (personal life vs. work), conflict of self (detachment due to stressful lifestyle) and conflict of relationships (no time to devote to people that we care about). This is absolutely true of the society we live in today, and these conflicts in turn make it very difficult to maintain internal stability, especially for someone who is struggling with an eating disorder.

Morrie's solution is a deceptively simple one: live everyday as if we have a little bird on our shoulder, constantly questioning and interpreting our actions, emotions, and sense of self. This is a task that can only be done with time, patience and understanding, things which often seem impossible to gather. But Morrie also argues that "we learn from what hurts us as much as what loves us." Through my long, ongoing journey towards recovery, I have realized how true this is. I had to learn how to turn anger and frustration into determination and persistence in order to become a stronger person. It is utterly impossible to love others if you cannot love yourself, and I truly believe that acceptance through understanding is the ultimate goal. Morrie put it perfectly: "Love is the only rational act."

For me, music has been the best way for me to interpret and understand my own personal struggles. The incredible thing about it, though, is that writing songs is also a form of what C. Wright Mills called the **sociological imagination**. Through attempting insight into my own problems, I have been able to better understand the conflicts that others experience because more often than not, our battles are of the same nature, just in different forms. Writing, playing and performing music is what fuels me every single day. It is what gives me a sense of purpose and it is how I want to make my own mark in bringing new meaning to the pursuit of beauty.

Music is, in many ways, also a way to engage in what Jürgen Habermas calls **communicative action**. This should be an extremely important component of culture because it is "through the act of communicating…that society actually operates and evolves." Not only is it a process of reaching a better understanding but it also involves "taking part in interactions through which [people] develop, confirm, and renew their memberships in social groups and their own identities" (Wallace and Wolf 181). It is a powerful feeling to be singing about battles in my own life to a room full of attentive people because I feel like I am connecting to them somehow.

What, then, is beauty? Is it a perfect body? A model with flawless features? Is it possible to ever attain or are we following an abstract ideal that can never be reached? I have been asking myself these questions for what seems like a lifetime. The days, months and years that I have spent obsessing is obscene to think about, and yet even now, after all the struggling and fighting, I can't quite seem to completely dispose of this need to pursue beauty. On some days, I still yearn for a perfect body and sit in a pit of dissatisfaction and frustration.

But I am gradually making progress and the good days are finally beginning to outnumber the bad. Self understanding is a lifelong process and the war on body image is one that cannot be conquered in a day. The only thing we can all do is to continue to connect, to relate to each other and to see beyond the obvious.

Perhaps there is not one solid definition of beauty; maybe it cannot be defined in simple, tangible terms. I think the truth is that the spectrum extends infinitely, rooted in the essential connection between the body and the mind. By writing essays like this one, it is possible to reach new understandings that can slowly and gradually help fill the mind/body gap that has been created by an eating disorder.

BIBLIOGRAPHY

DeSalvo, Louise. *Writing as a Way of Healing.* Boston: Beacon Press, 1999.

Farganis, James. *Readings in Social Theory.* 5th ed. New York, NY: McGraw-Hill, 2008.

Fredrickson, Barbara L., and Tomi-Ann Roberts. "Objectification Theory." *Psychology of Women Quarterly* 21 (1997): 173-206. Rutgers University. May 2008 <http://www.sanchezlab.com/pdfs/FredricksonRoberts.pdf>.

Knapp, Caroline. *Appetites: Why Women Want.* New York: Counterpoint, 2003.

M.D. "Body Image: A Clouded Reality." *Human Architecture,* Vol 2, Issue 2, Fall 2003/Spring 2004.

Morpheus, Neo. "The Drinking Matrix: A Symbolic Self Interaction." *Human Architecture,* Vol 2, Issue 1, Spring 2003.

Murray, Megan. "Treading Water: Self-Reflections on Generalized Anxiety Disorder." *Human Architecture,* Vol 2, Issue 1, Spring 2003.

University of North Carolina at Chapel Hill. "Three Out Of Four American Women Have Disordered Eating, Survey Suggests." *Science Daily* 23 April 2008. 7 May 2008 <\http://www.sciencedaily.com/releases/2008/04/080422202514.htm>.

Wallace, Ruth A., and Alison Wolf. *Contemporary Sociological Theory.* 6th ed. New Jersey: Pearson Education, Inc., 2006.

Choosing My Major and Career

A Sociological Inquiry

Jacquelyn Knoblock

University of Massachusetts Boston

jacquelyn.knobloc001@umb.edu

Abstract: In this paper, applying various micro- and macrosociological theories and concepts, I explore how I ended up taking Social Psychology as a major and what my options are for a future career. I use C. Wright Mills's notion of the sociological imagination as a way of looking at and interpreting the circumstances of my life and my feelings and reactions towards them. I explore how such circumstances and explorations have caused me to be where I am, and how they may influence where I am going. I find that it was mostly personal troubles that led me to this field, but it is public issues that keep me interested and that make me want to continue in this field as a profession. I understand the tension of opposites between what I want to do with my life and my time, and what I must do to "make it" in this society. I am coming to see that society's view of success and my own may differ and they don't have to be the same. I see that my learning of different sociological theories and perspectives has enriched my own viewpoints and that I desire to be able to extend these viewpoints to others through teaching and/or counseling, and that these fields would seem to suit me because I am comfortable in the realm of academia. I do not want to get stuck in the "rat race" of corporations and capitalism, but that I must do something to make money to be able to support myself and live a basic comfortable life. Therefore I must make a compromise between the opposites. I realize that it is my personal fears and insecurities as well as some of the larger and public institutions that do not make it easy for women to succeed both in the workplace and in the home. These may hold me back from making high goals for myself and following through. However, I also see that I should not worry too much about getting a job and what I am going to do, and should instead focus now on learning about what interests me in the hopes of finding a meaningful occupation.

My current major is Social Psychology. This means that half of my classes are under the sociology department and the other half under the psychology department. When I tell people about my major, the usual reply is "Oh, that's interesting, what are you planning on doing with it?" My response to this is always, "I'm going to go to Grad School." Then I'll usually give a list of a few fields of graduate studies that I am interest in, and four or five things that I may end up doing with my possible future degree. I can't limit myself to one choice, and I think I am afraid of choosing a goal, for fear of not achieving it.

I used to think that Sociology and Psychology were both just phony majors, full of information that was just commonsense,

Jacquelyn Knoblock is an undergraduate junior at UMass Boston, majoring in Social Psychology. She wrote this paper while enrolled in the course Soc. 341-1: "Elements of Sociological Theory," instructed by Mohammad H. Tamdgidi (Assistant Professor of Sociology at UMass Boston) during the Fall 2007 semester.

and for people who were bleeding hearts or who didn't know what they wanted to major in. Sociology and/or Psychology were never on my possible lists of majors during my senior year of high school. It was a difficult road that led me to this major, with the most salient events occurring in the past two years, but on a road that probably began as a young child.

In this paper, using a framework informed by various sociological theories, I am going to investigate what caused me to choose Social Psychology as a major and explore the possibilities for my future career goals. In order to do this I must employ what C. Wright Mills has termed the **Sociological Imagination**. By this notion, Mills means the ability to relate one's own life with society as a whole, and the capacity to shift views from the personal to the public (348).

To understand my choice in becoming a Social Psychology student and to determine my future educational goals, I will first look at what Mills has termed **"troubles."** Troubles happen "within the character of the individual and within the range of his immediate relations with others" (Mills 350). Troubles fall under the realm of **microsociology**, which is concerned with face-to-face encounters and the details of communication and human interaction (Wallace & Wolf 5). It is the microsociological connection that I will focus on during the first half of my paper.

To begin to question and comprehend the reasons I have chosen Social Psychology and what I plan to do with my degree, I must first **problematize** it from a **stranger's** point of view, and look at my life as though I have no preconceptions of what it should be, suspending my "learned cultural notions" (Wallace and Wolf 262). These approaches were used by Harold Garfinkel, the founder of **Ethnomethodology,** as ways of explaining his methods. Ethnomethodology is people's ways of making sense of their social world (Wallace & Wolf, 269). Ethnomethodology is a type of research associated with **Phenomenology.** Phenomenology is interested in things that can be directly understood by one's senses (Wallace & Wolf 263) and "asks us not to take the notions we have learned for granted, but...to question our way of looking at and our way of being in the world" (Wallace & Wolf 262). I need to look at my past experiences and problematize and understand them to question the place I am in today, and to figure out where I want to go in the future. I agree with the UMass Boston student Jennifer Kosmas that "If we can become aware of our own social constructions, then we can also break out of the self-destructive habits by knowing ourselves" (74). So if I gain awareness of what has led me here, I can begin to know myself and figure out the best path for my future.

In her article titled, "The Roots of Procrastination: A Sociological Inquiry into Why I Wait Until Tomorrow," Kosmas analyzed her relationship with her father in order to get to the bottom of her procrastination problem. Similarly, in her essay titled, "Accepting Myself: Negotiating Self-Esteem and Conformity in Light of Sociological Theories," Sheerin Hosseini discussed her relationship with her parents in terms of her issues dealing with self-esteem. I feel that both procrastination and self-esteem are issues for me as well. I try to avoid any type of conflict that will make me feel uncomfortable, and if I have an assignment that seems difficult or that I am not sure of, I will postpone it, perhaps in an effort to self-handicap.

This may also stem from my relationship with my parents, and my place in my family. The family is the **primary group** of **socialization**. This means that individuals in childhood come to identify emotionally with the significant others in their lives, and that they come to **internalize social norms** and **values** (Wallace & Wolf 290). My family was not rich, but we lived comfortably, and my parents did not force us to do chores or even to really clean up after ourselves very much. We were free to enjoy our childhood, which was nice at the time, but I think I may have overly internalized the role of a child, and that now this is why I have a fear of growing up, and can't imagine living on my own, and having all the extra responsibilities of paying my bills, cooking dinner, and cleaning up. My father is also what may be called a "helicopter dad" because he tends to hover over our affairs even now, so I feel like I have internalized a sense of helplessness in dealing with things that my dad usually takes care of. I think that part of the reason I may have chosen Social Psychology as a major is because it would take me a few extra years, since not many of the credits from my former major and my previous colleges transferred into this major. Also, I know that I will probably have to attend graduate school. Overall, these give me more years of schooling where I do not have to be a full "grown up."

Another reason that I have chosen to continue my education and become a social psychology major is because being a college student is important to me. In my family my parents did value education, and there was never any question that we could all go to college. When I started elementary school I moved into **secondary socialization**. Secondary socialization concerns more specific **roles** and is a later phase of socialization of an individual who has already been socialized (Wallace and Wolf 290). In these classes, I took on the role of the "brain" and sometimes the "teacher's pet." In her essay cited previously, Sheerin Hosseini states that, "As a child, it seemed I wanted more than anything to be noticed and admired by both my parents and my teachers. I wanted to be the one in class who got recognition from the teacher for being intelligent or artistic" (29). I felt very much the same way as Hosseini. Once I realized that I was good at learning, that it came easier to me than most of my classmates, I ran with it, and took this as an important part of my self-concept and self-esteem. This has led me to achieve, but also has given me an extreme fear of failure. I feel like educational success is such an important part of my self-concept that I generally avoid trying activities in which I may not succeed, or that I feel might be too hard, and that I won't fare well in.

I have taken my role as a student seriously for the most part. In the film *Twelve Angry Men*, the jurors come in with their own presupposed ideas of what a jury was supposed to do and what the role of a juror was. Some took the roles more seriously than others. They created their own arbitrary rules and followed them. This also connects to the work that Garfinkel did on jury deliberation. The jurors "knew how they were supposed to act" by drawing on knowledge from other sources (Wallace and Wolf 276). In the same way I, even as I entered elementary school for the first time, or entered college as a freshman, knew what was expected of me by drawing on knowledge from other sources, what I had learned from books, TV, movies, friends, and parents.

It's also important for me to be viewed by others as a college student. I work as a cashier, a job that isn't very prestigious, and I always feel the need to make it known that my first priority is being a student, and that I only work part time to pay tuition and have some spending money. I also feel that some of my reasons for wanting to earn a college degree are so that I can live "the good life." The film *Affluenza* illustrates how Americans are consumerists and that many of us equate happiness with spending, and the goods that we acquire. Although I don't feel that making money is the greatest factor in my decision to major in Social Psychology, I can't deny that the prospect of money is an important aspect of my choice to attend college.

The most significant event in my life that caused me to think about work in Social Services or Psychology and that led me to become a Social Psychology major was being involved in an abusive relationship. I became involved with a man who at first seemed like a good boyfriend, then he became controlling, emotionally abusive and finally physically abusive. It was during this time that I withdrew from my first college. I can analyze this time in my life by drawing on many of the concepts included in **Symbolic Interactionism** and **Rational Choice or Social Exchange Theory.**

Symbolic Interactionism is "a social-psychological perspective; it focuses primarily on the individual "with a self" and on the interaction between a person's internal thoughts and emotions and his or her behavior" (Wallace & Wolf 199). Rational Choice Theory assumes that people are rational and make their decisions and perform actions based on what they think are the most effective means to their goals. The theory presumes that people are constantly weighing alternatives.

There are many concepts in these theories that apply to my situation. At first I allowed a little bit of controlling in the relationship, because of self-esteem issues and my problems with making my own decisions, and because I enjoyed the **status** that came with having a boyfriend. In Social Exchange Theory, status is defined as common recognition by others of the amount of esteem or friendship that someone receives (Wallace and Wolf 340). Once my troubles got more serious, however, I tried to hide them, and refused to admit to anyone that anything was wrong. I engaged in what Goffman terms **impression management**; I tried to control how others viewed me. That I could do this may be explained best by Goffman's theory of **dramaturgy.** Dramaturgy is concerned with the lives of ordinary people, and how they act out their daily roles on the "stage" of life (Wallace & Wolf, 239). On the **front stage**, in the presence of an audience, I acted as if everything was perfectly fine, that I had it all together, and refused to let on that anything was wrong. It was only in the **back region**, hidden away from others, that any of the abuse took place and that I could cry and let myself feel the pain.

During this time, I felt like I was a **discreditable person**, that I carried this **stigma** of being an abused woman which could be discovered at any time. It seemed to me, if I didn't let anyone else know that something was wrong, then nothing was wrong; but if anyone else found out, I couldn't go on pretending anymore. Finally, when it became more than I could handle with a smile or an explanation, I began to isolate myself from my friends and family. I even lost my sense of **self.** George Herbert Mead describes the self not as a "passive receptacle" but as active and creative (Wallace & Wolf, 205). At this point in my life, I was not active and creative, I only felt that I was the "**me**" and not the "**I**," that I was acted upon, but not acting or reacting upon my own initiative. I left my first college at this time. Eventually my parents made me go to therapy (once again, I did not make my own decision). I ended up finally getting out of the relationship and moving to South Carolina.

It was in South Carolina that I decided that I wanted to take some classes in Psychology and Sociology to try to understand both myself and others better, and hopefully see what had happened to me. In this sense, I was sort of like Neo, the main character in the movie *The Matrix.* In one scene, Neo is given a choice whether to take a pill and gain an understanding of what the real world is, without any promises that such knowledge would make him happy, or that he could go back to living his artificially created life. Neo decided to take the pill that would cause him to learn about what was really happening. My choice was similar; I wanted to gain an understanding of myself and other people and the structures of society. I felt that maybe I could get some answers by studying Social Psychology. As Avery Gordon wrote in her book *Keeping Good Time*, "faced with our powerlessness to assert authoritatively what is going on, we only want to 'simply understand'" (9). Gordon was speaking about research during wartime, but I feel that her statement applies to most people who can't ascertain what is happening in their lives or feel powerless to take action to change it, but at least want to understand what is going on.

Now that I have gained some understanding, the question I have is about where I want to go once I get the degree. I have basically determined that graduate school will be necessary. But, I suffer from an anxiety that I won't be accepted, and I know I probably won't apply to any elite schools, because of the fear of being rejected. I have never been rejected by a college, and I think that since so much of my identity is tied up in my academic abilities, that rejection would hurt my sense of self. I limit myself to schools I know I will probably get into. I also limit myself in terms of being a woman and knowing that if I have children, I will want to have a job in which I can fit my family life and career.

In her essay titled "Altruism or Guilt: Applying My Sociological Imagination to Choosing a Helping Profession," Elizabeth McCauley writes about how her future career choices are based on ones that would have a mother-friendly schedule; but she also states that, "I am not certain I want to have children. I am positive that if I choose to I will be in my late thirties when I start, so why am I basing my career choices on non-priorities or priorities that are not my own?" (149).

Similarly, I am worried about experiencing the **bifurcation of consciousness**. This is experienced in women as they move between modes, generally work mode and home mode, and are constantly aware of and concerned about each, because there are ongoing pressing demands on women, especially those with children (Farganis, 375). Therefore, when I think about wanting to be a professor, I usually think about just being an adjunct so that I would still have time to have a family. Or, I may think I want to have my own counseling practice, but then I'll think that may take up too much time, or be too hard. Because of the expectation of women to raise a family, I think about it even when I am in no position to be having a family anytime in the near future. I do think that I want to help or at least educate others to avoid, or get out of the situation I was in when I was involved in the abusive relationship. But I struggle with this: is it this because I really want to help others, or is it just because I want to give meaning to this awful thing that I went through, the powerlessness that I felt? Avery Gordon states that,

> The struggle to achieve a certain independence from the ways we are imprisoned, with love in our hearts, preparing for a struggle that will change us, that will make us more free, more capable of sharing freedom than we were before: that's *the place we go to, the way we get there.* (66)

I feel like I have gone through a struggle, and part of the reason I chose Social Psychology was for understanding, and the other part was so that maybe in the future I would help others by sharing what happened, or just by being someone who could listen to their situations and not judge them. This is where I feel that I want to go, and I think that it is only my own apprehensions, and my interpretation of society's ideals, that can hold me back.

In his book, *The Sociological Imagination*, Mills wrote that distinguishing personal troubles—which I defined and explored in the first half of my paper—from ***public issues***" is an essential tool of the sociological imagination" (p.350). Public issues have to do with matters that involve the larger social structures and impact society as a whole. These are part of the realm of ***macrosociology.*** Macrosociological perspectives are concerned with analyzing "the large scale characteristics of social structures and roles" (Wallace & Wolf, p 5). Macrosociological theories include ***Functionalism***, ***Conflict Theory***, and ***Postmodernism***.

The functionalist perspective states that society is made up of different parts and that these parts must work together in order for society to function as a whole and maintain equilibrium (Wallace & Wolf 16). Conflict theory assumes that people have certain basic needs and wants, that power is at the core of social relationships, and that values and ideas are seen as weapons (Wallace & Wolf, 68). Postmodernism has come to mean different things, it has been associated with post-industrialism and post-Marxism, a movement in literary criticism. It deals with the Enlightenment promise of rationality and the relationship of knowledge to theory and power (Farganis 423).

We are influenced by the macrostructures of our society. In our socialization we are introduced to the prevailing ideas of our culture. Culture is defined as "the enduring behaviors, ideas, attitudes, and traditions shared by a large group of people and transmitted from one generation to the next" (Myers, 11). This transmission of culture through generations occurs through socialization and is part of what Emile Durkheim's called ***integration***, "the incorporation of individuals into the social order" (Wallace & Wolf, 20). Durkheim maintains that integration is important for the maintenance of social equilibrium.

An important part of integration is the understanding of the current ideologies comprising and shaping the culture. Conflict theorists, such as Karl Marx regarded established ***ideologies*** as beliefs that serve to legitimize the position of those who currently are in control (Wallace & Wolf, 87). I will discuss some of these ideas that have been internalized by most people and myself, and how they affect our perceptions of the world, and what we end up doing with our lives in it.

The contemporary U.S. society has an ideology that admires rugged-individualism. In our individualistic society identity is self-contained, becoming self-reliant is viewed as an important trait of a successful individual, and independence is valued (Myers, 41). As the division of labor increased in our society so did individualism, or at least the idea of it.

I have struggled with this individualism in my life. I have always been a dependent person; I haven't really lived on my own. I think that I want to at some point (or maybe just because society wants me to), but I don't have the money, and I think I am also scared of the prospect of not having my family or friends nearby for support. This will also limit my prospects for the graduate school and careers as I will probably stay in the Boston area or the South Shore. In the movie and book *Tuesdays with Morrie* by Mitch Albom, the late sociology professor from Brandeis Morrie Schwartz says that our culture tells us we should be ashamed of being dependent on others, but that "there is nothing innately shameful about being dependent." Most people in the U.S. would probably disagree with Morrie. I struggle with this because as I grew up I was allowed to be dependent, and never learned to do things for myself; but now at my age that dependency is deemed inappropriate, and I am told that I need to just grow up. Our society values independence and abhors dependency so much that we have pathologized dependency as a personality disorder in the DSM-IV.

The paradox is that we are not truly independent in our society; we are interdependent. We rely on what Durkheim termed ***organic solidarity***, which occurred as the division of labor and ***specialization*** increased. Organic Solidarity is "characterized by the interdependence of roles and a lack of self-sufficiency" (Wallace & Wolf, 20). Durkheim states that this is what holds people together when they are all performing separate tasks. Ferdinand Toennies first termed this type of modern industrial society a ***gesellschaft***, which is "characterized by a predominance of more impersonal or business-type relationships" (Wallace & Wolf, 30). The Functionalist Talcott Parsons labeled these types of relationships in the modern society ***instrumental***. The focus on independence, these formal relationships, as well as the exploitative relationships of ***Capitalism***, are what allow most people to justify their own or others' positions in life. They feel that whether one is on top of the chain or on the bottom is a product of their own hard-work. I used to feel this way, and still in part hold on to this ideology for myself; I think that if I just work hard enough and find my niche I will be able to support myself, make money, and live the "good-life." Of course there are many other factors that influence these outcomes other than "hard-work."

Although I still hold on to this ideology for myself, I do not expect it of others. A concern that has brought me into the Social Psychology field of study is that of the welfare of others. At my previous college I took business, accounting, management and economic classes. During these classes I never felt good about profits. I felt that leadership and management classes were all about manipulating people to get the most productive value out of them. In the book and movie *Tuesdays with Morrie*, Morrie was brought by his father to the factory where he worked making fur coats. Having seen and experienced the inside of the factory Morrie vowed that he would, "never do work that used people, that made money off their sweat and degraded them." I also felt this way; how could I be concerned about business and profits and not worry that the people in the company weren't happy or were being exploited or were in danger of losing their jobs?

Michael Moore tackles this issue in his film *The Big One*, where he illustrates some of the real casualties of capitalism. During one stop on his book tour, Moore visits Centralia, Illinois, home of the PayDay candy bar. When he arrives he finds that the workers in the PayDay plant just found out (through a video) that they were losing their jobs. The ironic part was that the workers had worked hard, done their jobs well, and made profits, ***surplus value***, for the company. The factory workers were termed ***proletarians*** by Marx, the laborers who don't own any of the means of production. These particular workers in Centralia are the embodiment of modern ***exploitation***. Karl Marx described exploitation as occurring when "any surplus value is appropriated by someone other than the worker" (Wallace & Wolf 84). Because of this exploitation, the company had the means to relocate their factory to another country with a cheaper labor force.

Of course at the PayDay factory in Centralia, and in many of the other corporations Moore visited during his book tour, the management and the owners and CEO (the ***bourgeoisie***, the people who "control capital and direct labor" [Wallace & Wolf 90]) did not seem to care about the people they had let go. To them it was all about "remaining competitive" and "having an obligation to the stockholders." Many of the CEOs had a corporate mentality in that in their thinking and conversation they became a spokesperson for the corporation and had the interests of the corporate structure in mind. This corporate mentality included what Max Weber called ***Zweckrational***, a "form of reasoning that breaks down all problems into a means-ends chain and entails rational calculation of costs incurred and benefits to be secured if a particular line of action is pursued" (Wallace & Wolf 93). The management and owners dis not seem to worry or care about what happened to their employees, because in the end they would be making profit by laying them off and downsizing to another country. This is a feature of capitalistic mentality and *Zweckrational*.

Weber came up with the term the latter term to describe ***bureaucracies***. Features of bureaucracies include highly specialized tasks that are arranged in a hierarchical order, depersonalization, routinization, and mechanical predictability. I do not want to end up working in a big bureaucracy or dealing with the ***alienation*** and impersonalization of a job like that. Therefore I am caught in what Morrie Schwartz termed, "a tension of opposites" and what Karl Marx termed ***dialectics***, a contradiction, or a "pattern of inner conflict" (Wallace & Wolf, 162). I, like most people, am stuck between what I want to do and what I must do. My dream would be to be a perpetual student, and I wish I could just learn for the sake of learning and not deal with the bureaucracy of deadlines and grades, etc. I enjoy learning and my ideal would be to be learn about everything and anything that I find interesting. Unfortunately, that is not the "real-world." I have already been in college for 5 years and can't afford to continue indefinitely. I need to finish college and earn a degree so that I can get a "real job."

Also the problem with school for me is that I am so concerned about grades and I really don't learn for the sake of learning. I wish I could, but I always know that the grade is what is really important. That is the way I can maintain scholarships. Graduate schools and future employers are going to judge me chiefly by my GPA.

Although grades are important to me, I hope to become more like the alternative students that Avery Gordon speaks about:

> By naming yourselves alternative students you have chosen a path, a perspective, that says you are part of a long tradition of those who see differently, who imagine and dream other possibilities than the normal ones taken for granted. You have aligned with those…who see the individual as part of the larger social fabric and consider this cooperative interaction freedom; with those who feel a source of pride and accomplishment in what is derided and undervalued; with those who don't take the easy way out but are willing to live and struggle with difficult contradictions. (72)

One of the reasons I decided to attend UMass Boston was because it is more of an alternative campus. There are students of all ages, from many different ethnicities, and ***socioeconomic statuses***. There are people who go to school full-time, and those who work full-time and take one or two classes a semester. I wanted to be exposed to a wider range of perspectives, which I get from the students around me and from being in a major that really consists of two majors.

I feel that the more I learn, the more I want to learn. I definitely feel that in the future I would like to counsel or teach. I think that it is important for people to be exposed to viewpoints that take them out of the conventional ways of thinking about the world. Rose Weitz writes about resistance in her article, "Women and Their Hair: Seeking Power Through Resistance and Accommodation." She states that we must define resistance as "actions that not only reject subordination *but do so by challenging the ideologies that support that subordination*" (137). In order for people to challenge and resist the social forces that act on their lives, they must first understand the ideologies that have put the forces in place and where they come from, to be able to think of different ways of thinking or living. By teaching or counseling I feel that I would have the means to at least expose people to some differing paradigms or ways of looking at life in hopes of allowing them opportunities to resist. Avery Gordon writes of this as dissent:

> The maintenance of an alternative perspective requires also an active commitment to dissent and to the insistence that dissent is never treason … to commit dissent is to commit self-governance, not administrative governance … social justice, not personal justice … to commit to learning and unlearning again and again, with pleasure. Dissent is the key link between education and democracy. (73)

I see that this dissent or resistance is important, and that education is a key factor in understanding systems and having the ability to resist them. I myself have always been a firm believer in learning and empiricism and the scientific method, but in reading about the Postmodern perspective I see that there is some room for subjective relativity concerning reality and that knowledge and science are bound with power. I feel that understanding this has made me less dogmatic in my views, and I feel that interaction with others on a face to face or a teaching level would continue to open my life towards new perspectives.

To conclude, while trying to understand what has brought me into the field of Social Psychology and what I am going to do with my life when I finish school, I see that it was mostly personal troubles that led me to this field, but it is public issues that keep me interested and that make me want to continue in this field as a profession. I understand the tension of opposites between what I want to do with my life and my time, and what I must do to "make it" in this society. I am coming to see that society's view of success and my own may differ and they don't have to be the same. I have seen that my learning of different sociological theories and perspectives has enriched my own viewpoints and that I desire to be able to extend these viewpoints to others through teaching and/ or counseling, and that these fields would seem to suit me because I am comfortable in the realm of academia.

I do not want to get stuck in the "rat race" of corporations and capitalism, but that I must do something to make money to be able to support myself and live a basic comfortable life. Therefore I must make a compromise between the opposites. I realize that it is my personal fears and insecurities as well as some of the larger and public institutions that do not make it easy for women to succeed both in the workplace and in the home. These may hold me back from making high goals for myself and following through. However, I also see that I should not worry too much about getting a job and what I am going to do, and should instead focus now on learning about what interests me in the hopes of finding a meaningful occupation.

References

Albom, Mitch. *Tuesdays with Morrie.* New York: Random House, 1997.

Farganis, James (ed.) *Readings in Social Theory: The Classical Tradition to Post-Modernism.* Fourth Edition. McGraw Hill College Division, 2004.

Gordon, Avery. *Keeping Good Time: Reflections on Knowledge, Power and People,* Paradigm Publishers, 2004.

Hosseini, Sheerin. "Accepting Myself: Negotiating Self-Esteem and Conformity in Light Of Sociological Theories." *Human Architecture: Journal of the Sociology Of Self Knowledge,* IV, 1&2 Fall 2005-Spring 2006, 29-44

Kosmas, Jennifer. "The Roots of Procrastination: A Sociological Inquiry into Why I Wait Until Tomorrow." *Human Architecture: Journal of the Sociology Of Self Knowledge.* Vol. II, No. 2, Fall 3003-Spring 2004. 74-81

McCauley, Elizabeth. "Altruism or Guilt: Applying My Sociological Imagination to Choosing a Helping Profession." *Human Architecture: Journal of the Sociology Of Self Knowledge,* IV, 1&2 Fall 2005-Spring 2006, 147-156.

Mills, C. Wright. The Sociological Imagination. Oxford: Oxford University Press, 2000.

Myers, David G. *Social Psychology.* Ninth Edition. McGraw Hill, 2008.

Wallace, Ruth A., and Alison Wolf. *Contemporary Sociological Theory: Expanding the Classical Tradition,* Sixth Edition. New Jersey: Prentice Hall, 2005.

Weitz, Rose. "Women and Their Hair: Seeking Power Through Resistance and Accommodation." *The Politics of Women's Bodies: Sexuality, Appearance, and Behavior.* Second Edition. Ed. Rose Weitz. New York: Oxford University Press: 2003.

Films:

Affluenza. Bullfrog Films, 1997.

The Big One. Miramax Home Entert., 1997.

The Matrix. Warner Brothers, 1999.

Tuesdays with Morrie. Touchstone, 1999.

Twelve Angry Men. MGM, 1957.

A Futile Struggle?

Power and Conformity in High School and the Society at Large

Eric Reed

University of Massachusetts Boston

eric.reed001@umb.edu

Abstract: George Simmel describes small groups as being "closely coherent," allowing "its individual members only a narrow field for the development of unique qualities and free self-responsible movements" (Simmel 2004:133). Throughout my experience in high school I witnessed this first hand, struggling with a need for acceptance, and being swept into conformity. What I have come to realize is that this drive acts as a sort of competition. In this competition a social ladder develops that is vague, but still significantly shapes the lives of the members of the group. It shaped my attitude toward other members of the group as well as myself, in turn influencing much of my actions. Moral judgments in turn were replaced by group-determined rules of what is acceptable, or "cool." I in turn explain how similar patterns possibly form in class-based stratifications, influencing the identity and attitudes taken among individuals of the lower, middle, and upper classes, where morality and trust may be set aside in order to promote one's status.

In his essay, "The Metropolis and Mental Life," George Simmel describes the nature of different sized groups. He states, "The earliest phase of social formations found in historical as well as in contemporary social structures is this: a relatively small circle firmly closed against neighboring, strange or in some way antagonistic circles. However, this circle is closely coherent and allows its individual members only a narrow field for the development of unique qualities and free self-responsible movements" (Simmel 133).

No statement so clearly reflects the structure of Natick High School. High schools like mine and those neighboring it are typically small, mine having about three hundred individuals in each class. Our school pitted against its neighboring schools by way of rivalry. And from the day I entered it I confronted a foggy yet specific view of what was considered cool.

Now, if one were to look back and examine the nature of coolness, he or she would find, like all cultural phenomena, that it is always changing. The finer points of coolness tend to also vary from group to group or from school to school. For example, after arriving in my freshman year at Northeastern University, I found many of

Eric Reed is an undergraduate sophomore at UMass Boston, double-majoring in Sociology and Psychology. He wrote this paper while enrolled in the course Soc. 341-2: "Elements of Sociological Theory," instructed by Mohammad H. Tamdgidi (Assistant Professor of Sociology at UMass Boston) during the Spring 2008 semester.

the same styles of dress and musical taste while finding others that varied greatly from those I had seen among students coming from New Jersey and LA. Even other students from neighboring towns in Massachusetts didn't match the expected schema of coolness that I was used to. Therefore, it is important to interpret my high school experience in terms of a confined, relatively small group structure that was influenced by all kinds of outside and internal factors. As a result, it may be regarded as a useful sample for the study of human interaction involving social hierarchies and competition for acceptance.

Upon entering the high school I noticed the system at work. Some individuals just seemed to be superior to others. These individuals comprised actually a very small percentage of my classmates, but they seemed to have a presence in the school which rewarded them more of my attention than others. This provided them with a higher **social capital**, which are "the aspects of social structures which make it easier for people to achieve things" (Wallace and Wolf 372). But how did these individuals come to have such **power**? George C. Homans describes power as the "ability to provide rewards that are valuable because they are scarce" (Wallace and Wolf 326).

These students seemed to possess little to offer the rest of us in terms of material rewards. Rather, what these students seemed to possess was the ability to punish rather than reward. Highly confident, these students were just a bit surer of themselves. In turn they had the ability to punish others for not meeting their set of criteria for normalness. If a fellow student found himself or herself on the radar to be put-down, their social status would in turn suffer. Their power, therefore, was based on their self-confidence and others' fear. These students were cool, more or less because they decided that they were. They could then dictate their norms to everyone, offering punishments if we did not comply. From the first day of high school, the so-called "popular kids" were more or less decided upon, and they continued to be so until graduation.

From that first day, I searched for various strategies in my head in order to gain acceptance and move up on the social ladder. While I found that my personal ideas of coolness was very much different from others, I would often compromise these ideas, seeking a balance between my own and other people's standards. As soon as I became used to acting a certain way I noticed that my own style and ideas changed to match those of others. The process perfectly demonstrated what Charles Horton Cooley called the "**looking-glass self,**" comprised of three elements which are "the imagination of our appearance to the other person; the imagination of his judgment of that appearance; and some sort of self-feeling, such as pride or mortification" (Wallace and Wolf 203). Cooley's concept describes well my strategy of behaving according to how I imagined others perceive being cool and acceptable, reinforcing my feelings of satisfaction with myself.

The reason I was doing this, however, was out of a drive to actively gain similar powers in relation to those above me. Such a strategy illustrates George Herbert Mead's view of "**the self**" and how it relates to Herbert Blumer's **Social Interactionist** perspective. In Mead's perspective "individuals are viewed as active constructors of their own action, rather than as passive beings who are impinged upon by outside forces" (Wallace and Wolf 199). Social interactionism stresses not only the influence of society on an individual, but also of the individual's own unique interpretations of social situations. My active pursuit of acceptance was done in search of a higher status, and was therefore also self-motivated. By making the achievement of this power a goal I utilized what I would later learn to be a basic proposition of the **Rational Choice Theory**, which states that "people are rational and base their actions on what they perceive to be the most effective means to their goals" (Wallace and Wolf 303). Even though I disliked and distrusted most of those students in power, by pursuing a similarly high status I acknowledged them to have the power to disapprove and halt my progress.

UMass Boston student Keyon Smith presents a similar situation in his article, "Lifting the Fog: Finding Freedom in Light of the Sociological Imagination" (2007). He states, "With the universality of these sequence of events, coupled with my humanity, I had my mind set on following that prescribed order, even if that meant not being myself, whatever myself was, whatever being myself meant" (74). Here Smith describes the way conformity controls us. People become more concerned with "getting ahead" and often don't consider what they are doing or changing about themselves in order to do so, simply because it is generally accepted as the correct, or in my high school's case, the cool, thing to do. Not doing so can be met with the anxiety that you are more or less falling behind.

The solution that Smith offers is for the individual to step back and examine the workings of society or the group as a whole and evaluate its problems, thus illustrating the value of what phenomenological sociologists encourage us to do. **Phenomenology** is "a method in philosophy that begins with the individual ad his own conscious experience and tries to avoid prior assumptions, prejudices and philosophical dogmas" (Wallace and Wolf 262). This method advocates adopting an unbiased view to look at each social situation, an attitude that I did not have during this time. Rather than questioning my practices, I took my past experiences for granted, automatically seeking what I viewed as being success in the past. **Ethnomethodology**, as defined by Harold Garfinkel, is the study of "how people make sense of everyday activities" (Wallace and Wolf 269). My efforts here in observing how I and my peers behaved and made sense of our everyday high school lives is an example of ethnomethodology at work. I had previously never really considered whether my strategies were making me happy or not; I just did them. Now I am making sense of them. I seemed to live a make-believe world before.

The film *The Matrix* describes a world which is in actuality an illusion. People go about their days living in a society they think is real and tangible. However, to the main character, Neo, it is revealed that his world is actually an illusion and a virtual reality, the real world being one in which people's bodies are used as an energy source to run a world ruled by robots. Once one awakens from the virtual world, the flaws in the real world are revealed and the domination of humans by robots become apparent. In a way *The Matrix* illustrates how the world is constructed in such a way that products of human activity become objectivated, taking a life of their own, coming in turn to rule human affairs instead. **Objectivation**, according to Peter Berger and Thomas Luckmann, involves "apprehending everyday life as an ordered pre-arranged reality that imposes itself upon but is seemingly independent of human beings" (Wallace and Wolf 288). In the same way my perceptions of what I should be as an "empowered cool student" were just examples of how I have interpreted others' **externalized** point-of-views as **objectivated** facts, and have **internalized** them, believing them to be my own illusory values, rather than considering them as what they actually are.

A similar strategy of reality altering is demonstrated in the documentary film, *Affluenza*. This film demonstrates the extraordinary degree of power the media and corporations have in controlling how we spend our money. As a result of affliction with affluenza, "the disease of consumerism," the average American spends way more on way too many products, far beyond what is necessary for survival. At the same time, through the same process of objectivation, individuals are led to believe that consumer goods are and have always been necessary for survival and what brings them real happiness. This is similar to how I perceived competition for power over peers and acceptance by them as my route to happiness in high school. Just as I feel only a short rush of excitement after buying a new pair of jeans and then move onto my next buy, with every moment of acceptance as a "cool kid," I only had the next to look forward to.

Though perhaps the ability to utilize my strategy in copying with my school environment may have come easier with maturity and experience, an understanding would have saved me much anxiety throughout my interactions in school. The extent to which I let others control my actions in most aspects of those four years is astounding.

It even reached a point where I would avoid bathroom breaks in order to not give anyone the impression that I was actually using the school's restrooms. This simple bodily function was surely performed by all my peers, and the toilets were specifically designed for such a function. However, I found some disgust in others' doing so. Erving Goffman's **dramaturgy**, or "the art of dramatic composition and theatrical representation" (Wallace and Wolf 238), comes to mind here. I could simply act as though these biological limits did not apply to me between 8 a.m. and 2 p.m. And this illustrates Erving Goffman's discussion of **impression management.**

Impression management is defined as "the ways in which an individual guides and controls the impressions form of him or her" (Wallace and Wolf 238). In other words, throughout my day at school I did my best to control how other people thought of me. Goffman identifies two concepts, the **front stage** and **back regions**. The front stage was were I interacted with others, including the part of me that I made visible to my peers. I would therefore try to present myself as being as acceptable to the rest of the group as possible, thereby securing my place. The back region consists of what I don't let people see, which not only included performing vital bodily functions but also any other practice that I felt would not improve my social stance.

Throughout high school I found just enough reinforcement for my actions to keep pursuing higher status. With every positive comment I was given an immediate dose of satisfaction. I found, however, that in conforming, these temporary highs were short, plentiful and addictive. Every time I felt I did the "coolest," most acceptable action possible, I felt happy and successful in my mission to become as "popular" and moving forward. However, I never went anywhere. Rather, I stayed in my position. The irony of this manner of competition was that as competing individuals many of my peers and I strove to distinguish ourselves through attempting to do what had in the passed proven successful. This meant meeting the norms of the few in power, in essence becoming more similar, and rather doing the opposite of what we set out to do.

Though not one of the elite, I don't consider myself to have been rejected socially either. Instead, I believe I was generally well like and accepted. I was able to interact casually with virtually everyone in my high school, but without the respect I saw possible. In my pursuit to move upward, however, I discovered myself having an unfortunate view of those I saw as less cool than myself. For whatever reason, perhaps these individuals did not concern themselves with the same standards that I did. Some found themselves having lower social status opposite to the few in power. I quickly acknowledged that these few were not of use to me in my quest for higher status. As a result I did as most other did, I paid little attention to them. What those in power felt over me, I did over these individuals. I did not want to be associated with them because they belonged to a class lower than my own, and association seemed to only bring me down to their level, which was the opposite of my goal.

This interaction between groups in power is not unlike that of the jurors in the motion picture *Twelve Angry Men*. This film features a group of twelve jurors deliberating on the guilt of a young man accused of killing his father. In the beginning of the film all but one juror votes the accused man guilty, as all the evidence seems to clearly point to him. However, as the lone juror, Juror #8, begins to break apart the evidence against him, it becomes more and more apparent that the evidence is foggier than the jurors had previously thought, and all eventually vote "not guilty." In fact, the reason that many thought the boy was guilty was more based on their prior prejudices against the boy since he was of lower class and lived in a rough neighborhood, This in turn rewarded him with a lawyer that for whatever reason put little effort in the case. In this situation the accused represents the lowest tier of social status. Several of the more vocal jurors against the accused show a similar status to those I recognized in my high school as the popular attention-getters. The rest of the jurors represent those in my situation, attempting to go along with the status quo, in order to maintain or improve their position. At first all but one wouldn't give the accused boy the time of day because he would have little to no impact on their position.

In his article, "Looking Inside Out: A Sociology of Knowledge and Ignorance of Geekness," UMass Boston student Johnny Yu provides a unique perspective highlighting individuals in the same lowest tier of social power. He uses the popular term "geek" to describe this group stating, "People who aren't in the 'in' group find it impossible to break out of their ascribed identities. They find they need to conform and change what their interests are to avoid being labeled as anti-social or geeky, even if the terms are not accurate descriptions of their characters" (Yu 42). Here, he describes the seemingly hopeless position for some not to be accepted, and describes the unfortunate steps one must undergo to move out of this position. He describes the importance of **coalition building**. This is a strategy in which individuals of similar interests form a sub-group in order to gain more power. In his case the sub-group is that of individuals who get together and play "Magic: The Gathering." If his characteristics put him on the bottom of the social latter, he seeks a new group that accepts him, thereby rejecting the culture as a whole that he is a part of. Though others may consider him to be a "geek" on the bottom of their social latter, he considers himself to be a part of some other latter, thereby escaping the negative popularized perspectives towards his behavior.

Johnny Yu's perspective gives me new insight into my own situation in high school. It is that determining one's identity as part of a group is very much also up to the individual. My desire to be part of Yu's "in" crowd made me a part of it. However, my acceptance to that group is completely up to the rest of the members. By following my interpretation of the norms of this particular group that made up my high school, I was participating, but not necessarily succeeding. Many other individuals, who may be seen as "geeks" or outcasts, are only so by that one specific group, while those individuals seen as "geeks" may see themselves as actually very powerful in the separate group to which they actively participate.

Upon leaving high school, I was slowly introduced to the universality of what could be considered cool. As I cringe at the dance music my friends from Los Angeles listen to, I remember how I forced myself to tolerate Jam Band music for the past four years. I agree with the perspective of Simmel, that in the presence of a larger group, I am more able to break out of my previously established ideas of coolness. But even in the largest of groups there are still societal norms to be grappled with as described by the SUNY-Oneonta student R.F.A. in his article "The Capitalist's Cuckoo's Nest." The question he means to answer is, "is it worth the problems I face not to conform to conventional customs and values?" (1). He uses the examples throughout the novels, *One Flew Over the Cuckoo's Nest*, and *Fight Club*, along with the life of John Lennon, displaying the struggle to be different as one in which the rewards are not immediately obvious. But behaving as such is what is required for exercising self-actualization. R.F.A. writes, "Society oppresses people like me, and therefore it is my job to be different" (6). As I now delve deeper into my college years, I fully agree, noticing that as one gets older norms are more general and obvious, but in some ways more imperative. This is because most of these norms start at the need to feed and shelter oneself, through the acquisition of resources. The most prevalent example of this is money and the power struggle that occurs between the haves, the have-somes and the have-nots.

The manner in which I have experienced social interactions as a middle class white male in the United States seems to fit well with Karl Marx's **conflict theory.** Wallace and Wolf generally define society from a conflict perspective as an "arena in which groups fight for power, and the control of conflict simply means that one group is able, temporarily, to suppress its rivals" (68). In other words, different groups, especially those of differing economic status (upper, middle, and lower classes) compete in order to gain power. **Power,** as defined by sociologist and social exchange theorist Peter Blau, is "the ability of persons or groups to impose their will on others despite resistance through deterrence either in the form of withholding regularly supplied rewards or in the form of punishment" (345). It is out of the collision of these competing interests that various forms of **social stratification** arises. I have witnessed this power struggle first hand, and not just in the confines of my high school, but in broader society as well.

In capitalist society, just like in my high school, the upper class attempts to maintain their status, while those lower attempt to better it. This creates the most prevalent conflict between the upper and lower classes, as the upper class seeks to maintain great control over the lower classes. This position of power is well exposed in the documentary *The Big One*, directed by Michael Moore. The film features the filmmaker on a book tour across theU.S., in which he makes stops at corporations seeking to expose inhumane business practices of outsourcing amid globalization and its effect on working class people. What he demonstrated was that throughout the United States, despite the fact that companies were turning record profits, they were closing factories and moving manufacturing to other countries where they could pay workers much lower wages, thus **exploiting** them. As a result the workers in the successful—yet apparently not quite successful enough—factories in the United States were losing their jobs.

From the standpoint of middle class individuals, the struggle is much more ambiguous, as they have both much to gain and much to lose. The middle class represents a group with perhaps the most prevalent **internal conflict** since, as Marx argued, it is often in the process of polarization between the two major classes of the bourgeoisie and the working class. With ready access to higher education, the middle class may much more readily than the working class move up into the upper class. At the same time, the upper class still holds control over the middle class, though having much control over their employment as well. As a result the middle class may mistrust the upper class, but generally suppresses conflict with the group they wish to join. As Wallace and Wolf suggest, "a general approach to organizations is to view them as arenas for struggle in which superiors try to control their subordinates" (Wallace and Wolf 144). The upper class seek to coerce the middle class to support them through illusions of necessity and the attractiveness of their situation, keeping the middle class docile in conflicts between the upper and lower class.

This is similar to my experience in high school; even though I ultimately didn't trust those I thought to be my superiors, I followed them out of the want of acceptance and a desire to be like them. Interactions between the middle and lower classes are similarly skewed. Just like I attempted not to be associated with the "out-crowd" in high school, members of the middle class wish to distinguish themselves from the lower class. And though they may ultimately mistrust the upper class more, their desire to be promoted into the upper class, redirects their conflict toward the lower class, similar to how I conformed with the upper echelon of my high school against the lower strata. As a result, much of the blame for problems in society gets placed on the lower class.

For example, many blame unemployment partially on illegal immigrants, who work for less money than citizens would. Less often demonstrated through the media is the effect outsourcing of jobs has on unemployment, as demonstrated in *The Big One*. Another example may be the blame middle class people put on lower class individuals on welfare for raising their taxes, while much of the unemployment which causes people to be on welfare may also come from not only outsourcing, but also from the taxes upper class people are more than able to pay, but don't. In my experience, I chose to ignore the atrocities of my envied peers, focusing on my distaste towards the unenvied.

The position of a middle class person is well described in the film *Tuesdays With Morrie*, through what the title character, Professor Morrie Schwartz of Brandeis University, described as "the tension of opposites." Morrie believed that the individual was constantly being pulled in two opposite directions, as if by a rubber band. In this case middle class people are often caught between both other groups. In his article, "Taking Sides and Constructing Identities: Reflections on Conflict Theory," for instance, Guenther Schlee provides a similar explanation for how individuals of a group, such as the middle class, take sides on an issue, and offers an equally relevant explanation. He states that one's reason for choosing a side "concerns the advantages and disadvantages that may arise…from the costs and benefits of taking sides" (136). His other reason relates to which class the middle class identifies with. He states, "Wishing to be or not to be something is not enough; one also needs a plausible claim to an identity or a plausible reason for rejecting it" (136). In other words, Schlee means to convey that individuals side with the group they find themselves to be more related to. He goes on to say, "If plausible reasons are lacking, one might be forced by one's own logic and the expectations of others to join the fight on a given side" (136). In the case of the middle class, individuals may not readily see themselves as being related to either the lower or the upper class, but side through conforming to what others in their group have done.

The problem with this trend is the way that the lower class stands alone in this struggle and must more or less fend for itself. An example of this can be pulled from the problems of impoverished people in California. In the edited book, *California's Social Problems*, author Robert Enoch Buck states that "the poverty rate among households headed by women with less than a high school diploma is more than three times the national family poverty rate"

(218). The reason given for this is that "most participants lack the human capital (education, training, and experience)" (Buck 218). Buck also describes the state of the wealthy stating, "a series of tax reforms in the 1980s...greatly reduced the federal income tax rate for the richest Americans" (184). He goes on to state that, "California has tended to rely on taxes that hit the poor and the middle class the hardest" (184). From these examples it becomes clear that the upper class is able to stay in power through the indirect support of their middle class rivals, who seem to act as a sort of buffer between the lower and upper class. As a result the lower class suffers greatly, as the poverty rate in California increased by 30% in the 1990s (Buck 189).

Another reason for the relative stability that is able to exist in this society despite the conflicting interests of the classes is that people generally often remain in the same class. In their book *Using Conflict Theory* Otomar J. Bartos and Paul Wehr describe injustice as both, "if we receive less reward than is appropriate" (33), and as "what we get now as to what we were getting in the past" (34). Though people in the lower class may feel exploited by harsh labor jobs, through the **reproduction** of social class these conditions become all they know and all their parents knew, making social mobility less possible. This also keeps the upper classes reluctant to budge some of their wealth, as they may feel it would be an injustice for them to make less than they have always made. This presents a huge problem for the upper class if a depression was ever to hit again, as a dissolution of the middle class to the lower class could potentially lead to the call for mass reforms, perhaps even an upheaval of the capitalist system, as illustrated by the experience of t the "New Deal" amid the great depression.

The above macro model similarly sheds light on my experience in the high school. As a former mid-level cool kid and an always middle class individual, I found myself inept when choosing which side to place my support. Throughout high school I sought after emulating my own oppressors, hoping to receive acceptance and what now seeming insignificant power. At the same time I devoted myself to countless hours of television, watching ads, and purchasing unneeded products, many of which were made outside the United States by factory workers in downsized factories abroad.

So what is the remedy for this? The first step is simply to be self-aware. Being self-aware means to ask myself "Am I only doing this for myself" and if so "is it doing damage to others." Self-awareness is even more needed in the presence of **social density**, a term coined by the sociologist Randall Collins, that indicates "The greater the degree of **mutual surveillance**, the more [people] accept the culture of the group and expect precise conformity of others" (Wallace and Wolf 151); it is easy to lose oneself to the crowd to which one belongs. In high school my support gave power to the upper tier, thereby damaging those who would get picked on or neglected. Everyday I do these things without even considering if what I am doing is supporting the very literal pain and suffering of others. I therefore held what Marx described as a **false consciousness**, when I falsely thought and behaved in such a way that actually ran against my own interests in terms of the need to develop my own identify as an individual and side with those not following the crowd. Similarly, in class terms, I behaved in such a way that violated the interests of those being exploited by the capitalist economy. Now, things as simple as deciding to purchase a pair of sneakers seems exacerbating to me, since I realize that by doing so, I am contributing to what Michael Moore showed in his film to be the downsizing by companies like Nike, involving the outsourcing of huge numbers of American jobs.

The next step is to keep myself well informed about the world. It is not always clear who is the "good guy" in a conflict. It is up to me to get to know both sides and come up to a position. This is sometimes difficult to recognize this easily in society, however, since media outlets often present biased information to back special interests. Cultural droning is also a danger given the way pop-culture can itself be a means of manipulation, as Max Horkheimer points out in his **Critique of Mass Culture** (Wallace and Wolf 105). However, varied enough information are around to look at all sides of any issue, and I should therefore do my best to develop a well-rounded view of what is going on in the world.

Once I have made a decision about how not to follow the crowd as usual, the final step for me would be to always and continually re-evaluate my position, never letting my stance succumb to what Pierre Bourdieu calls **habitus,** "a system of durably acquired schemes of perception, thought, and action, engendered by objective conditions, but tending to persist even after an alteration of those conditions" (Wallace and Wolf 115-116). What is important is to be take sides based on knowledge of issues than attachment to groups.

As illustrated well by the movie *The Matrix*, our views of the world and our selves may appear to be truthful but are in fact illusions. Like Neo when deciding to take the red or the blue pill, I can make a choice to know myself and the world as I really am rather than as how others in high school or society at large think I am and how the world should be.

BIBLIOGRAPHY

A.F.R. 2002. "The Capitalist Cuckoo's Nest." *Human Architecture: Journal of the Sociology of Self-Knowledge* I(1): 1-8.

Bartos, Otomar J., and Paul Wehr. 2002. *Using Conflict Theory.* New York: Cambridge University Press.

Buck, Robert Enoch. 2002. "Inequality and Poverty." Pp. 177-202 in *California's Social Problems* edited by Charles H. Hohm, and James A. Glynn. Thousand Oaks: Pine Forge Press.

Buck, Robert Enoch. 2002. "Poverty and Welfare Reform." Pp. 203-234 in *California's Social Problems* edited by Charles H. Hohm, and James A. Glynn. Thousand Oaks: Pine Forge Press.

Schlee, Guenther. 2004. "Taking Sides and Constructing Identities: Reflections on Conflict Theory." *Journal of the Royal Anthropological Institute* 10: 135-156.

Simmel, George. 2004. "The Metropolis and Mental Life." Pp. 129-137 in *Readings in Social Theory: The Classical Tradition to Post Modernism*. Ed. James Farganis. McGraw Hill College Division.

Smith, Keyon. 2007. "Lifting the Fog: Finding Freedom in Light of the Sociological Imagination." *Human Architecture: Journal of the Sociology of Self-Knowledge* V(2): 73-82.

Wallace, Ruth A., and Alison Wolf. 2005. *Contemporary Sociological Theory: Expanding the Classical Tradition.* Sixth Edition. New Jersey: Prentice Hall

Yu, Johnny. 2007. "Looking Inside Out: A Sociology of Knowledge and Ignorance of Geekness." *Human Architecture: Journal of the Sociology of Self-Knowledge* I(2): 41-50

Films:

Affluenza. Bullfrog Films, 1997.

The Big One. Miramax Home Entert., 1997.

The Matrix. Warner Brothers, 1999.

Tuesdays with Morrie. Touchstone, 1999.

Twelve Angry Men. MGM, 1957.

What Do I Want to Be?

A Sociological Exploration in Choosing A Career

Joel Bartlett

University of Massachusetts Boston

joelb84@yahoo.com

Abstract: This is an effort at applying some sociological theories and concept to the question of my major and career in life. I left this class everyday with my mind completely racing and had all these thoughts I wanted to put together. I guess it goes back to the importance of putting C. Wright Mill's *Sociological Imagination* on the back of the syllabus. Nothing really quite clicked until I could understand my place in today's society. It was about being able to understand the bigger picture and myself at the same time, and how world and I related to each other. I took things like my high school experience and there was a whole different view of what had gone on. I can now understand why people harp on high school being so influential on kids at that time. They absorb so much interaction and mold into whatever is going on at that time, but they don't notice it when they are in it. They kind of get lost within the environment that they are in. The sociological imagination makes us go back and consciously study our past and our present in a broader social context, so we can better shape our future.

When it was first brought to my attention that we should be coming up with an issue or problem in our lives as a topic to write this paper, many topics came to my mind. I wanted to do something that I really felt could give me a better perspective on life. It took me around a week, but I finally came to the topic about what I want to do with my life. This remains a constant question for every 18 to 20-something years old, and considering the fact that one are in college, one should know, right?

What about the people that do not know the answer to that question? They have talked to numerous career counselors, taken "career tests," and spoken with other professionals to get a feel for it. But what if that doesn't do it for them? I have found myself trying to answer that question for the past five or six years, and this paper has finally forced me to come to terms with it.

I know that when you start out in college it's acceptable to not really be sure what you want to do, but after a while people start asking more questions. "How do you not know what you want do?" "You have no idea what you want to do?" "What are you good at?" "What do you like to do?" The questions just keep coming, and most of the time they don't make things any easier. If anything, it seems like the more people ask, the more confused I get as I try to scramble for things I could possibly do with my life.

Joel Bartlett is an undergraduate junior at UMass Boston, majoring in Sociology. He wrote this paper while enrolled in the course Soc. 341-1: "Elements of Sociological Theory," instructed by Mohammad H. Tamdgidi (Assistant Professor of Sociology at UMass Boston) during the Fall 2007 semester.

While trying to find and understand terms or concepts that I could possibly relate to my exploration, I came across the notion of the "rational choice." **Rational choice theory** assumes "that people are rational and base their actions on what they perceive to be the most effective means to their goals. In a world of scarce resources this means constantly weighing alternative means to alternative ends and choosing between them" (Wallace & Wolf, pg. 303). Up until now I had never really took the time to completely submerge myself into possible "rational" career choices. I had just assumed that what I had been doing and what I was settling for were completely rational. I had been going to school and "trying" to figure out what the best route for me was. Little did I know that there is much more out there.

GROWING UP

When I was growing up, I was blessed to have a mother, father and three sisters in our household. I am the third oldest of four, and I have always considered where I grew up in to be a stable environment. My parents worked very hard to provide my sisters and me with the things we needed throughout my childhood. I remember being in elementary school and never having any problems with school, always maintaining good grades.

When I moved on to middle school, many changes came about that I never considered until now. As a child, you tend to neglect how much money your family has or what type of "status" your family holds. However, I remember middle school was the time when I began to gain some **class consciousness.** The term class consciousness is derived from Karl Marx's view in **conflict theory**. Marx believed that the "degree to which members of a class recognize their interests will depend on [their] level of class consciousness" (Wallace & Wolf, pg. 85). Realizing that my family had four children, and to provide them with expensive clothes would have been very difficult, it was not too hard to determine that my family was of lower-middle class. That idea became very apparent to me when I thought about the movie we watched in the beginning of the year, ***Affluenza***, which portrayed how people try to keep up with the new fads. I now realize more how ridiculous such "disease of consumerism" really is.

I had a hard time meeting new people and never really attached myself to friends or a group. This led me to further examine my issue and I came across George Herbert Mead's notions of the **self,** and the **I/Me**:

> ...the "me," is something that is, so to speak, responding to a social situation which is within the experience of the individual. It is the answer which the individual provides to the attitude which others take toward him when he assumes an attitude toward them. Now, the attitudes he is taking toward them will contain a novel element. The "I" gives the sense of freedom, of initiative. (Wallace & Wolf, pg. 207)

I had gone through school without ever really meeting or keeping friends, but I began to grow up more when in the later years of middle school I took the *initiative* to meet new friends. I still had little notion of what I was planning to do with my life, but who does at 13 years old? When I went to high school, however, I felt things would be different and I would have a fresh new start.

HIGH SCHOOL

Previously I had not put much emphasis on my years spent in high school, mainly because I did not think they had that big of an impact on my life. Little did I know that I was so very wrong with that assumption. I had been playing football for around four years at this time, and during the summer I knew it was going to be an excellent season. Between football and trying to meet new people, school was completely on the back burner at this point. Like middle school, meeting new people and desiring acceptance among peers was very important to me. I had lost weight and made some friends in middle school so I felt pretty good about going to Lowell High School. It is only now, however, that I realize how relevant Erving Goffman's concept of **impression management** is to how I behaved during my years at that school. Goffman studied "the way individuals in their everyday lives present themselves and their activities to others," and used the term "impression management" for "the ways in which the individual guides and controls the impressions others form of him or her" (Wallace & Wolf, pg. 238). There were 1,000 entering freshman in the school that year, and there were many different people that I wanted to meet. So I had a particular need for impression management to make sure I appeared to others in such a way that I could make some friends.

My cousin and I often talk about the way we make friends and the extent to which we are real or fake with one or another of them. My cousin, who is also a sociology major, also brought to my attention Goffman's notion of **dramaturgy**, "the art of dramatic composition and theatrical representation" (Wallace & Wolf, pg. 238), and how everyday life can be like a theater and we its actors. I thought about how I had constantly acted throughout high school in a way that now seems to be completely stupid, always acting in order to please people. I could see how in the **front stage** we acted out particular selves, while in the **back stage** we acted differently. The more I thought about this, the more I could not believe that for four years at Lowell High I was preoccupied with acting out. Rarely did I ever study, and instead I acted stupid with the teachers I had, and when it came down to it, begged them to pass me at the end. I continued this acting throughout and told my friends how easy the class was, and how I never did anything, and so on. The **back stage** roles I also played rarely seemed to come out, however.

I almost failed my senior year and barely graduated from high school, not thinking of what I was actually going to do for college. I took an easier route and went to a community college.

COLLEGE

I had decided against applying to any colleges fearing rejection, so I enrolled in Middlesex Community College in Lowell which was not far from my house. I thought that I could easily finish in two years and then decide what it is that I wanted to do with my life. Again, I had been so very wrong in thinking that I would just cruise through this and come to an answer for what career I would pursue. The difference between **value vs. interest** that we read in class this semester was something that really caught my attention when I decided to explore this stage of my life in this paper. In their book *Contemporary Sociological Theory*, Ruth Wallace and Alison Wolf describe how various sociological perspectives can be classified in terms of how they regard people's motives for acting in society. Functionalists consider values as what motivates people to behave in society; from this perspective, what shaped and motivated my behavior in school was the shared "value" of going to school so as to find a job and move on to better things. However, conflict theorists "emphasize interests which include being free from subjugation, and being in a position of power" (Wallace & Wolf, pg. 8) as motivating factors influencing people's behavior. I guess you can say that I really was also motivated by seeking to find a job where I could be more in control of my life rather than following other people's orders.

Robert K. Merton's theory of **deviance**

was brought to my attention in class as well, though I did not quite understand it at first. After reading the material and discussing it further with my cousin, I felt that it also applies to the way I was feeling at the time. Deviance involves a "dissociation between culturally instilled goals and institutionally provided means of realization; discrepancy between ends and means is seen as a basic social source of individual frustration" (Wallace & Wolf, pg. 54). I felt I could relate to this concept of deviance because I felt frustrated that there was a separation between the culturally defined goals I wanted to reach (having a good job) and the problems I was having in my school work, which was for me a means to achieve that end.

My parents have always wanted me to finish school and keep the options open for whatever it is that I felt was right for me. My father has always wanted me to finish because he dropped out of college at my age and has regretted it ever since. My mother, on the other hand, has received her bachelor's and master's degree in the area of human services and finally this year has secured a job that she deserves because of all the work she has put in finding it. I'm not sure if I was angry and frustrated at the fact that in either situation my parents felt like they were being held back. In *Keeping Good Time*, author Avery Gordon writes, "Undergraduate education, as we all know, only tangentially produces disciplined students looking to enter the profession as equals" (pg. 77). But, even assuming that undergraduate education produced equally prepared students, why should you be an equal in employment? You have a degree in college; shouldn't that be more than just an equal? Why would I spend time studying and earning a college degree if my degree wouldn't earn me a better chance at a job compared to those who didn't go to college? These are some of the thoughts I had when faced with what I wanted to do after school.

I switched back and forth between Liberal Arts and Psychology and really could not concentrate on Psychology so I switched to Liberal Arts again. The problem wasn't that I didn't have the means to do well in school and find a career path; the problem was that of not knowing what I wanted to do after I I went through a four-year school. "What do you want to be?" I told myself. I took much time when I was finishing Middlesex Community College to really analyze who I really am and what I wanted to do and be. I can now see that the concept "**looking-glass self"** explains some of my thoughts and perceptions at the time. According to Charles Horton Cooley the looking glass self has three elements, "the imagination of our appearance to the other person; the imagination of his judgement of that appearance; and some sort of self-feeling, such as pride or mortification" (Wallace & Wolf 203). It basically means your sense of self is influenced by how you imagine others see and judge you. At the time, I imagined that people thought sI was a nice guy who otherwise doesn't really know what it is that he wants to do. When I think back about that time, I realize that that isn't the type of person I want be in other people's eyes.

The group of people I hung out with knew I had lived in Lowell my entire life. They knew that I like to slack off, and that I had been dancing around the fact of actually getting my bachelor's degree. I felt at the time that I was trapped in this carousel of doing the same things everyday, and acting in this way because people liked it. But I wanted to change what I was doing and to finish Middlesex Community College and hopefully turn things around for myself.

I graduated from Middlesex after four years, and decided that things were going to be different. After reading the article titled "Exiting the Self-Destructive Highway: A Sociological Path Back to A Future Career" by UMass Boston student Paul Connor, I compared how he decided to listen to his father with that of my attitude toward my parents. He questioned what he had taken for granted in his everyday life in terms of slacking off in school and considered how his father had wanted the best for him. Similarly, I thought not taking school seriously and going out and partying every night were not good. I needed a goal and a path to something, a change of environment seemed like just the right thing to do.

TIME TO GROW UP?

I applied to UMass Boston about two weeks after I graduated from Middlesex Community College, and was very excited when I received my acceptance letter. Finally I could go to a school that would make me realize that I had to do better things with my life and find something that I could do later on. When it came to picking a major, I went out on a limb and selected sociology. Considering that I had only taken Introduction to Sociology, I really didn't know all that much about it. My lack of knowledge in other professional fields, and fear of commitment, had me think twice about my decision but it proved to be one that I most definitely do not regret.

Since I had only taken Introduction to Sociology, I was pretty much only aware of what I know now to be **microsociology** which is "concerned with person-to person encounters and the details of human interaction and communication" (Wallace & Wolf pg. 5). This was about all I had known up to the beginning of this class, and then was introduced to a very important concept, C. Wright Mills's notion of the **sociological imagination**. At first this was a very hard concept to grasp, but then I began to understand it more as the topic of this paper began to take shape in my mind. It made me look at my "personal troubles" in terms of how they relate to the bigger picture of social life, to the "public issues" about which I learned more when we studied the **macrosociological** theories. The way I was able to more clearly comprehend how C. Wright Mills defined this term was when I re-read the excerpt from his book copied in the back of the syllabus. Having a sociological imagination, Mills wrote, involves "the capacity to shift from one perspective to another—from the political to the psychological; it is the capacity to range from the most impersonal and remote transformations to the most intimate features of the human self and to see the relations between the two" (Mills, pg. 7).

The first time I read through the excerpt, I really had to try and make sense of what was in front of me. I just did not understand it. But increasingly I left class sessions with new ideas in my head and just basically began to think more about what really goes on in this world. I thought about things like poverty and how many people are poor in this country. The thought of racism and how some people today are still so racist and the stereotyping that goes along with it preoccupied my mind. Then it all seemed to make sense after we watched Michael Moore's film, *The Big One*. It may sound rather odd that this movie made this much sense to me, but it just completely clicked in my head. All those things I had been thinking about seemed to fit into what we were learning about macrosociological theories in class. I left classes thinking that I finally understood some of the problems that I had faced both as an individual and as a member of society.

Michael Moore's film focused specifically on the nature of **capitalism** in the U.S., and how business owners and CEOs of companies use their workers to maximize their profits and then, when they have reached a certain point, downsize their companies by moving to other countries with cheaper labor costs in order to "remain profitable." In our society people commonly associate the "American Dream" with capitalism. It made me so incredibly angry at the way many of the workers portrayed in the film were treated, the time and effort they gave for these companies and then how, in Marx's terms, they were **ex-**

ploited.

Early on in the year we watched a part of the film *The Matrix*, and the first time I watched it, I did not really get much out of it other than that it was a good film to watch. The thought that we could all just be essentially an energy source for robots and being controlled by them didn't seem to make sense. After we reached to the macrosociological theories, however, the film started to make more sense to me. In my philosophy class, we also had a discussion about *The Matrix* and how it related to ***Rene Descartes's*** saying "I think therefore I am..." Everyone seems to be going through the motions every day, and I seemed to be taking the same approach. When the quote from Descartes was brought to my attention it really got me thinking differently about the difference between being a robot, or merely an energy source to be exploited by others, and being a human that thinks, critically.

Halfway through the semester I started to brainstorm about what I would write when exploring my "issue" on the macro-level and what I wanted to do with my life. I read about the different theories like ***Functionalism*** which involves "The analysis of social and cultural phenomena in terms of the functions they perform in a sociocultural system. Society is conceived as a system of interrelated pasts in which no part can be understood in isolation from the whole" (Wallace & Wolf, pg. 17). I noticed that the understanding I had of my issue was that I was taking such a distant approach to it. I needed to mix myself into what I was studying and find out for myself what the real issue was. This particular quote made me think of the way some sociologists have approached their study using different theories. Robert K. Merton wrote that "Some sociologists are seeking a single unified theory—a generalized body of explanations as to what cements society together, how institutions fit into a social framework, how discrepant values arise and work their changes upon a society, and so on" (Wallace & Wolf, pg. 46). However, he thought it is better to study society not on too abstract a level, but at mid-range, trying to theorize about specific issues of social concern.

I also brushed upon different concepts such as ***Communicative Action,*** and like most other terms we had gone over, I did not grasp it at first. This concept was presented by Jurgen Habermas, a German critical social theorist who argued that communicative action, "is not only a process of reaching understanding; communicative actions are at the same time processes of social integration and of socialization" (Wallace & Wolf, pg. 181). This again brought me to the point that I had not tried to fully understand what my parents and advisors were telling and teaching me, and had not fully used the school in a way that I could communicate with and understand others and the world. That is one reason why I felt **alienated** from my environment—a stupid kid who just took life as it came to him.

As I have slowly begun to learn more about the society at large and realize my place in it, this has also brought on other concerns that I have had throughout my whole life. I have always believed, and still believe, that you should be doing things that you like to do and that you enjoy doing. However, in America today, the focus of capitalism is about how to get richer and richer. Earlier in the semester we received a copy of an article as a hand out in class, titled "Would You Be Happier If You Were Richer?: A Focusing Illusion" (Kahneman et al). To be completely honest, I read over it briefly and then put it to the side. As I was approaching the final paper deadline and my constant concern to hopefully one day be able to provide for a family of my own I pulled it back out. The authors conclude that "people exaggerate the contribution of income to happiness because they focus, in part, on conventional achievements when evaluating their life or the lives of others." They refer to this concept as ***the focusing illusion***. It is understandable to a point how unimportant money can be, but as I read the paper I thought in reality how happy are you if your child is screaming because he/ she is hungry and your wife complaining that you should get a better paying job? It seems easy to justify always wanting to do things you love and enjoy, but looking at the bigger picture in America, how many people are poor? I do not look at this as being selfish or that I want to be a millionaire, because it's far from that.

Then, recently I was reminded of the meaning of the term ***reification***, which we covered early on in the semester as part of the readings in Phenomenology. What I took from that term was that a person does not question what goes on in their world or society and just takes things for what they are—completely forgetting the fact that we as a society have created this and therefore things don't have to remain as they are. I then made the comparison, in the movie ***Twelve Angry Men***, to the man who has tickets to the Mets game and just goes with the flow so he can get out of there. This is what I had been essentially doing with my own life, taking things and my own situation for granted, not realizing that I could influence and shape my own life and future.

When I was buying books for this class early in the semester, I picked up ***Tuesdays with Morrie***, by Mitch Albom. Before this semester I had not read a single book since I was around 15 or 16 unless it had been required for school. Although this book was required, the syllabus had said that it be read before the end of the semester. On the cover of the book it said, "An old man, a young man, and life's greatest lesson." I was completely drawn into the book and become extremely curious. Considering the fact that I hadn't read a book for a while, and that it was required reading, I said to myself that I would give it a shot. Not only did it bring attention to so many aspects of my life, but it really summed up some of the questions I needed answered. Morrie has so many great things to say throughout the book, and this was the first time I actually read a book before seeing the movie. I finished the book, and now I can understand why people say the book is always better than the film.

The conversations Morrie has with Mitch, the way he forces Mitch to really dig deep into what he is doing with his life really struck a chord with me. I thought about my own life and what I had been doing all these years, and what I could be doing. Morrie mentions the idea of never working at his father's workplace, where they will exploit the workers and doing something you really love. One of Morrie's quotes that I really enjoyed and reassured me that what I thought was right was, "Devote yourself to loving others, devote yourself to the community around you and devote yourself to creating something that gives you purpose and meaning" (Albom, pg.127).

I had a conversation with an older woman earlier this year, and we got into the subject of what is really important. We had thrown ideas at each other the whole time and we both agreed that the most important thing in this world is love. We discussed in class about love being the only real "rational act." Morrie says that everyone needs to be loved, from the day you're born until the day you die. He tells Mitch about how people lose sight of the nurturing we need and the barriers that we put up to protect ourselves. There is just so much going on in the world that we tend to forget the one universal thing that everyone can relate to in life, the need to be loved by someone.

CONCLUSION

As I talked with another student in the class about the paper that we were both stressed about, we both found ourselves sharing the same feeling that it was worthwhile to explore our majors and career choices. This was a topic that we both liked to do because of the fact that in today's society people get trapped with majors and careers they did not consciously choose and being to question them when they have already been submerged into them. I had been so reluctant to choose a major or career path because I did not want to be stuck in something that eventually I would not enjoy or realize that it was not for me.

Throughout this semester a constant question for me to answer has been, what do you want to be? As the semester has progressed and I've learned so much; I have come to realize that I really just want to help people. I met recently with an advisor and she asked me about my classes and how they went. After analyzing each one we got to "Elements of Sociological Theory." She asked me how it was going and I replied that there was a lot of information to learn and apply, and that I really did learn a lot—not only about different theorists and their concepts, but about me as well.

I left this class everyday with my mind completely racing and had all these thoughts I wanted to put together. I guess it goes back to the importance of putting C. Wright Mill's *Sociological Imagination* on the back of the syllabus. Nothing really quite clicked until I could understand my place in today's society. It was about being able to understand the bigger picture and myself at the same time, and how world and I related to each other. I took things like my high school experience and there was a whole different view of what had gone on. I can now understand why people harp on high school being so influential on kids at that time. They absorb so much interaction and mold into whatever is going on at that time, but they don't notice it when they are in it. They kind of get lost within the environment that they are in. The sociological imagination makes us go back and consciously study our past and our present in a broader social context, so we can better shape our future.

Going back and studying the feeling of being lost in the past five or six years and not having anything to really base them on opened up a new light for me and my journey in finding a career path. I haven't totally decided what it is that I want to do when I get out of college, but I know that I will follow sociology as a major and wherever it may take me from there I think will be great for me. Not too many papers that I have written really changed the way I have perceived myself, and forced me to confront an issue that I have been dealing with for quite some time.

I just want to do something that people will benefit from. That will be great. With all the people and potential that this world holds, I just don't want to miss out on anything or let it pass me by.

BIBLIOGRAPHY

Albom, Mitch. 2002. *Tuesdays with Morrie.* Broadway.

Connor, Paul. 2007. "Exiting the Self-Destructive Highway: A Sociological Path Back to A Future Career." *Human Architecture: Journal of the Sociology of Self-Knowledge*, V, 2, Spring, pp. 139-146.

Farganis, James. 2008. *Readings in Social Theory: The Classic Tradition to Post-Modernism.* McGraw Hill Publishing,.

Gordon, Avery F. 2004. *Keeping Good Time: Reflections on Knowledge, Power and People.* Boulder, CO and London: Paradigm Publishers.

Kahneman, Daniel, Allen B. Krueger, David Schkade, Norbert Schwarz, and Arthur A. Stone. 2006. "Would You Be Happier If You Were Richer? A Focusing Illusion." *Science* Vol. 312, 30 June: 1908-1910.

Mills, C. Wright. 2000 [1959]. *The Sociological Imagination.* Oxford University Press.

Shipka, Thomas A. Arthur J. Minton. 2004. *Philosophy: Paradox and Discovery.* McGraw-Hill Publishing. New York, NY.

Wallace, Ruth A. Allison Wolf. 2006. *Contemporary Sociological Theory: Expanding the Classical Tradition.* 6th Edition. Upper Saddle River, NJ: Pearson Prentice Hall Publishing.

A Lifetime of Labor
A Sociological Imagination of Work as Life

Stacey Melchin

University of Massachusetts Boston

staceymelchin@gmail.com

Abstract: This is a critical reflection on the sociology of work in a sociological imagination framework, involving an effort to critically analyze my work experience in light of various sociological perspectives, concepts, and literature. Each day, nearly every person on this planet works. This may be hard to believe when thinking in terms of the basic, more stereotypical idea of work: getting woken up early in the morning by an alarm clock and rushing off to a paid job. However, work is actually a very complex system which everybody takes part in. If we, as members of a capitalist culture, begin to see life every day as a group of tasks—some we enjoy and maybe some not so much—we may begin to value each person in society equally for the work they do and become more "human" in the process. The idea of blurring the lines between paid and unpaid work in a positive way and valuing each of them is something that might slow the progressive dehumanization of our capitalist culture. Over the years, we have become programmed as a society to only value money, and in turn, to value the idea of hard work and constant work as the norm. This idea has caused irreparable damage which has shown itself in increased divorce rates, less empathy and sympathy for our common person, high rates of crime including murder and the senseless war we are currently involved in. If we as a society and as humans could begin to value care work and see the significance of unpaid work in general, maybe we could begin to correct many of the mistakes we have made.

Each day, nearly every person on this planet works. This may be hard to believe when thinking in terms of the basic, more stereotypical idea of work: getting woken up early in the morning by an alarm clock and rushing off to a paid job. However, work is actually a very complex system which everybody takes part in. If we, as members of a capitalist culture, begin to see life every day as a group of tasks—some we enjoy and maybe some not so much—we may begin to value each person in society equally for the work they do and become more "human" in the process.

Work can be broken down into two different categories: **paid** and **unpaid**. Each of these has subcategories. Unpaid work is work that everyone does daily, but tends to be undervalued in our society. We all live in **households**, or "the organizational units for working people the world over" (p. 28), according to Dickinson and Schaeffer as stated in their book *Fast Forward* (2001). A household can be comprised of any number of people and can span the globe. Households are basically groups of people

Stacey Melchin is an undergraduate junior at UMass Boston, majoring in Psychology with a minor in Women's Studies. She wrote this paper while enrolled in the course Soc. 232: "Sociology of Work," instructed by Anna Beckwith (Lecturer of Sociology at UMass Boston) during the Spring 2008 semester.

who pool their various resources together in order to survive.

In our households, we do **subsistence work**, a form of unpaid work. Household members work together to do laundry, cook, clean, and possibly attend school. My current household is comprised of me, my boyfriend, my sister, her boyfriend and one other roommate. We each take turns cleaning the common areas of the house and my boyfriend and I take turns doing other chores, like laundry and cooking, depending on whether he is working late or I have to spend time doing homework. I would say that up to five hours of my time is devoted to subsistence work daily.

Another time of unpaid work is **share work**, where according to Dickinson and Schaeffer, "people share, swap and barter with relatives, friends and neighbors" (p. 24). Because this work is usually done outside of the household, it can be called **intra-household** share work. For example, in exchange for visiting and providing emotional support a few times a month to my parents, both of whom are ill, my sister and I are provided with dinner. In this way, we are swapping with each other things that we need. My parents receive love, attention and companionship and my sister and I receive a free meal, which is beneficial to the both of us since neither of us earns much at our paying jobs.

On the other hand, there is also **paid work**, which is what people commonly think of when "work" is mentioned. One type of paid work is **enterprising**, where people "rent rooms, repair houses, make and recycle goods and sell goods and services from their home…" (Dickinson and Schaeffer, p. 24). To make extra money, my sister creates and sells artwork through a community gallery near our home. She earns about $150 extra a month on average and does not report this income on her taxes. She feels the need to earn extra money this way because the money she earns at her paid job, when doing **wage work**, does not pay enough for her to have a decent quality of life.

Wage work is the stereotypical, more valued kind of work because it is the more standardized, regulated or **formal economy** type where people earn a set wage for selling their **labor power**, according to Braverman in his book *Labor and Monopoly Capital* (1974). This form of work is a newer form which has evolved in human societies over the course of hundreds of years. Our capitalist society is completely dependent on this set up for its survival.

Braverman states that work is an activity that alters nature to change its form to make it more suitable for our needs. We differ from other animals in that our work, human work, is not instinctive. We have the ability to imagine how something should be and then we can physically go through the motions to bring this idea into existence. As Braverman states, humans have turned work into a "purposive action" (p. 49). If this is the case, why has it become important to turn the majority of working people into mindless machines, like worker bees buzzing in the hive?

It seems that those in power are more interested in maintaining working class conditions in such a state that they would increase their own profits. Our world of work is divided up by institutions such as race, gender, and social status. This is called **the social division of labor.** Once industrialization began, people realized that if a job is divided into many parts, with each person specializing in each step, production increases and more money can be made. This idea of chopping up a job into many parts that a number of people can work on is called **detail division of labor**, according to Braverman. This system totally destroys the skills of many workers to benefit the corporation. It completely polarizes society so that the majority of people are performing menial, repetitive tasks for the "greater good" of humanity. Meanwhile, the people who come up with the ideas earn large salaries along with the ability to sit back and relax, while the people doing the more physical work earn sometimes pennies an hour. Braverman calls this the **capitalist division of labor**. This system takes advantage of many people on a global level.

Some of the people taken advantage of by our capitalist economy are the working poor, people of color and immigrants who are expected to work long hours doing grueling jobs while earning sometimes less than minimum wage. At the very least, the amount earned is usually never equal to the cost of living. In "Household Structure and Family Ideologies: The Case of Vietnamese Refugees" (1998), Kibria discusses the difficulty of entering this world of paid work for Vietnamese immigrants in Philadelphia in the late 1970s and early 1980s. The jobs that were most easily accessible to these immigrants were, "low level, service sector positions, such as waitressing and cleaning, which tended to be poorly paid, part time, unstable and devoid of benefits and opportunities for advancement (p. 58). Another example of the exploitation that happens in our society is described in the book *Fast Food Nation* (2006) by Eric Schlosser. In this book, Schlosser gives many examples of workers being mistreated at meat processing plants in America where the **conditions of work** tend to be horrible. Because the workers are expected to work at such a fast pace while not being provided adequate training or safety equipment, some are injured while using machines and knives to cut the meat and some are injured from inhaling toxic products used to clean the plants. Most of the people interviewed by Schlosser for his book were Mexican immigrants, but other people are also taken advantage of in similar ways.

One group of people who are rarely thought of during the discussion of work are the people with disabilities. Recently, I shifted positions within the non-profit agency I work for and have gotten to see first hand how people with mental retardation and other disabilities are completely taken advantage of. Currently, I work as a job coach in a vocational training program. It would seem from the title that it would be my job to teach skills to people with disabilities and provide them the support they need to hold a job. In actuality, I am nothing more than one of the dreaded supervisors, similar to the ones I loathed in the Charlie Chaplin movie, *Modern Times*.

There are different sections in the factory building set up to allow workers to do different kinds of tasks. Some of the jobs the lower skilled workers have are building boxes, stuffing envelopes and sorting nuts and bolts. Some of these jobs are provided by the agency I work for in order to keep people busy. On average, a worker will earn one penny per piece that they complete. This is often referred to as **piecework.** Some of the higher skilled workers package nails. This job is broken up exactly as how work was broken up in *Modern Times* and is similarly done in real life factories complete with conveyer belt. It is a good example of **Taylorism** in that the entire workshop is set up for mass production and in the way that incentives are built into the job. The disabled people I work with know that the faster they work and the fewer breaks they take means they will earn more money for that pay period. I have just begun working in this department and I already feel completely **alienated** from the job; I feel I have no control over the work going on around me and I cannot relate to the process as a whole. I am sure the people I work with feel this way as well.

Besides alienation, another problem caused by the Taylorist model is the anger and hostility that I have observed among the workers. In the nail room, the entire flow of work is dependent upon every worker in the room. If one person is working too slowly or takes breaks too frequently, everyone else suffers. Workers get angry at one another, and there is bickering and constant arguments that need to be attended to by the supervisors. It really is not a nice environment to work in. The entire time I am there, I want to interact more with the people around me and I oftentimes talk and joke with the disabled people I am supposed to be "bossing." Some of my colleagues do not seem to appreciate this very much and seem much more concerned with producing more and more goods. My **role** as a supervisor, or what is expected of me, is to simply make sure that the workers stay focused on their work and to keep tabs on the people I am responsible for.

Another way these disabled workers are exploited directly relates to Schlosser's book, *Fast Food Nation*. Before the nail room, I worked in the vinyl room where approximately 25 disabled people manufacture different kinds of medical bags commonly used in hospitals, such as feeding bags and colostomy bags. These workers are constantly exposed to sheets of vinyl being heat sealed and to a chemical called methyl ethyl ketone, both of which are known to be toxic to the human body. On top of this, no chemical masks are provided for the workers, even the supervisors. While talking with a supervisor who had worked in the vinyl room for twenty consecutive years, I learned that up until recently, ventilating fans were only allowed to be turned on for one hour a day, a half hour in the morning and a half hour in the afternoon. Fans were set up and connected to a timer. Only after months of protests by the staff did the company allow for the fans to be kept on all day. The supervisor I spoke to, who happens to be a Chinese immigrant, said to me that she would not be surprised if she was diagnosed with lung cancer in a few years but since she does not possess many **hard skills**, she felt that she had no option but to stay with the agency.

In the article "The Evolution of the New Industrial Technology" (2006) Meyer argues that Ford's ideal worker would be "...the mindless automaton who applied himself constantly and consistently with little thought" (p. 39). Since people who are mentally retarded do not fully understand their situation—the health risks they encounter, the low wages or poor working conditions, they seem to be the ideal Taylorist or Fordist worker. Ironically, worker's rights posters are placed in every workshop but only a handful of workers have the ability to read them. This population is definitely being exploited under the guise of job training by a non profit agency.

I can relate to resorting to a low wage job, although I am not an immigrant or disabled person forced to do physical labor. Because I do not yet have a college degree, I am seen as unskilled by potential employers. The job I had up until recently, which required much **emotional labor**—or labor that turns emotions into a commodity, according to Hochschild in the article "The Managed Heart" (2006)—among a variety of other skills, only paid $12.38 an hour. At this job, which was in a day habilitation center, my title was "Community Training Specialist," or CTS. This title was a way for the agency I worked for to try and make me feel that my job was important. Really, I was a glorified direct care worker. It was my duty to teach life skills to developmentally delayed adults. Hidden in the description of my job were additional tasks to teach communication skills, sign language, run an art therapy group, work as a lunch monitor, plan out curriculum, do paperwork and generally be a parent, friend, psychologist and behaviorist for the thirteen people whose well being I was personally responsible for. I had to do all of this every day with a smile on my face, hiding all of the emotions I may have been feeling that day.

Some mornings I might have woken up with a headache, or maybe I did not get enough sleep. Maybe I had an argument with my boyfriend or just did not feel like going to work that day. For whatever reason, sometimes I am in a bad mood because

I am a human being, but it was my job to sell my emotional labor for a wage. It was my job to be friendly, caring and responsible at all times. By training people to force warmth, happiness and spontaneity, we are training people to be able to turn their emotions on and off. This is something that affects the world on a macro level as well. As a society, we are less apt to sympathize with people. If we do, we turn around and complain about the person we were just sympathizing with. We have become angry as a society and the emotional labor brought about by capitalism plays a part in this.

While I complain about the inequality that is a direct result of the capitalist idea of work, I also directly benefit from it. Living in the United States, low paying job or not, I am a child of privilege. After the Industrial Revolution in America, roughly from 1860-1900, we quickly out-earned our Western European counterparts. This **global inequality** is still a trend today. As Glenn Firebaugh states in his article "The New Geography of Global Income Inequality" (2006), "...the growth has disproportionately benefited different regions of the world, with richer regions generally benefiting more than poorer regions" (172). Firebaugh goes on to present data using **nations as a unit of analysis** which show the dramatic difference between the United States and other countries around the world. In the U.S., Income per Capita grew by about $16,000 from 1820 to 1990. In contrast, income in Africa (all countries combined) only rose by $4,500 approximately (according to data from Angus Maddison). So just by living in this country at this particular time, I am earning more than most people around the world, even though I still lack a college degree. If viewed from this global angle alone, I am definitely a privileged person.

Another way that I benefit from this system of inequality is my ethnicity. According to Deirdre Royster in her article "Race and the Invisible Hand" (2006), "While blacks did experience significant educational and occupational gains during the 1970's, their upward trajectory appears to have tapered off in the 1980s and 1990s" (285). Royster goes on to say that white men always have the upper hand because of certain community contacts they might have. Usually, blacks are lacking these contacts. What is completely unfair is that, even though the men she mentions in her article went to the same school and had the same teachers, these teachers, who seemed to have been all white, only acted as job connectors and contacts for the white students. The black students searching for employment were left to fend for themselves while white students like me could take advantage of **white privilege**.

Again, although I can recognize the unfairness of this situation, I still did use this white privilege and system of contacts to find work at the agency I am still employed by. Four years ago, my father was doing construction work in front of the building I work in now. He became friendly with some of the staff who would come outside for smoke breaks and while talking to them, found that the company was hiring. When my father came home to tell me about the opportunity, I was instructed to write on the top of my application that I was directly referred by one of the workers my dad met earlier. I got the job and, although I do feel that I am completely qualified for the position, had I not had contacts I never would have even known the company was hiring. I have a strong system of contacts and in this way, I am directly benefiting from inequality in this country.

Much of the oppression and inequality felt in the workplace comes directly from bosses and managers. I can completely relate to this. According to Philip Moss and Chris Tilly in their article, "Stories Employers Tell" (2006), a manager's perception of different workers, especially based on their race, has a huge influence on how they rate the skill level of the worker, and, it seems, how they treat their workers. Moss and Tilly go on to say, "To what extent are assessments of various racial or ethnic groups as 'more focused,' 'friendlier,' 'faster,' or 'not as dependable' shaped by stereotypes?" (235). My previous boss had strong opinions regarding me and my coworkers which seemed to be based on **racial stereotypes**, ageism and sexism and he was not afraid to express them. He is a very hostile person and most of the staff is afraid of him. His overt prejudice, sexism and general harassment would probably make anybody uncomfortable, to say the least.

Last summer, MassHealth was coming in to do an audit of our day program. Because my boss had not been keeping up with the requirements we were supposed to meet, our company was in deep trouble. One of the case managers, who were responsible for doing much of the paperwork and documentation for the program, had left months before and a replacement had not been found. As an act of desperation, my boss decided to put three direct care staff in charge of doing the mountains of paperwork that would need to be done in order for us to have a chance of passing the audit. He elected his three "best" staff, as he said. The staff he elected included me and two other coworkers. One is white and the other is Filipino. Over the course of that summer, my boss was uncharacteristically nice to the three of us, speaking to us like human beings, asking for our opinions and even throwing us a breakfast which the other staff were invited to. All of this attention, especially the breakfast, made me incredibly uncomfortable because I knew my boss was not being genuine and because it was causing the other staff to feel even more undervalued than before. It was obvious to me that we were considered the "best" staff because of our race and age. I knew this because of the way I had seen him treat my other coworkers on a day to day basis.

During meetings, my old boss, who is a middle aged black man originally from Trinidad, deliberately berated and humiliated staff whose first language is not English. One of my coworkers, a young man originally from Kenya, got the brunt of this abuse. My boss would ask a confusing, wordy question and they ask my coworker to rephrase the question to make sure he understood it and then he would have to answer it. My boss would smirk at my coworker and the rest of us would stare down at the table. My coworker always mumbled and averted his eyes, and his hands usually began to shake. Since I still work in the same building and am close with many of the people I used to work with, I know this boss still treats other people in a minority position horribly. He once made the program RN cry. She is a 70 year old woman who has worked for the agency for fifteen years. As she cried he clicked his teeth and said, "You've been around for such a long time. I thought you were tougher than that." Because I am white, young and have decent writing skills I am perceived to be smart and my boss left me alone, until some of my coworkers and I started to file complaints about him with upper management.

When my boss realized that I was actually the first person to go up and complain about his behavior and encouraged my coworkers to do so, he began actively ignoring me and bullying me at work. One day fairly recently, he shouted at me in front of my group of clients and I began crying. After four years of dealing with abuse from my boss, I decided to do something about it. My response was to go upstairs to the vice president of the company while still crying to thoroughly express the way this man constantly made me feel. I was sent home for the rest of the week with pay by the vice president and my boss was disciplined. I wrote a formal complaint letter about him and requested to be transferred to a different department within the agency, but a few days later I gave my notice to the agency. Also, after I gave my notice, the case manager from my old department was inspired to give her notice as well. Her reason for leaving was my old boss as well. Hopefully our acts of resistance will force the company to finally take some action against this abusive man who seemed to deliberately mistreat workers who are already underpaid and undervalued.

According to Braverman in his book *Labor and Monopoly Capital,* some of the outcomes or symptoms of this way of breaking up paid work and treating the workers poorly are "...dissatisfaction as expressed in high turnover rates, absenteeism, resistance to the prescribed work pace, indifference, neglect and overt hostility towards management" (p. 141). This angry mindset of the paid worker, which seems to be a common one, not only affects the worker and his life within the corporation, he or she also takes home those feelings of anger, hostility, alienation or indifference and they, in turn, affect the worker's household and the quality of unpaid work done by these workers in the home. In this way, the lines of paid and unpaid work definitely do cross.

Another outcome of this evolution of paid work is the complete devaluation of unpaid work. People rarely see the complex list of chores they do daily at home as being work. Nor do they see **care work** for elderly or sick relatives or visiting friends to maintain relationships as actual work. In actuality, everything we do on a day to day basis is work, not just the part we get paid for. We do find joy in work as a way of life, just as many animals do; we just have forgotten to consider all of the kinds of work that are done. The devaluation of most work is what separates humans, especially those who live in a capitalist society, from animals. According to England and Folbre in their article "Capitalism and the Erosion of Care" (2006), "care that goes unrewarded is likely to diminish over time" (p. 497). This directly relates to women feeling that "just being a mother" is not enough. Nowadays it is common for women to do a full load of unpaid work at home while simultaneously holding down a full time paid job. What often ends up happening is women are forced to choose between one or the other and since we no longer seem to value unpaid work as a society, paid work wins out. In turn, we end up paying other people to do our care work for us.

In my work utopia, one of the main things I want is a combination of and equal value for paid and unpaid work I will be doing. Because child care workers earn just above minimum wage while agencies collect huge fees from parents, I know that it is hard to find highly skilled workers to take care of children. Also, I do not share the typical view of child rearing that most people in Western society have. To me, this form of unpaid work is absolutely on the same level as any paying job I could ever have. Because our capitalist society only values flexible workers, it is hard for women of childbearing age to find work. Most employers are reluctant to hire younger women or mothers because of the common opinion that mothers are not committed to their jobs. On the other hand, if a woman does decide to have a baby and leave work, the family is faced with financial hardship because of the need for a **dual earner** household, or a household in which both parents work. The problem is that women are simultaneously in their prime years to have children and start a career. I know I will be faced with this dilemma in a few years, but I plan on finding a company who will allow me to bring my baby to work with me.

According to Carla Moquin from the Parenting in the Workplace Institute, babies in the workplace tend to be very content and cry less than other babies. When the mother responds quickly to the baby, the crying rate is drastically lowered. Babies up to eight months of age sleep for most of the day and the frequent social interaction and stimulation keeps them very happy and entertained. Since I want to be a counselor for teen mothers and teach parenting skills in general, this combination of paid and unpaid jobs would work out beautifully. This idea of a combination of paid and unpaid work was seen in the movie *Patch Adams*, in which a medical student provided care work to sickly people, essentially for free. While the real life Patch Adams went on to run his own hospital which receives grants, the people who work at his hospital earn an average of $2,000 to $3,000 a year, but are provided room and board at the hospital. This was a perfect example of my idea of a work utopia.

Another part of my work utopia that ties in with valuing unpaid work is the issue of time. In her article "The (Even More) Overworked American" (2003), Juliet Schor gives data on the annual number of hours Americans spend working. She states, "What the data show is that from 1973 to 2000, the average American worker added an additional 1999 hours to his or her annual schedule—or nearly five additional weeks or work per year" (p. 7). This data supports Schor's argument that, contrary to popular belief, leisure time in America is decreasing. According to Schor, married dual earner couples now only have an average of twelve minutes per day to sit down and talk to each other. No wonder the divorce rate in the United States is at an all time high! People are actually being alienated from their unpaid work. It is now a status symbol in this country to pay others to cook our meals, care for our children, mow our lawns and clean our houses because we are all simply too busy earning money to do these things ourselves.

It is also a status symbol to force this constantly busy way of life on our children. In the article "Overscheduled Kids, Underconnected Families," Doherty and Carlson argue that enrolling kids in too many extracurricular activities, such as sports, drama, dance and acting classes, combined with the chronic overwork of the parents, is tearing families apart. The **standard of living** may be increasing, because of the amount of things we all have and the number of things our children participate in, but the **quality of life** is actually rapidly declining. In my work utopia, I would work a 30 hour work week; six hours per day five days per week, or if it could be done, in four days per week. This is not an insane, out of reach idea. Most workers in Europe follow this schedule now, in addition to having four to six weeks of paid vacation each year. As Schor states in "The (Even More) Overworked American" (2003), "From today's vantage point, a time-surplus society may seem utopian, almost unnatural. But that's only because we've been going at 24/7 for too many years and have lost sight of other possibilities" (p.11). It does seem "utopian," but it is completely possible and I plan on living my life according to this standard as I progress with my career.

The idea of blurring the lines between paid and unpaid work in a positive way and valuing each of them is something that might slow the progressive dehumanization of our capitalist culture. Over the years, we have become programmed as a society to only value money, and in turn, to value the idea of hard work and constant work as the norm. This idea has caused irreparable damage which has shown itself in increased divorce rates, less empathy and sympathy for our common person, high rates of crime including murder and the senseless war we are currently involved in. If we as a society and as humans could begin to value care work and see the significance of unpaid work in general, maybe we could begin to correct many of the mistakes we have made.

REFERENCES

Brandt, Barbara. 2003. "An Issue For Everybody." Pp. 12-19 in *Take Back Your Time: Fighting Overwork and Time Poverty in America*, edited by John DeGraaf. Berrett Koehler Publishing, Inc.

Braverman, Harry. 1974. *Labor and Monopoly Capital: The Degradation of Work in the 20th Century*. Monthly Review.

Dickinson, Torry and Robert Schaeffer. 2001. *Fast Forward*. Rowman and Littlefield Publishing, Inc.

Doherty, William and Barbara Carlson. 2003. "Overscheduled Kids, Underconnected Families." Pp. 38-43 in *Take Back Your Time: Fighting Overwork and Time Poverty in America*, edited by John DeGraaf. Berrett Koehler Publishing, Inc.

England, Paula and Nancy Folbre. 2006. "Capitalism and the Erosion of Care." Pp. 496-510 in *Working In America*, edited by Amy Wharton. New York, New York: McGraw Hill Companies, Inc.

Firebaugh, Glenn. 2006. "The New Geography of Global Inequality." Pp. 170-178 in *Working In America*, edited by Amy Wharton. New York, New York: McGraw Hill Companies, Inc.

Herring, Cedric. 2006. "Is Job Discrimination Dead?" Pp. 277-284 in *Race and Ethnicity in Society: The Changing Landscape*, edited by Higgenbotham and Anderson. Thompson Wadsworth.

Hochschild, Arlie Russell. 2006. "The Managed Heart." Pp. 69-78 in *Working In America*, edited by Amy Wharton. New York, New York: McGraw Hill Companies, Inc.

Kibria, Nazli. 1998. "Household Structure and Family Ideologies: The Case of Vietnamese Refugees." Pp. 55-68 in *Families in the US: Kinship and Domestic Politics*, edited by Karen V. Hansen Anita Ilta Garey. Temple University Press.

Meyer, Stephen III. 2006. "The Evolution of the New Industrial Technology." Pp. 31-44 In *Working In America*, edited by Amy Wharton. New York, New York: McGraw Hill Companies, Inc.

Moss, Phillip and Chris Tilly. 2006. "Stories Employers Tell: Employer Perceptions of Race and Skill." Pp. 235-261 in *Working In America*, edited by Amy Wharton. New York, New York: McGraw Hill Companies, Inc.

Robinson, Joe. 2003. "The Incredible Shrinking Vacation." Pp. 20-27 in *Working in America*, edited by Amy Wharton. New York, New York: McGraw Hill Companies, Inc.

Royster, Deirdre. 2006. "Race and The Invisible Hand: How White Networks Exclude Black Men From Blue Collar Jobs." Pp. 285-294 in *Race And Ethnicity in Society*, edited by Higginbotham and Margaret L. Andersen. Thomas Wadsworth.

Schlosser, Eric. 2006. "Fast Food Nation: The Most Dangerous Job." Pp. 330-345 in *Working In America*, edited by Amy Wharton. New York, New York: McGraw Hill Companies, Inc.

Schor, Juliet. 2003. "The (Even More) Overworked American." Pp. 6-11 in *Take Back Your Time: Fighting Overwork and Time Poverty in America*, edited by John DeGraaf. Berrett Koehler Publishing, Inc.

http://www.babiesatwork.org

Films:

Patch Adams. Universal Studies, 1998.

Modern Times. Image Entertainment, 1936.

Finding My Work Utopia

Examining My Work Experiences and Position in Society

Laura Zuzevich

University of Massachusetts Boston

laura.zuzevich001@umb.edu

Abstract: As I prepare to search for a job after graduation, there are many factors that I must consider in order to understand exactly what I am looking for in a job. I am unsure at this time as to what type of job I am looking for. However, I do realize that I can draw on my past paid and unpaid work and labor experiences to recognize what characteristics I find desirable and undesirable in a job or work environment. Furthermore, through examining my position in society in relation to my social class, gender, and race, I will be able to understand what advantages or obstacles I may face in finding and maintaining a job in relation to others. Through examining these factors in this paper I hope to reach a greater understanding of what I am searching for.

In his book, *The Sociological Imagination*, C. Wright Mills states that an individual can understand his life's experiences better if he or she takes the time to examine them in relation to what goes on in the world (349). As I am preparing to graduate and find a job other than the minimum-wage jobs that I have held until now, I believe that by examining my past work experiences, both paid and nonpaid, I will gain a greater understanding of what I am looking for in a potential work. Furthermore, by examining the experiences of others I will be able to decide if my preferences are feasible for me to achieve in competition with others who have similar goals to mine.

As of right now, I am unsure of what course I would like to take in regards to my occupation. I am considering taking a year off of school and then attending law school at night. However I want to take time off to make sure law school is really what I want before I make any major decisions. As for jobs that I will have in the meantime, I am unsure of what type of work I am looking for. Over the course of this paper I seek to examine the aspects of my previous work experiences that I value and which characteristics I would prefer to avoid in future work. In addition, for future unpaid work I assume that I will one day have a family and children, so in imagining my work utopia I will take into consideration how children will affect my future jobs.

Over the course of my life, I have participated in many different types of **wage and nonwage work** that have helped prepare me for future jobs through learning what characteristics I would preferably want and not want out of a job and to iden-

Laura Zuzevich graduated from UMass Boston in May 2008, double-majoring in Sociology and Criminal Justice. She wrote this paper while enrolled in the course, Soc. 232: "Sociology of Work," instructed by Anna Beckwith (Lecturer of Sociology at UMass Boston) during the Spring 2008 semester.

tify the characteristics that would form my ideal job. Wage work is commonly referred to as "real" work, as it is done in exchange for monetary wages and is organized by an employer (Dickinson & Schaeffer 30). Wage work that I have done includes being a crew member at Dunkin Donuts, a salesperson at the clothing store American Eagle, and an intern at the State House where I completed various administrative duties. On the other hand, nonwage work is just as important but is often held in low esteem as it is work that is often done to assist the individual, family or friends and is rewarded with thanks, a gift, or cash (Dickinson & Schaeffer 30). The nonwage work that I have done includes being a student, babysitting for friends and family, and completing chores at home. While wage and nonwage work is often very different, both play instrumental roles in developing myself as a worker and to identify what I want in a job.

One thing that may make finding a job difficult is the current state of the economy. Dickinson and Schaeffer in *Fast Forward* address some of the new features of the **changing world of the economy and work**. Initially, the authors address the issue of the **immiseration of workers**, which refers to the decline in worker resources and income (Dickinson & Schaeffer 42). It is believed that between 1979 and 1995 the median income in the United States fell by approximately $100 a month. As a result, there were many workers that were making less money in 1995 than workers made in 1955. Another feature of the new economy is the reduction of state benefits. Through this reduction, benefit to groups of workers such as women on welfare and immigrants have been largely reduced, resulting in diminished income and quality of life for these individuals (Dickinson & Schaeffer 37). Dickinson and Schaeffer also discuss how income from both wages and benefits has declined, while the cost of basic necessities has increased. For instance, workers are now expected to pay for the increased cost of housing, utilities, healthcare, education, taxes and rising gas prices on a lesser income (Dickinson 39).

In "The Capitalist Firm in the Twenty-First Century," Walter Powell discusses how unemployment is currently at a thirty year low, as one feature of the new economy. Even though there are more people working, this has not meant there has been a steady or increase in job security. Workers constantly have to worry about becoming unemployed due to **downsizing** (Powell 81). Downsizing has affected the workplace and economy because it has reduced the number of jobs available since workers are employed only when there is a pressing need for assistance within a company. As a result of downsizing there are many individuals who are temporary workers, only being employed when needed by the company. This practice leaves individuals that are permanently employed by the company to have greater job responsibility, since they are compensating for job positions that have been consolidated into fewer positions (Powell 81). This knowledge of downsizing makes me apprehensive about my immediate future as a recent college graduate with very little experience or skills in the paid workforce. To find a job only to be a victim of downsizing would be disappointing and something to worry about in terms of maintaining a steady income.

As graduation and the responsibility of finding a job approaches, I find that I am placing my faith in the **achievement ideology**. I believe that as long as I have the proper education and training, the ability to demonstrate my skills, and that I work hard and have the proper attitude, I will have no problem in finding a job and being successful (Royster 288). However I do have to realize that there are always going to be people who may be more qualified, interview better, or have more personal contacts, than I do. The idea of using personal connections to have greater access to jobs is known as the **contacts ideology** (Royster 288). Up until recently, I never really believed in this ideology since I deeply believed that hard work and level of skill was the determining factor in finding employment. However, at my internship at the State House I have heard countless times that getting a job is "all about who you know." My supervisors at the State House have instructed me to use them as contacts for whatever position I may apply for, since having the recommendation of a State Representative almost guarantees an applicant an interview and certainly works in favor of the applicant.

I am part of a household that financially supports me in the form of paying for my education, bills and other necessities. As a result, while I have been in school I have had the ability to work solely to make extra money to spend on myself. Furthermore, as I have only worked in minimum wage paying jobs for extra money, I never really recognized the importance of having contacts. However, I received my first two jobs as a result of my contacts. When I first applied to Dunkin Donuts, I applied in the town next to where I live, but never received a call back. A few weeks later, a friend from school who knew the manager at the Dunkin Donuts in my town got me a job. I had no job experience at this time and I think that the only reason I was given the job was based on the fact that my friend knew the manager. I have seen this type of situation many times throughout my years at Dunkin Donuts. Whenever a new application was submitted to the manager, our manager would ask the employees if they knew who the applicant was. There were many times that an applicant would be disliked by an employee and subsequently that applicant would not be considered for a job. Furthermore, there were even more times when an applicant was a friend of a current employee and subsequently put on the top of the application pile and that individual was interviewed first and usually given the job.

I also experienced hiring as a result of contacts at American Eagle. I received my job because my mother is friends with the assistant manager's mother. Had I not had this connection, I am positive I would not have even been considered for the job. Almost every employee at American Eagle knew each other before starting to work there. It was common practice to hire from friends of employees. For example, during my time at the store there were many instances when the manager would make an announcement that the store was hiring and that employees should inform their family and friends. As a result, it almost became a joke when an "outsider" whom no one knew came in and asked for an application. These individuals were almost never considered for a position. Of particular interest, in many instances the friends that were hired were valued for their physical attractiveness. These employees had no skills and were poor workers in regard to their work ethic; however these employees were never reprimanded. This behavior disproves the achievement ideology, as people who had no particular skill or personality were given a job because they were attractive and knew employees who already worked there. These hiring practices assist in maintaining racial boundaries in the area in which American Eagle is located. The area is predominantly white and all of the employees that work at the store are white as well. Furthermore, the friends of employees are white as well, presumably because of the location in which they live. Since friends of employees take precedent over other applicants, white individuals are the only people employed by the particular store in which I worked.

Ideally, I want to work what is referred to as the **fixed day**, which requires that at least half of the hours worked during the majority of the work week fall between the hours of 8 A.M. and 4 P.M. Also, I would prefer to work during the week days and to avoid working **nonstandard days**, which are Saturdays and Sundays (Presser 459). Furthermore, to be able to have **flexitime** would be optimal, as I would be able to choose what time I want to begin and end my work day, "within a few hours of the core schedule" (Presser 459). By being able to flexibly work during the week would be ideal in the future when I decide to have a family. By being able to work standard hours during the week, I would be able to have the time to spend with my family and to take care of family obligations. This schedule would match up with the schedule of children attending school, so I would be able to be at home while my children are at home. Furthermore, by being able to be flexible with my hours I would be able to accommodate anything unexpected such as taking children to doctor's appointments or taking care of any family emergencies that may arise.

While I want to work hard and contribute as much as possible to my future work environment, I want to avoid working excessively or to the point of constantly worrying about work. The first step I would ideologically avoid taking would be to not frequently work **overtime**, which is work that is completed in addition to a worker's scheduled hours (Golden 29). While working overtime sparingly would not be a problem, I would want to avoid frequently working extra hours as it is very important for me to have time and space away from work. This time away would help me relax and concern myself with things that happen outside of the work environment. Allowing myself to separate work and time away from work will prevent me from working more than is physically or mentally healthy, which is known as **overworking** (Golden 30). It is important to avoid this as it is known that employees who consistently work overtime are twice as likely to consider their jobs to be very stressful and are at risk for high blood pressure, heart attack and occupational burnout (Golden 30). The worst situation that I could potentially have would be **overemployment**, which occurs when employees are willing to reduce the amount of time they spend earning an income but are unable to due to constrictions of the job (Golden 30).

I have learned from past experience that overtime and overwork can lead to undesirable consequences. At one time during my college career, I was working upwards of forty hours a week at two jobs, while attending class for fifteen hours a week, and then completing countless hours of schoolwork. This was obviously more hours than a standard forty hour workweek and I was unable to reduce the amount of time I spent on this work, no matter how much I wanted to. I clearly could not cut back on my class attendance or schoolwork as I would do poorly in my classes. I was also unable to cut back at my two jobs as I worked part time at both, and both jobs required me to work a minimum of twenty hours a week in order to continue working there. I wanted to keep both of these jobs and as a result I continued to work. The longer this schedule went on I became very anxious and stressed out. I became tired all the time and my grades and class attendance began to fall. It was only after I forced myself to cut back on one job that I began to feel better, both physically and in relation to the work I had to complete. As a result, I believe I must learn from my past experiences and do my best to prevent working too much overtime, becoming overworked, overemployed, and burnt out at future jobs.

In addition to the actual time spent at work, I also have to be aware of the **emotional labor** that I do and how I manage it. The work of emotional labor requires an individual to suppress any felt emotion, in order to present themselves to the customer a certain way. This was described in Hochschild's "The Managed Heart" in the example of a flight attendant who was expected to serve customers and be polite and friendly at all times during the flight. The flight attendant was not afforded a break as she constantly had to maintain an appearance of being relaxed, friendly and willing to serve. In the reading, Marx is consulted to note that work can become dangerous because the worker can become detached from the object of himself that is used to do work, either physically or emotionally. For instance, a factory worker can become detached from his arm, which is vital to his work and the flight attendant may become detached from her emotions, which are vital to her work (Hochschild 70).

Drawing on this notion of the worker becoming detached or alienated from body or emotion, Marx suggests that it may be vital to survival for the worker to become mentally detached from himself (Hochschild 70). While I hope to avoid a job that forces me to rely heavily on emotional labor, I hope that I am able to separate my emotions from work with my life away from work. I am aware that I tend to become completely dedicated and engulfed in the work that I do, to the point that it consumes my thoughts day and night. While that increases my work productivity and makes me a hard worker, I need to constantly be aware that there is a line that I should not cross between being emotionally invested in my work and being overly invested in what I do. As a result, I am aware of this potential and I must continuously watch myself from becoming too involved.

While it may seem as if there are large differences between types of jobs, especially between minimum wage paying and corporate, I think that it is important to realize that there are no real great differences in the core structure of work. I believe that this knowledge will help me in knowing what to expect out of a job and to successfully acclimate myself to future jobs. For instance, most work organizations utilize **centralization and decentralization of authority**. In using this system, there is a concentration of power at the top of organizations, generally in the form of a CEO or manager. From there, the individual or group in power generally pushes responsibility for profits and decisions "as far down the organizational line as possible" (Jackall 406). In relaying responsibility down to the less powerful within an organization, superiors are able to rid themselves of the pressure and obligation of having to deal with minute details and tasks, which is referred to as the **pushing down of details** (Jackall 408). However, even though superiors do not do the actual tedious work themselves, they often receive all of the credit for any profit, ideas, or recognition that is produced by inferior employees, which is known as the **pulling up of credit** (Jackall 409).

Through the centralization of power, pushing down of details, and pulling up of power in both corporate and minimum wage positions, employees are used to assist and benefit superiors. Employees assist their managers and bosses with little or no gratification, in terms of receiving credit for work done or monetary compensation. In my experience, I have found that this occurs everywhere. At Dunkin Donuts and American Eagle, there would be certain quotas of product that employees were required to sell during the day. If we did not meet this quota, employee hours were often reduced as a punishment. However, on days that employees did exceed the quotas, we received little acknowledgment, even if our superiors received bonuses or company recognition.

This was also evident, though in less harsh of a manner during my internship at the State House. In the House of Representatives, each Representative has at least one legislative aide to assist them in the massive amounts of work that needs to be done. These aides complete the work behind the scenes and put in all of the effort to make things happen; however the Representatives receive all the credit. While Representatives are very appreciative and generous towards their legislative aides, the aides receive no recognition by any other individuals such as constituents, the media, or other legislatures. In my situation, even though it may seem unfair for someone to get full credit when others have done the hard work, I realize that I am employed to benefit the organization I work for and I am an integral part of that company, whether or not I receive credit for what I have produced.

My past work experiences have definitely helped me to become more aware of what I want and don't want out of a future job. Initially, by working at Dunkin Donuts I realized that of utmost importance in having a job, aside from making more money than minimum wage, is to have more freedom and variety in a job. My work at Dunkin Donuts was primarily characterized by **standardization**, which orders that there be strict and detailed control over both work routines and product quality (Leidner 347). This was enforced in every aspect of the job, from the way in which every employee was required to wear the uniform of company issued pants, visor, and whatever color shirt was being worn on a particular day, to the way in which every breakfast sandwich that was ordered was not considered done until the egg, meat and cheese had been assembled in proper order and the wrapping had been folded around the sandwich in the proper technique.

The quality of employee behavior was managed through **routinization**, which achieves expected behavior through repetition of work habits until they become essentially programmed into the employee. At Dunkin Donuts, this occurs with everything from the way orders are taken, to the manner in which they are processed and distributed (Leidner 351). For instance, in Robin Leidner's "Over the Counter," the way in which McDonald's routinizes behavior is exemplified by the way employees complete every task through the use of a computer instructing them what to do. The employees press buttons according to what food is ordered, allowing the employee to never have to remember anything or write anything down. The characteristics of standardization and routinization are definitely what I want to avoid in future work, as there is no room for individuality, creativity or freedom to think or act on ones own.

To avoid high levels of social control in my future employment, I hope to convey that I view supervision as social control of the worker. Due to standardization and concern over employee and product quality, supervision at American Eagle and especially Dunkin Donuts was extremely high. This characteristic is exemplified in the film *Modern Times*, which illustrates the notion that someone, namely a boss or manager, is always supervising employees to make sure that all tasks are completed and that employees are producing maximum output for the company. In the film, when Charlie Chaplin enters the bathroom to smoke a cigarette he is seen on film by his boss who promptly tells him to "quit stalling and get back to work" (*Modern Times* 1936). By monitoring employees it can be ensured that company time and funds will not be wasted by idle employees.

Every morning at Dunkin Donuts, a manager would review the amount of time it took for employees to complete their transactions from the previous day. This would be done through review of receipts or of security tapes. If the manager noticed that, for example, between 4:00-5:00 p.m. one day there were multiple customers that were not served in sufficient time (usually in under one minute), the manager would find out who had worked that shift and reprimand the employees. Furthermore, managers would watch large portions of the film to monitor employee behavior and to make sure that everything was being done properly and efficiently. If an employee did not complete all cleaning duties or did something else deemed undesirable, they would be written up.

The excessive level of supervision often made work an uncomfortable environment. There were many trivial reasons an employee could be reprimanded, such as leaving the drive-thru window open on hot days to allow air into the building. Supervision was vital to the manager's role in **getting workers to work**. For instance, Robin Leidner in "Over the Counter," mentions the McDonald's slogan of "[i]f there's time to lean, there's time to clean" (354). This slogan is utilized, most of the time word for word, by Dunkin Donuts. Weekend mornings at our Dunkin Donuts were very hectic. There were frequent periods of endless customers, followed by a few short minutes of relative quiet. During that time, employees would sometimes stop and rest for a second to collect themselves in anticipation for the next rush. When this occurred, managers would see this happening on the camera in their office, and promptly rush out to tell us "i[f] you can lean, you can clean." This also happened on days where there were few customers and the store was completely clean, leaving employees to wonder what else we could possibly clean. As a result, many would pick up a wash cloth and stand scrubbing the same area as to appear busy.

Through these experiences I have learned that I desire relative freedom in future work in the sense that I will not have constant supervision from a supervisor. However, as uncomfortable as the supervision may have been at times, I am glad that I experienced it at such a young age. I believe that because the level of supervision made employees constantly accountable and nervous, I was able to strengthen my work ethic. Since I knew that I was being observed, I wanted to make sure that everything I did was of the best quality possible. As a result, everything I did was of high quality, which is something that I naturally strive for today.

There are additional incentives that encouraged employees to put forth great effort at Dunkin Donuts and American Eagle that I hope to continue using as a motivator in successfully completing future work. One of the major motivators that is echoed in Leidner's article is that many workers view the work they produce as a team effort and therefore work hard in order to earn and maintain the respect of their coworkers (354). In my situation this was true because if one individual did not complete a certain task, another worker would have to complete that task in addition to their own tasks. Therefore, working as a team to assist coworkers and to keep coworkers from growing angry with one another was a great motivator in getting work done.

Another major motivator was the customer. Leidner states that at McDonald's "the sea of expectant faces provided a great deal of pressure to keep moving" (354). This provided significant motivation at my jobs because employees knew that the faster we worked, the faster we could finish the customer transactions and shorten lines. Furthermore, employees knew that effort had to go into every customer transaction or it would just prolong the amount of time the customer was in the store. For instance, at Dunkin Donuts if an employee did not toast a customer's bagel properly, that customer would keep coming back until the bagel was properly toasted. As an employee I am extremely motivated by these factors to please everyone I work for, whether it is a company, my coworkers, or customers.

One issue that I should be aware of potentially occurring in future work is gender discrimination. As a result of my brief participation in part time, unpaid and minimum wage work I do not feel that I have been granted any privilege or dealt any disadvantage in relation to finding work or the pay I receive based on my gender; however I do have to realize that this may be an issue in the future. I am aware that discrimination exists in some form in the paid workplace environment, as I believe I have faced particular instances. My experience can relate to the article "Gender Inequality of Work," where authors Cotter, Hermsen and Vanneman state that in terms of inequality, men and women have made great advances; however there is still "a long way to go" (Cotter 190). This was proven to be true during my time at American Eagle.

One particular instance of discrimination is illustrated in the opportunities that were afforded to males over females. About once a month new clothes would arrive at the store, which had to be placed on the sales floor. When new clothes arrive, new decorative cases and structures to hold the clothes would also arrive. These new cases require assembly with the use of tools. Every month, managers would request that one or two employees come in on a Sunday after the store closed to assemble these structures. This was desirable because on Sundays we were able to earn time and a half.

The issue with this request, however, was that managers would only allow male employees to assemble these structures. In the two years that I was employed at the store, never once was a girl allowed the opportunity. Girls would frequently request that they be allowed the opportunity to no avail. In one particular instance I remember asking my assistant manager why the boys were always chosen to assemble the cases. She replied that boys are "stronger and better at that stuff" and suggested that girls "don't really want to do that kind of work anyways." Not only did she make a huge generalization, she confirmed through those statements that management only chose boys because it was assumed that they would do a better job with the assembly, based solely on the fact that they were male.

The issue of concern with this particular situation is in reference to the **dissimilarity index**, which is a tool that is used to "describe patterns of gender segregation" by interpreting the percentage of employees that would have to change occupations in order to make each occupation equally represented by male and female workers (Cotter 191). On this scale, my retail work at American Eagle was categorized as evenly represented by male and female employees (Cotter 191). This information has prepared me to expect discrimination because if discrimination can occur in a field that is commonly evenly represented by male and female workers, I have to assume that discrimination also occurs in jobs that are unequally represented by females.

While I may have experienced discrimination based on my gender, I believe that I have and will continue to be privileged based on my location in relation to my race as one of the **mechanisms by which racial discrimination around jobs take place**. In his article "Is Job Discrimination Dead?" Cedric Herring discusses the way in which companies alienate and discriminate against African Americans. Herring cites that discrimination occurs when companies locate themselves away from urban areas where large numbers of African Americans reside. The author furthers that public transportation to desirable job sites is not available from the inner city, where many blacks live, to the central city or suburban jobs (Herring 282). This lack of access forces black inner city residents to remain isolated and to find meager jobs located in their own communities.

In relation to Herring's information, I feel that I am privileged based on the fact that I am white and live in a predominantly white, suburban, middle class community. While African Americans are disadvantaged due to their lack of good jobs and locations, I believe that I benefit. The areas that I work and live in are the same areas that are inaccessible to inner city jobseekers. I am able to access these jobs as a result of my proximity to a variety of companies and because of this, I believe I have privilege over minority groups. I also benefit in getting to a job based on my easy access to transportation. Based on where I live and the income levels of my area, I am able to afford a car to take me to and from work. However, even if I did not have access to a car, every town in my area has access to a commuter rail which connects all of the towns to one another. However, the benefit of public transportation in my area would not be a viable option to inner city residents. The train system in my area connects towns to one another and in order to reach inner city areas, a rider must change trains and trolleys multiple times. In addition, fare costs eight dollars each way to ride. While this is discouraging for poor individuals that reside in inner city areas, this will continue to benefit me as I seek for a job after graduation.

While my past paid work experiences have definitely helped to prepare me for future work by giving me experience and an idea of what I am looking for from a paid work position, my non-paid work has also given me considerable skill. To begin with, I have benefited from my work to assist the **household as an institution of the economy**. A household consists of separate workers that are engaged in different types of work. These workers pool resources and income in order to survive and sustain themselves economically (Dickinson & Schaeffer 28). Initially, I have learned from my experiences in lending myself as a resource to my cousin through subsistence work, which is unpaid work that is done to share resources or assist family members (Dickinson & Schaeffer 23). My cousin has a ten-month old baby and was having financial difficulties as she was unable to work, since she could not afford childcare for her daughter. I attend school relatively close to where she lives, so I volunteered to baby-sit two nights a week so that my cousin could work. Other family members have since offered their services as well, so my cousin is able to work around thirty hours a week. Due to this experience, I have learned not only the value of assisting someone in order for them to survive, but I have gained experience for future unpaid work I will do. Before my cousin's baby was born, I had relatively little experience around babies. By babysitting, I have learned all of the basics of childcare, such as feeding, diaper changing, bathing and interaction with the child, all of which will benefit me when I decide to raise my own children.

Also, Goldscheider and Waite in "Children's Share in Household Tasks" discuss the **gender and age division of household labor**. Some characteristics of this division include that girls are more apt to spend almost twice as much time on housework than boys. It is believed that this pattern parallels the amount of work participation in by their mothers and fathers, with mothers contributing more work than fathers. (Goldscheider & Waite 809). Furthermore, as children grow older they become more involved with both indoor and outdoor chores. Families with teenagers also break down the types and amounts of chores by gender, with teenage girls completing five times more tasks than teenage boys. For instance, when there are chores that are completed by multiple children, teenage girls are often expected to complete three quarters of an entire task, for instance by doing most of the laundry or most of the dishes. Furthermore, girls contribute more in the area of dishes, cleaning and sharing in cooking and laundry (Goldscheider & Waite 813).

This is relevant to my experiences since when I was younger and still to this day I complete more work than my brother. When we were younger, we were both expected to complete equal amounts of work, however my brother frequently chose not to complete the tasks assigned to him. As a result, as time passed my parents grew tired of constantly asking him to participate in chores to no avail and began asking me instead. I believe that my experiences in participating in non-paid, household work has increased my work ethic and made me realize that in order to succeed, work must be completed. I did not receive any significant pay for this work; however I knew that it was my responsibility to complete my chores. I knew that if I chose not to complete a certain chore, that chore would simply not get done. As a result, from a young age I realized that I need to work hard at things that I sometimes may not want to do, in order to get things done. Also, when I was younger if I did not do a satisfactory job in something like folding laundry or dusting, my parents would make me redo the chore. This helped me to learn that I might as well give one hundred percent the first time, or I will have to waste more time by redoing the task until it is done properly. Since I have always been responsible for completing my own work and work that does not get done by others, I have become used to completing work on my own without help from others. This could present a problem in my future as I may have a difficult time allowing others in my family to complete a share of the work that needs to be completed, as I am so used to doing everything myself. This is something that I need to be conscious of and fix if the problem presents itself.

The desire to work hard is also echoed in my unpaid work as a student. I took the initiative to attend college and from the first day I realized that all of the work was my responsibility. No one was going to pick up my slack if I decided not to pass in a paper or study properly for a test. As a result, I needed to develop a strong work ethic to get all my work done. Furthermore, I was aware of the necessity to put all of my effort into every piece of work that I did, or my grades would not be as good as possible. My work as a student has helped me develop a strong work ethic and sense of responsibility as I was able to manage taking many classes and completing all assignments. By also giving all of my efforts to my assignments I ensured that I was successful in achieving the highest grade possible, which will benefit me in finding paid work after graduation.

One of the major reasons I want to be successful in getting a job after graduation is to benefit my family. In "Household Structure and Family Ideologies," Nazli Kibria describes how the **ideology of family collectivism** is the belief that family is of much greater importance than the individual. Furthermore, there is the trust that the family is the most significant and reliable source of support in an individual's life. The family is considered such a vital support system, that Kibria believes that individuals in a household should share resources and assistance to survive. This type of support can come in the form of economic or emotional assistance (Kibria 64). The information that Kibria provides is one of the most prominent factors to why I want to be successful in future work.

In conclusion, my family has financially supported me my entire life from buying me a car, to paying my college tuition and bills, and giving me money during times that I have been unable to work. Over time, they have spent a significant amount of their money on me, which they could have easily spent elsewhere. In addition, my family has always been there to assist me physically, mentally, and emotionally as well, whether I needed help after my car broke down or if I needed help deciding a difficult decision. As a result of this level of support and commitment from my family, I want to do well in my work in order to make them proud of me and so that I can begin to pay them back for all of the money they have given me over time. As I prepare to find a job after graduation, there are many factors that I must consider before I make any decisions. I realize what type of environments I enjoy working in and what types of environments are not well suited for me. Furthermore, I have realized that there are many individuals that have similar goals to mine; however I may be more privileged or disadvantaged than them based solely on gender, race, or location. In taking what I have learned, I will be able to analyze my past experiences and my future goals before making an informed decision on what job I will choose.

REFERENCES

Braverman, Harry. 1974. *Labor and Monopoly Capital: The Degradation of Work in the Twentieth Century* (Monthly Review).

Cotter, David A., Hermsen, Joan M. and Vanneman, Reeve. 2006. "Gender Inequality at Work." Pp. 189-201 in *Working in America: Continuity, Conflict, and Change* Third Edition, edited by Amy S. Wharton. Boston: McGraw Hill.

Dickinson, Torry and Schaeffer, Robert. 2001. *Fast Forward*. Rowman and Littlefield Publications Inc.

Golden, Lonnie. 2003. "Forced Overtime in the Land of the Free." Pp 28-36 in *Take Back Your Time: Fighting Overwork and Time Poverty in America*, edited by John de Graaf. Berrett Koehler Publications Inc.

Goldscheider, Frances K. and Waite, Linda J. 1998. "Children's Share in Household Tasks." Pp. 809-817 in *Families in the U.S.: Kinship and Domestic Politics*, edited by Anita Ilta Garey and Karen V. Hansen. Temple University Press.

Herring, Cedric. 2006. "Is Job Discrimination Dead?." Pp.277-285 in *Race and Ethnicity in Society: The Changing Landscape*, edited by Elizabeth Higginbotham and Margaret L. Andersen. Thomson-Wadsworth.

Hochschild, Arlie Russell.1983. "The Managed Heart." Pp. 69-78 in *Working in America: Continuity, Conflict, and Change* Third Edition, edited by Amy S. Wharton. Boston: McGraw Hill.

Jackall, Robert. 2006. "The Social Structure of Managerial Work." Pp. 406-418 in *Working in America: Continuity, Conflict, and Change* Third Edition, edited by Amy S. Wharton. Boston: McGraw Hill.

Kibria, Nazli. 1998. "Household Structure and Family Ideologies: The Case of Vietnamese Refugees." Pp. 55-68 in Karen V. Hansen and Anita Ilta Garey; *Families in the U.S.: Kinship and Domestic Politics*. Temple University Press.

Leidner, Robin. 2006. "Over the Counter." Pp. 345-357 in *Working in America: Continuity, Conflict, and Change* Third Edition, edited by Amy S. Wharton. Boston: McGraw Hill.

Marx, Karl. 2006. "Alienated Labour." Pp. 44-51 in *Working in America: Continuity, Conflict, and Change* Third Edition, edited by Amy S. Wharton. Boston: McGraw Hill.

Meyer, Stephen III. 2006. "The Evolution of the New Industrial Technology." Pp. 31-43 in *Working in America: Continuity, Conflict, and Change* Third Edition, edited by Amy S. Wharton. Boston: McGraw Hill.

Mills, C. Wright. 1959. *The Sociological Imagination*. Oxford University Press, Inc.

Powell, Walter W. 2006. "The Capitalist Firm in the Twenty-First Century: Emerging Patterns in Western Enterprise." Pp. 80-94 in *Working in America: Continuity, Conflict, and Change* Third Edition, edited by Amy S. Wharton. Boston: McGraw Hill.

Presser, Harriet B. 2006. "Toward a 24 Hour Economy: The U.S. Experience and Implications for the Family." Pp. 459-465 in *Working in America: Continuity, Conflict, and Change* Third Edition, edited by Amy S. Wharton. Boston: McGraw Hill.

Royster, Deirdre A. 2006. "Race and the Invisible Hand: How White Networks Exclude Black Men from Blue Collar Jobs." Pp. 255-294 in *Race and Ethnicity in Society*, edited by Elizabeth Higginbotham and Margaret L. Andersen. Thomson-Wadsworth.

Zuboff, Shoshana. 2006. "In the Age of the Smart Machine." Pp.113-125 in *Working in America: Continuity, Conflict, and Change* Third Edition, edited by Amy S. Wharton. Boston: McGraw Hill.

Film:

Modern Times. 1936. Directed by Charlie Chaplin, Charles Chaplin Productions.

"Patching" My Life
Sociological Lessons for a Joyful Work

Michelle Tanney

University of Massachusetts Boston

shell21035@aol.com

Abstract: Various literature in the sociology of work, and my own past work experience, are consulted to develop a general philosophy of my work utopia. There are so many different types of work in this world. Some people may define work by their job or their career, most people only consider work to be paid work where they receive a wage for their labor, the time that they spend and the tasks they accomplish within their job. Some people love their work; they find joy in going to their job everyday. I want to do something I enjoy, something that even if I hate my boss I'll still keep working because I love my job. I want to help people. I am studying alcohol and substance abuse in school right now. I plan to get my masters in counseling so that I can help people who can't help themselves. I love working with people. Almost all of my jobs that have involved dealing with people have been my favorites. I like being a part of someone's life, and by being a counselor I will be able to do this. In a way I want to be like the character Patch in the film *Patch Adams*, he wanted to help people with their problems, give them the hope that someone is there to listen to them. And so do I.

There are so many different types of **work** in this world. Some people may define work by their job or their career, most people only consider work to be paid work where they receive a wage for their labor, the time that they spend and the tasks they accomplish within their job. Some people love their work; they find joy in going to their job everyday. For others, however, they consider their work simply a job.

There is no denying that there are many different kinds of work that people do every day. In their *Fast Forward*, the authors Torry Dickinson and Robert Schaeffer state "when most people think of work, they imagine men with paid jobs. But this is the narrowest possible conception of work. What is left out of this picture is the fact that different kinds of people work, not just men, and they do different kinds of work. Actually, women do most of the world's work, and it is unpaid" (Dickinson& Schaeffer, 23).

Work can be done outside or inside the house throughout the daily **tasks** that members of a household complete. These tasks are vital to the household; if they were not done then their household may not run smoothly. This work is known as **subsistence work**. It is a type of work that is necessary to allow the household to survive. Work is divided up by society as a whole.

Michelle Tanney is an undergraduate junior at UMass Boston, majoring in Sociology with a concentration on Alcohol and Substance Abuse. She wrote this paper while enrolled in the course Soc. 232: "Sociology of Work," instructed by Anna Beckwith (Lecturer of Sociology at UMass Boston) during the Spring 2008 semester.

Each person does some sort of work to help society produce. About this **division of labor** Harry Braverman writes, "each individual of the human species cannot alone produce in accordance with the standards unknown to any animal, but the species as a whole finds it possible to do this, in part through the social division of labor" (Braverman, 72).

There are even every day activities that we do that might not be realized are actually work. A mother soothing her child when he or she is crying is actually work; it is **emotional work**. Every person in this world does some type of work at one point in their lives, and they say that once a person starts working, they never stop.

More and more Americans are working harder and harder to survive. They are working longer hours to try and keep their jobs. In her article "The (Even More) Overworked American" Juliet Schor finds data proving that "the average American worker added an additional 199 hours to his or her annual schedule—or nearly five additional weeks of work per year (assuming a forty hour week)" (Schor, 7). People are spending less time with their families and more time at their jobs. They are losing out on leisure time and relaxation. Harry Braverman defines the work cycle very well. He says "the transformation of working humanity into a labor force as a factor of production and instrument of capital, is an incessant and unending process" (Braverman, 139). Some people feel like they are on a track, where they just keep on going and keep on working until one day they can reach retirement. They hope that at retirement they'll eventually be able to slow down.

Growing up I lived with my mother and sister although the composition of our household was constantly changing. When I was seven years old my parents separated and we moved in with my elderly grandmother. Living in this house there was a sense of **patchworking,** everyone contributed to the household in some way. Nazli Kibria refers to the term patchworking in her article, "Household Structure and Family Ideologies: The Case of Vietnamese Refugees." She says that "patchworking [is]... the bringing together and sharing of diverse resources" (Kibria, 60). My mother was the only person who worked outside of the home. She brought in the main source of income with her paycheck from her job. Her role was definitely that of the **breadwinner**. My sister and I helped around the house by doing chores like laundry, washing dishes, light cooking, ironing, cleaning, and grocery shopping. In "Children's Share in Household Tasks," authors Frances K. Goldscheider and Linda J. Waite report that "children who live in a mother-only family play a key role in the household economy: they share more overall and they share in every single task" (Goldscheider & Waite, 814). My sister and I lived in a single parent home and we did do more tasks and chores than some other children may have done. We also helped through **care work** by taking care of my elderly grandmother.

While living with my grandmother we received benefits from the state. My grandmother, because she was elderly, received social security. Because my mother was a **single-headed household** we also received welfare from the state. All these benefits being used together within the household along with the wages from outside labor, and the unpaid work going on inside of the household, made us a **collectivist family**. We all combined our wages and helped out around the house so that everyone within the family could reap the benefits.

As a student studying I do **unpaid** work all the time. Most people may not consider school a place of work, but students are actually doing unpaid labor. Harry Braverman, in his *Labor and Monopoly Capital,* states that "to seize upon the materials of nature ready made is not work, work is an activity that alters these materials from their natural state to improve their usefulness" (Braverman, 45). In addition I am working for my degree, a degree which will eventually help bring in some sort of income into my household. In school I study to get a degree so that I can get a good job that I like, a job that can realize my work utopia. I hope to work as a counselor that works with alcohol and substance abuse patients which is a type of care work that involves emotional labor. This type of work is often **gender divided**. Women are often thought of as counselors because the work entails care, and deals with emotions.

When I was younger my mother would work Saturdays at her job for **overtime**. She would work her normal five days a week and then work a sixth day for extra pay. At this time my mother was a travel agent. My sister and I would go to work with her. We accompanied her mostly because there was no one else to watch over us, but also because we "worked" there as well. One year the brochure magazine for the company had a misprint. My sister and I worked on an **assembly line**, where I would open the brochure to the page, pass it to her and she would place a sticker over the incorrect information; we would then move on to the next brochure. We didn't get paid much, but it was nice to have a little money and not be bored throughout the day.

When I was about thirteen I worked for my aunt at her job for about two or three weeks. My aunt works in a life insurance agency. She started there when she was twenty years old and has been working there for almost forty three years. Today not many people stay at their job for life; however, in the past it was pretty common for workers to stay in one workplace for long periods of time, or even for their entire career life. Henry Ford helped to allow workers a job for life. In his "Evolution of the New Industrial Technology," Stephen Meyer III writes about how Ford was able to keep employees for life. Ford was able to provide benefits for his employees and guarantee them a job for life. I, however, did not last in my work as long as my aunt did at MassMutual. I was merely a **contingent worker**. I was there to do a job and once that job was finished I was done at MassMutual. I photocopied all of the agents files for the new general agent. It was great because at the time I was about thirteen or fourteen, just going into high school and enjoyed the extra spending money. I felt a little **alienated** doing the job though. I was in the photocopier room all day long by myself and didn't really get to talk to other employees; I was just left with a machine all day.

My next job was working at a concert center. The work was almost like **piece work**, where workers are paid by each product they create. The case at the concert center was a little different, though; I got paid by how many concerts I worked at. I worked as an usher, which held some **status** and **power**. I could try and get my friends into better seats than they had paid for. I really disliked this job, however. No one wanted to listen to a sixteen year old girl tell them they couldn't sit where they wanted to sit. The patrons were often drunk and did not listen to what I had to say. My last concert that I worked for was a Dido concert. Dido didn't show up for the show; the patrons arrived at 7:30 p.m. to watch her perform; and no one told them until 10:30 that she wasn't going to be making an appearance. People were extremely angry and of course took it out on us as workers. I was so fed up that I took off my uniform and just started walking to the main entrance. I never went back.

Work is a place that someone should feel comfortable in. It is a place where you should be judged on your job performance and your character. No one should have to worry about their race, age, or gender while working. They should also not have to feel like their race, age, or gender will prevent them from getting a job. **Discrimination** and **inequality** are supposed to be two subjects that should no longer be associated with the world of work. Unfortunately that is not the case. Discrimination and inequality still exist in our working culture. Although many people may believe that they no longer exist, they do. What has changed concerning discrimination and inequality is the way employers act. Even though there are many laws that try and prevent discrimination and inequality, employers are still able to discriminate if they so wish. Employers and companies are able to hide their attitudes towards a certain race, age, or gender in many new ways so that they do not get caught in the act (Herring).

I have worked a few different jobs over the years. I never really felt at the time that I was discriminated against, but I did feel people took advantage of me, whether it was because of my age or gender.

I worked in a summer camp for three years every summer. I worked three days a week from 8 to 4 and on one day a week I had my late day, where I would stay until 5. I was hired by the city, so the actual camp didn't pay; Mayor Menino's city program The Red Shirts did. The pay for the time and type of work was actually pretty good. We were paid a little more than minimum wage and were able to go to a beach or lake everyday and work on our tans. I was paid $326 every other week, which now doesn't seem like much, but at fourteen, fifteen, and sixteen, with no bills it seemed like more than enough. I know that in many countries adults wouldn't even get paid this much. Glenn Firebaugh in his article, "The New Geography of Global Income Inequality," says that there has been a huge growth in the world's average income, but that the growth has "disproportionately benefited different regions of the world, with richer regions generally benefiting much more than poorer regions" (Firebaugh, 172). Maybe due to this unequal distribution of wealth I was able to get paid a higher wage than most youth my age.

Each morning there were certain things that had to be done. We had to pack the children's lunches, snacks, juices, and milks; we had to pack sports equipment for the day, arts and crafts equipment; we had to fill two five gallon jugs of water and pack them; and we had to pack or refill the first aid kit, and then watch the children in gym until the bus arrived. Once the bus did arrive we had to pack all the supplies, food, and equipment into the back of the bus and the camp's van. Here's where the problem was.

I always arrived early to camp. Out of all the tasks that had to be done in the morning each worker was supposed to do only one or two of them. I always ended up doing more. Every day I would pack the milk and juices first and then maybe the lunches and snacks. After that I should have been more than done; however, that was not the case. My other red shirt coworkers ended up arriving late or would go straight into the gym so I would have to do all the work. I also was told "oh you're the only one who can get all the milks to fit in perfectly." It was not fair that we were all getting paid the same and I had to do more work than the others. Then when it came time to pack up the bus I would end up carrying mostly everything into the bus. I was mad that I had to do so much work and that no one was helping me.

In her article, *"Lives on the Line,"* author Ruth Buchanan discusses the difference between **talent** and **skill**. She writes that the work that requires the development and use of interpersonal skills are often understood as personal development rather than actual job skills (Buchanan, 369). Buchanan continues on to say that "work typically performed by men is identified as requiring certain kinds of skills, what women do is more often described as the utilization of their so-called natural talents" (Buchanan, 369). Perhaps that was the case at the Little House Summer Camp. Perhaps this ability to load the milks in "perfectly" was connected with the fact that I was a female, and it was just one of my "natural" talents.

When we arrived at whatever beach or lake we were going to for the day it was time to unpack the bus and van. Again I was left to do most of the work. Not only did I have to do most of the work, but the bags were heavy. We had just between fifty and seventy five campers in the camp each week. Fifty to seventy five lunches are very heavy to carry. Being a girl who was just at five feet tall at the time, the bag definitely outweighed me. There was no way I could carry it by myself. The red shirt boys would walk right past me and not offer to help. I was very angry, but because I was new, I was young, I was a girl, and I was one of the only white employees there, I said nothing. I was also from another neighborhood than most of the workers there. Most of my co-workers had grown up together and knew each other their whole lives. The older counselors who were more superior to us as red shirts were not going to yell at their friends for not helping me.

In their article, "Gender Inequality at Work," David A. Cotter, Joan M. Hermsen, and Reeve Vanneman discuss how some females are more likely to be discriminated at work than males. They say that females make less money than males, and are hired less than males. At the camp we were all paid the same, which I'm sure was due to the fact that we were being paid by the city. However there were more males working than there were females. There were about three female red shirt workers, while there were about six male red shirt workers. That shows that males were hired twice as many as the females were. I don't believe that this job was a sex typed occupation (a job that is commonly preformed by one gender or the other); instead, I feel that there were simply more males that were friends with other previous years' staff and had heard about the job through **social networking**.

One day I overheard some of the older counselors talking, saying that the boy red shirts were the best workers. I was so mad; just because they were the same color, same gender, from the same neighborhood, and they were friends, did not mean that they were better workers than I. In *Race and the Invisible Hand*, Deidre Royster writes about blue-collar African-American men not being able to utilize their personal contacts to further themselves in a career. The author says that blacks do not rely on their personal contacts to ignite their careers. However, at my job that was not the case; all of my co workers knew one another. They had either gone to the camp as children, had a family member working there, or went to the community center growing up. When it came time to get summer jobs they relied on their contacts to help them get the jobs they needed. I still knew I wasn't going to say anything though; after all there were more of them than there were of me. Then I heard one of the older, Spanish, male counselors saying, "are you kidding me? Michelle's the best worker here. She's here before everyone else. She does all the work. She carries everything everyday and she doesn't complain about it." Little did he know that I was complaining about it, not to my co workers but to my friends and family. Still, nothing changed. People still acted the same towards me and my co-workers still acted the same.

In their article "Stories Employers Tell," Philip Moss and Chris Tilly write about how employers say that certain races may be better workers than others. In one case a white employer said that black workers might work a little harder and are more willing to stay a little longer, but that whites were more friendly and more dependent, more educated, but maybe not the best workers. Based on my experience at the camp, I would have to disagree with this Atlanta fast food restaurant owner. Your race doesn't necessarily determine that you are going to work a certain way. If that was the case my black coworkers would have been doing most of the work.

However in my situation that was not the case. I was doing most of the work. I believe that you have to base your opinions on a person by person situation, not on color, because being a certain race doesn't mean you'll work in a certain way. I believe that my supervisors may have believed that because I was quiet and a quick worker it was easier to ask me to just do the tasks that needed to be done rather than asking the males to do it—because they knew it would be a struggle to get them to actually do what they wanted.

One day we pulled up to a lake parking lot. I got out of the bus and walked to the back door to open it and start unloading. One of the older, white, male counselors was standing near by. Technically it was not the older counselor's responsibility to help us carry the stuff down to the lake; we were supposed to do it. As I lifted out one of the five gallon jugs of water he came over and said "Put that down." I did what he said, then he handed me a bag of balls. He said "Here take these down to the lake, and don't come back up for anything else." I looked at him like what's going on? I did what he said but I kind of hung back a little to see what was going to happen. The person that was calling out to them was John. John grew up in the neighborhood; he had also been working at the camp since he was old enough, and before he was old enough to work there he went to the camp as a child. He was also well over six feet, close to six five. This was someone that these boys had known and respected for quite some time and he definitely held authority over them with not only his position at the camp, but also his size.

Before I walked down the hill I heard him call a few of the boys over to him. He told them that I and the other two girls I worked with weren't going to be carrying any of the heavy stuff anymore, that they were going to take the lunch bags everyday and the jugs with the five gallons of water. None of the boys told him no. They didn't say much and I could sense that they were not pleased with what he was saying, but what could they do. They certainly were not going to say no to John. He outranked them within their jobs, but was also older and bigger than him. I'm sure they were afraid about what would happen if they didn't do as he said.

After that day not much changed. John was the leader of the oldest group in camp and so he took his groups to places most days that were different than where we went to. I think the boys thought if John wasn't around they didn't have to listen to what he told them; when he was in the camp, though, I never carried the lunch bag or the jugs of water. I still had to do most of the work in the morning; that didn't change.

Although not much changed I was glad that John had stood up for me because due to my position within the camp I wouldn't have defended my self. I still carried the lunch bag every once in a while but one of the other girls and I would take it together and we would take our time bringing it down to the beach or lake, in hopes that when we got back to the bus most of the stuff would have already been brought down. So maybe I carried that bag a few more times than I should have, but at least I saw that someone else recognized the work that I was doing and someone else saw that it wasn't fair that I was doing most of the work with little to no help from my coworkers.

This may not be the most extreme case of inequality, it may not even be inequality in some people's eyes, and I realize that most people are affected a lot more than I was through discrimination. At the time, however, it seemed like I was being singled out and that no one was on my side except for the one time John stepped in. I feel like that is what it feels like to be a part of an unequal job. When you feel you can't rely on someone else to help you or confide in, that is what I imagine to be a part of an unequal job.

Since I was seventeen I have worked in a local convenience store. The store sells coffee, tea, pastries, donuts, muffins, and groceries, but caters to the Irish community in the area. I recently quit my job, and it was the first job I had actually ever quit. The concert hall was used on a per show basis and I had just never signed up for any more shows. At the store I was a hard worker. At one time I worked five mornings a week at five a.m. I was a store clerk. My job basically included ringing up customers' purchases. Because I worked there for almost four years I really got to know the customers, as most of the workers do. I knew about their lives; what they did for a living; if they were married, or had a boyfriend or girlfriend; and if they had children and so forth. Most importantly I knew their coffee order; I knew how they preferred their tea or coffee, what they liked to eat in the morning, and how many packets of ketchup or napkins they liked with their breakfast.

When I first started working mornings I was amazed at the other girls' knowledge of all these different orders; but after a while I began to describe people by what kind of muffin they got, what time of the morning they came in, or what type of cigarettes they smoked. People like that in the morning, they enjoy the fact that they don't have to speak, that someone is there to make their lives a little easier, and get them exactly what they want without a word.

I loved most of the customers there. In fact I became friends with many of them. I knew about most of their lives' stories just from their coming in and some of them even got to know mine. Some people even gave us Christmas presents to show us how much they appreciated us. The problem was my boss. I know that this is a common occurrence for so many workers in society. So many people have issues with their bosses. They may enjoy the work, but simply can not work for the boss. When you can't get along with your boss or you know that he or she do not respect you, that makes it extremely difficult to do your job. Even if it is one that you love, you hate going to your work place and eventually resent the things you once loved.

My boss did not appreciate his workers. If he needed something we were just supposed to give up our own time out of our lives for him with no questions asked. This would be fine if there were any thanks involved after the favor was done, but with him there were no thanks. It was also hard to deal with because I was getting paid minimum wage. I never received a pay raise in the almost four years I was there, except for when the state increased the minimum wage and my boss was forced by law to pay us more.

In his article entitled "On the Digital Assembly Line," Simon Head looks at various call centers through out the U.S. He listens to the complaints of the workers there. Head finds that the workers really disliked the set up of the call center. The arrangement of the desks and floor plans made it such that their supervisors could watch the employees while they were working. Just like in *Nickel & Dimed*, Barbara Ehrenreich discusses how so many workers felt like they were being watched by their employers. When Barbara went out into the low wage workforce she was surprised to see that so many employees were treated very badly; she was especially surprised that she would have to be drug-tested. Returning to Heads study, the workers were also set up with computers that monitored the amount of calls being taken in. If a worker was not getting enough calls in then the supervisor could do a number of things to the employees. Some employees were given special classes to try and be able to meet the target times, while others were immediately dismissed. This shows to me that favoritism was going on.

I definitely saw this within the store. My boss Gerard let some employees get away with anything while he reprimanded, and even worse, fired some for those exact things that others got away with. One employee in a Teletech establishment did not comply with the company's footwear policy. One day she wore one sandal with a sock and one shoe to work, because she suffered from a swollen toe. Managers made her take off her sandal and sock to prove that the toe was black and blue (Head, 128).

We also had a similar policy where we were not able to wear sandals. Many people, however, disregarded this policy and wore sandals anyway. Most people got away with it. Gerard didn't say anything to them (including his daughter); others however weren't as lucky. One employee, Emily, wore flip flop sandals to work one day. Gerard had a problem with Emily for quite some time previous to this day. Emily had sold cigarettes to a minor without knowing and caused Gerard to be fined, and also received a punishment of not being able to sell cigarettes for a few days. Gerard happened to notice that Emily was wearing sandals that day. He made her sign a piece of paper that stated that if any thing had happened to her because she was wearing flip flops, that she could not sue him. No one else had ever had to sign this piece of paper when wearing sandals, but because Gerard disliked Emily he made her sign the paper.

Returning to the Head's study, another incident occurred at a call center in Arizona. A pregnant woman was criticized by her supervisors because she got sick in the bathroom. Her employers told her that she should throw up in the trash basket beside her desk. While we were working, we did not receive breaks, even though that is against the law. Gerard told us that if anyone had a complaint with that then he would have to have us start reporting our tips. We had tip cups on the counter; we hardly made enough for tips to be reported. Although some days you might make good tips, there were other days were you might only get five dollars. If I worked a long day time shift like 9 to 4, I would most likely eat something at lunch time. I usually went across the street to a deli and picked up a sandwich. Gerard would get mad that not only were we eating on the job, but also that we weren't eating food from the restaurant he owned right next door. He would stare at you and watch you eat and make you feel very uncomfortable, so much so that you wouldn't even want to eat anymore. Or, he would purposely bump into you while eating. Better yet, he would actually put his face right next to your food, ask you what it was, where you got it, and how much it was. Then he would inform you that he makes the same thing, for a lot cheaper.

Today I currently work at that same life insurance that I did contingent work, when I was thirteen years old. I am a part time worker there and also a part time worker at a credit union.

In the past few years, in the industrial capitalist nations, the field of work has been referred to as "**Post-Fordism**." The Post-Fordist ideology is discussed in "The Transformation of Work Revisited" written by Steven P. Vallas and John P. Beck. This term refers to "large scale shifts [which] have occurred in product markets and process technologies. Large corporations are letting go their more traditional dependence on fundamental bureaucracy and routinized tasks, and instead are using a new group of work arrangements. These arrangements are often referred to as flexible specialization, the post-hierarchical workplace, and the learning organization" (Vallas and Beck, 136-137). Basically employers are trying to use more technology instead of humans to do the work to produce their product, and when they need an actual human to do the work they hire part-time, seasonal, or temporary workers.

One of my jobs is working in a life insurance agency. The agency consists of two general agents; they are more or less the biggest bosses, although there are higher ups in the home office. These two men have contracts with agents who go out and sell life insurance. The agents have no set work hours or days but must produce a certain amount each year in order to remain a full time agent. In order to get the applications of those applying for life insurance (the people that the agent went out and found), the general agents hire staff to get the applications in and maintain a smoothly run office. I am a staff member; but in reality I don't do much. I answer phones, photocopy, file away completed applications, and help out where needed. My fellow staff workers, though, have a lot of work to do.

The two women I work with input the applications into the system. I would say that their work is **skilled mental work**. They do not use their body to produce something, instead they use their mind. They know the computer system and know which questions must be answered. They know how the system works and they know how to "fix" the system. They know that there are certain answers to life application questions that will most likely guarantee a "no" answer; they also know that there are some questions that must be answered, therefore if they are not answered then they return the application until the questions are answered. The agents, I believe, are more **practical skill** workers. They use their clients' body language in order to determine whether or not they believe they will purchase life insurance. They also debate and list the reasons as to why a person should buy insurance; therefore they are relying heavily on their body.

I believe that this work is telling of the Post Fordism era. The two general agents have no problem firing those agents who do not make their contract. They are also constantly looking for ways to have computers do the work for humans, like using programs like QuickBook or Microsoft Excel within the accounting department. Most importantly they try to get work that should be done by more than one out of one worker.

All these jobs are just for the present time. They are jobs that I am doing to support myself through school, and give me a little spending money. However, my work utopia is to help people. I don't want a job, I want a **career**. I want to be able to move up within a company and do something that I love which what I believe is the difference between a job and a career. I want to do something I enjoy, something that even if I hate my boss I'll still keep working because I love my job. I want to help people. I am studying alcohol and substance abuse in school right now. I plan to get my masters in counseling so that I can help people who can't help themselves. I love working with people. Almost all of my jobs that have involved dealing with people have been my favorites. I like being a part of someone's life, and by being a counselor I will be able to do this. In a way I want to be like the character Patch in the film *Patch Adams*; he wanted to help people with their problems, give them the hope that someone is there to listen to them. And so do I.

REFERENCES

Braverman, Harry. *Labor & Monopoly Capital: The Degradation of Work in the Twentieth Century* (Monthly Review, 1974), Pg. 45.

Buchanan, Ruth. "Lives on the Line," in Amy S. Wharton (ed.), *Working in America: Continuity, Conflict, and Change, Third Edition* (McGraw-Hill, 2006), Pg. 369.

Cotter, David A., Joan M. Hermsen, and Reeve Vanneman. "Gender Inequality at Work," in Amy S. Wharton (ed.), *Working in America: Continuity, Conflict, and Change, Third Edition* (McGraw-Hill, 2006).

Dickinson, Torry D. and Robert K. Schaeffer. *Fast Forward: Work, Gender, and Protest in a Changing World* (Rowman & Littlefield, 2001), Pg. 23.

Ehrenreich, Barbara, *Nickel and Dimed: On (Not) Getting By in America* (Owl Books, 2001).

Firebaugh, Glenn. "The New Geography of Global Income Inequality," in Amy S. Wharton (ed.), *Working in America: Continuity, Conflict, and Change, Third Edition* (McGraw-Hill, 2006), Pg. 172.

Goldscheider, Frances K. and Linda J. Waite. "Children's Share in Household Tasks," in Karen V. Hansen and Anita Ilta Garey (eds.), *Families in the U.S.: Kinship and Domestic Politics* (University of Temple Press, 1998), Pg.814.

Head, Simon. "On the Digital Assembly Line," in Amy S. Wharton (ed.), *Working in America: Continuity, Conflict, and Change, Third Edition* (McGraw-Hill, 2006).

Herring, Cedric. "Is Job Discrimination Dead?," in Elizabeth Higginbotham and Margaret L. Andersen (eds.), *Race and Ethnicity in Society: The Changing Landscape* (Thomson Wadsworth, 2006).

Kibria, Nazli. "Household Structure and Family Ideologies: The Case of Vietnamese Refugees," in Karen V. Hansen and Anita Ilta Garey (eds.), *Families in the U.S.: Kinship and Domestic Politics* (University of Temple Press, 1998), Pg. 60.

Meyer III, Stephen. "The Evolution of the New Industrial Technology," in Amy S. Wharton (ed.), *Working in America: Continuity, Conflict, and Change, Third Edition* (McGraw-Hill, 2006).

Moss, Philip and Chris Tilly. "Stories Employers Tell," in Amy S. Wharton (ed.), *Working in America: Continuity, Conflict, and Change, Third Edition* (McGraw-Hill, 2006).

Royster, Deidre. "Race and the Invisible Hand," in Elizabeth Higginbotham and Margaret L. Andersen (eds.), *Race and Ethnicity in Society: The Changing Landscape* (Thomson Wadsworth, 2006).

Schor, Juliet. "The (Even More) Overworked American. "in John de Graaf (ed.), *Take Back Your Time: Fighting Overwork and Time Poverty in America* (Berrett-Koehler Publications, Inc., 2003), Pg. 7.

Vallas, Steven P. and John P. Beck. "The Transformation of Work Revisited" in Amy S. Wharton (ed.), *Working in America: Continuity, Conflict, and Change, Third Edition* (McGraw-Hill, 2006), Pg. 136.

Films

Patch Adams. Universal Pictures. 1998.

Life is Change

"My Adolescent State of Mind"

Lora Aurise

University of Massachusetts Boston

laurise@deweyleboeuf.com

Abstract: Using a variety of literature, theories, and concepts in the sociology of youth and adolescence, in this paper I reflect back on my own "adolescent state of mind" to draw needed lessons for my further growth as a person. Adolescence is a very important time for growth and exploration. Cote and Allahar state that "[t]he average person looks upon adolescence as a natural and necessary stage of development and feels that this period is the time of one's life when one is carefree and has no responsibilities" (107). However, since an adolescent suddenly moves beyond the certainties of childhood and is suddenly exposed to many different perspectives, certain experiences can become incredibly overwhelming and difficult. In thinking back to my adolescence, I wish I was more open-minded when my father and I moved to Florida. I spent a good portion of my adolescence trying to prove a point instead of enjoying my life and my freedom. Leaving my father was the first time I experienced a broken heart and I will never forget how much that affected my identity development. Many days I was unsure of who I was without him because he was my provider and caretaker. In conclusion, my decision to leave him and move back to Massachusetts made me a stronger person. I learned the concept of responsibility and independence which are two very important qualities when entering adulthood.

Adolescence is a very important time for growth and exploration. Cote and Allahar state that "[t]he average person looks upon adolescence as a natural and necessary stage of development and feels that this period is the time of one's life when one is carefree and has no responsibilities" (107). However, since an adolescent suddenly moves beyond the certainties of childhood and is suddenly exposed to many different perspectives, certain experiences can become incredibly overwhelming and difficult.

My father raised me because my parents were never married. I was born when they were teenagers and my father happened to be the more responsible one of the two. When I was fifteen years old, my father decided to move from Springfield, Massachusetts, to St. Petersburg, Florida, the summer before my junior year of high school. He told me it would be an experience of a lifetime so I tried to look at our move as an adventure instead of an obligation. In many ways, I was frightened of this drastic change, primarily because I was

Lora Aurise is an undergraduate senior at UMass Boston, majoring in Psychology. She wrote this paper while enrolled in the course, Soc. 201: "Youth and Society," instructed by Anna Beckwith (Lecturer of Sociology at UMass Boston) during the Spring 2008 semester.

leaving behind the only life I knew including my mother, my younger sister, and *all* of my friends. The feeling was so powerful and overwhelming that it is indescribable.

Erik Erikson believed that "**adolescence** is the period in the human life cycle during which the individual must establish a sense of personal identity and avoid the dangers of role diffusion and identity confusion" (Muuss: 51). In order for individuals to develop stable identities, they must be able to understand themselves. As Erikson describes, "**identity** must be acquired through sustained individual effort. Unwillingness to work actively on one's identity formation carries with it the danger of **role diffusion**, which may result in alienation and a sense of isolation and confusion" (Muuss: 51).

In understanding the process of identity development, I feel as though my identity formation was delayed as I began to lose myself during the moving process. Once we left Massachusetts, I felt very lost and insecure in my surroundings. I felt as though I was alone in Florida and unable to identify with anyone socially. I could not maintain what Erik Erikson calls **psychosocial reciprocity**, because nobody seemed to empathize with my recent experience.

Erik Erikson considered the idea of adolescence as a Moratorium. **Identity moratorium**, as described by Cote & Allahar, "can provide young people with opportunities to experiment with roles, ideas, beliefs, and life-styles and can set them on a life course that is rich and rewarding" (74). In understanding this adolescent process, I feel as though I never experienced such an important stage of identity development until much later in life. As everything about my identity seemed unknown to me at that time, I only found myself experimenting with *ideas* and *beliefs* of happiness because I did not know how to express myself through my current state of mind. I just wanted to be happy and feel like a teenager again, instead of a lost soul in a new state.

Cote and Allahar also write, "From Erikson's perspective, these identity moratoria should function to lessen the trauma of the identity transformation experience during the coming-of-age period" (74). As I was experiencing my own form of psychological trauma, I did not have the opportunity to lessen the trauma of identity transformation. At the time of the move, I was not experiencing identity transformation because I was trying to understand and relate to my new identity in Florida. Once I moved, I was a stranger to myself and in many ways I felt as though I had left my identity in Massachusetts.

As my time in Florida progressed, I felt worse. However, I tried to look on the bright side of the situation. I was looking forward to the first day of school because it was my chance to meet new people and focus on my school work instead of my sadness. "A sense of rolelessness in current society is one of the most common dilemmas of adolescents. This is particularly problematic given fundamental needs among youth to find a sense of belonging and to feel a sense of worth as a person" (Herr: 48). My lack of self-worth led to much bigger problems than feeling alone. Unfortunately at this point, I was so emotionally traumatized from my sadness that I stopped eating regularly. Eating became a chore and within three weeks of not eating properly I lost about fifteen pounds, which was a lot for me at the time. As my first day of school approached, I weighed about ninety-two pounds. I had always been on the smaller side, but from my clothing and how it fit my body I noticed that I lost a lot of weight.

On the first day of school I felt very out of place. For the first time in my life I knew what it felt like to be the minority. My school was primarily black, which was not an issue to me but I felt like an outcast with my pale skin and undernourished slender figure. It was apparent at lunch time that there was a lack of racial diversity within the high school community. The few white students were always together, while the black students had their own social group. There was very little racial integration, and the only form of integration that I witnessed was in the classroom when certain groups were assigned to work with one another.

Beverly Daniel Tatum writes, "The search for personal identity that intensifies in adolescence can involve several dimensions of an adolescent's life .... The process of exploration may vary across these identity domains" (53). At my school in Florida, I could sense that the black students felt a stronger connection with one another while they were experiencing their identity formation process. I did not care if all of my friends were black, white, or Asian—I just wanted to feel a connection with some kind of peer group so I could start to find myself again. A **peer group** is "an association of self-selected equals who coalesce around common interests, tastes, preferences, and beliefs" (Bensman & Rosenberg: 80). In many ways, adolescents form peer groups to create a sense of belonging with other youths. I was in desperate need of friendship and acceptance at this school and for some reason it was impossible for me to meet the need.

Proveda writes, "Relationships with other adolescents are the major influences in adolescent identity development" (42). My constant need to connect with my peers and maintain psychosocial reciprocity is not limited to a peer group. Adolescence is "a period of a great need for peer group recognition and almost compulsive peer group involvement" (Muuss: 52). My constant need to find a peer group was reflective of my efforts in developing a new identity in Florida. While I was not interacting with other adolescents on a social level, I was not experiencing any positive aspects throughout this process of my identity development. I began to lose faith in forming new adolescent relationships. Alas, these high school students did not seem very open to new students and I did not have the energy to try to fit into that social environment.

A few weeks into the school year, I wore a pair of gaucho pants that tied in front with a cream colored top that fell as a midriff. In the middle of my day, I was walking to science class and two black girls stopped in the hallway, looked at me and said, "She is so disgusting, she is anorexic." I ran to the bathroom and cried for the rest of the afternoon. I did not know if they were verbally attacking me because of my appearance, or because of the fact that I was white. Either way, I constantly felt out of place and unwelcome.

In many ways, that day was a turning point in my moving experience. I became negative and refused to build my life in Florida. I started eating regularly as I quickly became unsatisfied with my appearance. I was addicted to peanut M&M's and started eating them by the pound because they made me feel happy. After three months I managed to gain twenty pounds. I looked great; but my father could tell that I was emotionally unstable and he began to worry since I was not adjusting well to my life in Florida. After a very long conversation, my father gave me the opportunity to move back to Massachusetts and live with my mother. Two weeks later, my father drove me back to Massachusetts.

Throughout my drive back to Massachusetts, I feared the consequences of my decision to leave my father. I wondered if my home was ultimately with my father because he had always been my provider. Alas, I felt the need to spread my wings to obtain a feeling of independence and individualism in regard to finding myself again. **Individualism** is defined as "an encouragement to be yourself, develop your own talent, and try your best" (Sawyer: 196). In my current state of mind, I wanted to simply feel like myself and learn the importance of independence as I never felt I understood the concept of independence while growing up with my father. In regard to the genital stage of development, "Freud said that from puberty onward the individual's great task is 'freeing himself from the parents.' ... [F]or the daughter, the tasks are the same, she too must separate from the parents and establish a life of her own" (Crain: 258). My father was my caretaker, and I never had anything to worry about when I lived with him. I always felt a sense of security, but in the back of my mind I always feared that I would be dependent on him throughout my life if I did not leave him and go back to Massachusetts.

While part of me was excited for my new adventure, I had a lot of doubt in the back of my mind. I never had a close relationship with my mother for a number of different reasons. Throughout my childhood, she proved herself to be a very irresponsible and confrontational individual; I was not raised to be that way with my father. I was raised in a very peaceful and loving environment which is something my mother had never experienced in her lifetime. When I moved back from Florida, my mother was residing in Chicopee, Massachusetts, in a one bedroom apartment with my five year old sister, Elana. All three of us shared a one bedroom apartment because that was all my mother could afford at the time. We had very little privacy from one another which made our living situation very difficult at times. Most days I felt an incredible amount of regret for the decision I had made, but I promised myself I would make the best of the situation because I was back home in Massachusetts where I wanted to be.

Once again, I had to start fresh at another new school. This was my second high school in three months. Overall, it was the fourth high school I had attended during my high school career. At some point, I began to doubt the idea of stability as I had not experienced *any* kind of stability in months. As a junior in high school, I always imagined myself having many friends and feeling a sense of acceptance amongst my peers. Unfortunately, I found myself in my junior year of high school entering another new school without a single friend in the hallway. The first two weeks of school, I ate my lunch in front of my locker while pretending to read a book. The cafeteria was too overwhelming for me and I was more comfortable by myself as I was trying to maintain a positive frame of mind.

As weeks passed, I was still eating lunch alone in front of my locker. I felt insecure about myself and I began analyzing all of my thoughts and feelings because I did not have anyone to talk to. Crain writes that "[o]ur thoughts and feelings can be as important as actual events" (247). I started to wonder if I was meant to spend my life alone because this was the first time in my life that I actually felt any kind of permanent sadness and loneliness. Regardless of my positive thoughts, I found it more difficult to get out of bed in the morning. Many days I would go to school for two hours and leave because I could not bear the idea of eating alone in front of my locker. Luckily, I was able to find a deli down the street from my school. I bought a journal and started to write down all of my thoughts and feelings. It was helpful as I found myself engaging with a lot of my problems.

Chicopee High school, as an educational institution, was a very welcoming environment for individuals who had the potential to feel out of place in life. For example, the majority of students were white but there was a racial/ethnic diversity of groups which I had not witnessed in Florida. In regard to the concept of institutional violence toward students that represent a minority, such as gay and lesbian students, I did not see any violence related to institutional practices or socio-cultural beliefs and values. The act of **violence** is defined as "any act or situation in which a person is injured, whether physically or psychologically" (Herr: 241). To my knowledge, I cannot remember witnessing any arguments where faculty and staff did not break it up and solve the issue with students and parents behind closed doors.

I remember attending school with two lesbian girls and I will always remember them as being very popular for their personality and their athletic abilities. Chicopee High School was very competitive in regards to sports teams and most of the athletes were popular amongst the student body. However, in understanding the concept of **institutional violence**—defined as violence that is "submerged from view so that its forms are almost completely invisible. Violence at this level includes harmful actions by societal institutions and their various organizational units that obstruct the spontaneous unfolding of human potential" (Herr: 241). While the most commonly reported acts of violence are aimed towards gay and lesbians, it is possible that many students at Chicopee High School were blinded by institutional violence because of its invisibility to outsiders. To my knowledge, Chicopee High School was a very well organized institution that accepted individuals for their academic abilities and skills, not to their inherent sexuality which should never be categorized as a problem.

In many ways, Chicopee High School was *very* different compared to any other high school I had ever attended. It was a small town school which consisted of many **cliques**—small groups that are derived from the neighborhood or a school setting (Bensman & Rosenberg: 80). As many of my peers grew up in the same town for most of their lives, they knew each other from the neighborhood and / or school. I was interested in trying out for the cheerleading squad but I was afraid of rejection. I knew I had the physical skills to be a cheerleader, but I was not sure if I had the social skills for the cheerleaders.

After attending Chicopee High School for two weeks I was able to pick up on the sexual double standard between the cheerleaders and the jocks. The sexual **double standard** is defined as the idea that women are disgraced by sex outside marriage; the sex transforms them into 'sluts' (Tanenbaum: 76). A **slut** is usually a girl that is categorized as promiscuous amongst her peers. "Roughly half of American teenagers are sexually active—in 1997, 49 percent of boys aged fifteen to nineteen reported ever having sexual intercourse, compared at the 48 percent of girls—but when it comes to moralizing about the excesses of teenage sexuality, girls alone are ridiculed and made to feel cheap" (Tanenbaum: 77). As I witnessed the degradation certain cheerleaders experienced in the hallways, I feared the idea of becoming visible to the jocks. The **jocks** were the popular guys in high school that embraced sports as part of their high school career (Eckert: 7). As part of me wanted to utilize my physical skills towards cheerleading, I did not want to gain a negative reputation in a new school environment, so I started to look into other after school clubs. Participation in after school clubs would be categorized as part of the jock environment as it is within the acceptance of the institution.

I always considered myself to be a good girl, so I was adamant about protecting my reputation as I was already trying to deal with overcoming the idea of being the new girl. Leora Tanenbaum believes a **good girl** is usually considered the smart girl that does not allow her peers to become involved in the morality that she maintains as a sense of self (91). I always had the tendency to mind my own business and stand clear of rumors. In some ways I desired belonging to a clique, but in other ways I was content in my own world of adjustment. In considering myself a good girl, the recent changes in my life forced me to feel insecure about myself and my surroundings. While I was still interested in meeting people, I decided to join a calligraphy club after school so I could focus some of my negative energy on something positive and rewarding. One of my teachers suggested it because she thought I had a nice handwriting. At this point in my life, I was not about to turn down compliments or opportunities to engage my creative skills into something positive for my self-esteem.

It was a very slow process, but as time progressed I began to feel more comfortable in my school surrounding. As the comfort settled in, so did my ambition for independence. I decided that I was ready to find an after school job. I wanted to focus on my independence, such as obtaining a job, getting my license, and buying my first car. I knew that an after school job would allow me to accomplish these goals, as well as give me more opportunities to meet new people in my town.

I needed something to keep me productive as I decided to give myself space from my familial environment. My mother did not seem to appreciate bringing me places, so I was determined to find my own form of transportation by earning my license. In my job search, I sensed a lot of hesitation from management as they were afraid to hire a young individual, such as myself, because they wanted to feel as though they could depend on their employees. This was something I never thought I'd experience in my job search. As Susan Willis describes, "In an open market of casual youth employment, many teens move from one job to the next in a matter of weeks" (351).

This lack of faith in teenagers made me want to prove society wrong. I knew I was dependable, and I also knew I could handle the responsibility of a job. I promised myself that I would be as reliable as possible in my employment environment. Part of me was striving to become a producer so I could prove to myself and to society that I had self-worth and the capabilities to be independent and responsible. Cote and Allahar refer to maturity and responsibility as qualities acquired through experience and practice (108). Two interviews later, I received my first job position as a sales associate in a small clothing chain called Cherry & Webb. I was hired at minimum wage, which was $5.25 at the time. Cote and Allahar believe that "young people lack power, rights, and legitimacy, which means they are disenfranchised. At the same time they are crucial to the economic system because they constitute a source of cheap labor, as well as a massive consumer market" (109). I knew I was underpaid as well as every other teenager due to the lack of experience and education obtained while searching for a first job. I was, however, excited for my new role as a sales associate and I was going to use this position as a stepping stone to gain the necessary experience to make me successful in the retail business.

In thinking back to obtaining my first job, I found myself able to relate to *Ghost World* in many different ways. Rebecca was the one character in the movie that I was able to identify with myself. She graduated from high school with aspirations of finding a job and moving out of her parent's house and into an apartment with her friend, Enid. She was successful in her job as she understood that the customer came before her thoughts, which is something that Enid was never able to understand when she began working. Rebecca's independence became apparently clear when she decided to look for an apartment by herself because she slowly came to realize that Enid was financially unstable as she was not able to maintain a steady job position. I feel as though my responsibility and need for independence was very much like Rebecca's character.

While my personal life was starting to pick up in social aspects, such as my newly founded after school activity and my new job, my home life was suffering tremendously as my mother and I did not have a very strong connection with one another. Everything I did was wrong. I was not clean enough for her around the house and neither was my younger sister. She was always complaining about something and it was starting to make me feel unwelcome, just as I felt when I was attending school in Florida. Nancy Lesko describes this as "youth as probation." **Youth as a probation** is a period experienced during youth in which everything they do is monitored through some kind of adult supervision. Lesko elaborates on youth as "always 'becoming,' a situation that provoked endless watching, monitoring, and evaluating ... careful development in-time was identified as the safest path" (111). As my mother continued to express her impatience with me, I started to spend less time at home and more time in school and at work to give myself distance from my familial situation. I kept telling myself it was only temporary and regardless of my mother's personality, she was still my mother and I was determined to make things work because this was *my* decision.

Throughout my junior year of high school, I spent most of my time working, studying, and participating in the calligraphy club. I was very devoted to my job because it was enabling me to feel a sense of responsibility and independence. In March of 1997, I enrolled in a driving school in Ludlow, Massachusetts. During this time, my mother was not making an effort to participate in my after school activities, such as work, school, and driving lessons so I had to find an alternative form of transportation. I never had these problems when I lived with my father as he would make arrangements for me regardless of his schedule. He always went out of his way to accommodate my needs and my mother was not like that at all. Luckily, my father was still a large part of my life and he was able to make arrangements with one of his friends in Massachusetts to transport me to my driving lessons every Wednesday after school. After three months of driving lessons, I earned my license.

I reached my first rite of passage at the age of sixteen and it was the first time in months that I felt good about myself and my accomplishments. The **rite of passage** is defined as the "experience that changes the status of the individual from within and without, in the eyes of other community members." Such rites require the participation, guidance, and witness of adults in the society (Hoffman: 59). ... the acquisition of a driver's license is perhaps one of the earliest and most universally accepted benchmarks of pending adult status unrelated to school in our society" (Hoffman: 60). This rite of passage was so important to me and the fact that I had accomplished it made me feel as though I could accomplish anything. At this point, I had been saving all of my money for months and I finally had enough money in the bank to purchase my first car. I viewed owning a car as my ticket to freedom; freedom from adolescence and freedom from my living environment.

As my senior year approached, I started to feel a sense of personal identity again. I was working diligently at work and school and I felt as though I was on the road to success because I was finally determined. At the end of the summer, I saved $1,600 from my part time job. I worked so hard and saved every penny so I could officially call myself a consumer. My first large purchase, bought with my own savings, was my first car. It was a 1990 Ford Tempo with automatic seat belts. "Young people are enthusiastic consumers of many products of the mass media" (Cote & Allahar: 146). It was the first time in my life I had actually worked for something that I really wanted, and I felt a sense of accomplishment. My father raised me to save and spend money responsibly; however, my dependence on him throughout my childhood never gave me the opportunity to experience how it felt to earn my own money and save it for something that I truly desired. Fortunately, his sense of responsibility encouraged me to spend my money on things I needed, and not necessarily the things everyone else had. I was never distracted by media sales through magazines or television. My sense of responsibility was an amazing feeling and for the first time in my life, I felt like a responsible adult.

It was finally my senior year of high school. As I began the last year of my high school career, I actually felt a sense of comfort in my educational environment. I was not exactly happy, but I was content with my life choices. On the first day back to school, I met someone who managed to imprint my life, my beliefs, and my values. His name was Michael[1]. I met him in English class and we became friends due to our competitiveness with each other in regards to our grades. At some point during our senior year, I felt a close bond forming between us and it felt so good to have that connection with someone again. I knew he was growing attached to me as well because I could feel it.

One day after school, we met at the park because he said he needed to talk to me about something important. I remember it was a ninety degree day in April. The sun felt so good on my face. As I sat under the tree next to Michael, he looked at me and told me he was in love with me. At that moment, I was not sure whether I should kiss him or slap him for potentially ruining our comfortable friendship. I decided to just stay calm because part of me felt like I was in love with him too; but most times I was so out of touch with my feelings after the trauma I just experienced in the last year. After that day, we started meeting up at night to watch movies or lie on the ground in the park and look at the stars. He would bring me flowers and he had this amazing ability to make me feel so loved and beautiful. By the end of our senior year, we were inseparable. Our friendship grew stronger; but not strong enough to turn our friendship into a romance. For some reason, I could not bring myself to kiss him because I was afraid to lose myself to someone else.

As I began to separate myself from him, he became bitter and very upset with me. On the day of our high school graduation, he would not look at me or speak to me—it was heartbreaking. Six months after graduation, he finally called me and asked me to have lunch with him because he needed to talk to me. The last time he told me he needed to talk to me, he told me he was in love with me so I knew this was important. During our lunch, he apologized for his behavior in regards to our graduation and his lack of response to my phone calls prior to graduation. He told me that he was so upset about our distance with each other that he tried to kill himself. He had a large red scar on his neck and I felt as though he was blaming me for his attempted suicide. Mike Males writes, "Suicide is a pure indicator of high risk behavior, of self-destructive intent" (229).

By definition, suicide is always self-inflicted and always fatal" (Males: 229). As Michael did not succeed with his attempted suicide, he still found a way to place blame on me. Some days I still feel as though his actions in 1998 still affect my life decisions when it comes to dating and opening my life to another individual. "Suicides, like all forms of crime, are becoming more and more precocious. In these days children leave their marbles and tops to commit suicide, tired of life almost before they have tasted it" (Males: 231). The more I analyzed the situation involving Michael and me I realized that his reasons for attempted suicide were simplistic. He was a high risk teenager and his decision to potentially end his life had nothing to do with me as much as he wanted to blame me. Throughout the years, we have remained good friends. In many ways, I consider him to be my first true love because at times I felt he was my true soul-mate.

[1] Name has been changed to respect privacy.

In thinking back to my adolescence, I wish I was more open-minded when my father and I moved to Florida. I spent a good portion of my adolescence trying to prove a point instead of enjoying my life and my freedom. Leaving my father was the first time I experienced a broken heart and I will never forget how much that affected my identity development. Many days I was unsure of who I was without him because he was my provider and caretaker. In conclusion, my decision to leave him and move back to Massachusetts made me a stronger person. I learned the concept of responsibility and independence which are two very important qualities when entering adulthood.

WORKS CITED

Bensman, Joseph & Bernard Rosenberg. "The Peer Group," in *Socialization and the Life Cycle*, edited by Peter I. Rose (St. Martins Press, 1979), pp. 79-96.

Cote, James E. & Anton L. Allahar. *Generation on Hold: Coming of Age in the Late Twentieth Century*, New York: New York University Press, 1995.

Crain, William. *Theories of Development: Concepts and Applications*, Forth Edition (Prentice Hall, 2000)/

Eckert, Penelope. *Jocks & Burnouts: Social Categories and Identity in the High School* (Teachers College Press, 1989).

Herr, Katheryn. "Institutional Violence in the Everyday Practice of School: The Narrative of a Young Lesbian," in Lee D. Baker, *Life in America: Identity and Everyday Experience* (Blackwell Publishing, 2004), pg. 240-247.

Herr, Katheryn. "Problematizing the 'Problem' Teen: Reconceptualizing Adolescent Development," in Shirley Steinberg, Priya Parmar, and Birgit Richard (eds.), *Contemporary Youth Culture, Volume I* (Greenwood Press, 2006), pp. 47-56.

Hoffmann, Lynn M. "Adolescent Rites of Passage," in Shirley Steinberg, Priya Parmar, and Birgit Richard (eds.), *Contemporary Youth Culture, Volume I* (Greenwood Press, 2006), pp. 58-63.

Lesko, Nancy. *Act your Age! The Cultural Construction of Adolescence.* Routledge Falmer, 2001.

Males, Mike. "Growing up Referred," in *The Scapegoat Generation* (Common Courage Press, 1996), pp. 219 -253.

Muuss, Rolf. "Erik Erikson's Theory of Identity Development" in *Theories of Adolescence*, Sixth Edition (McGraw Hill, 1996), pp. 42-57.

Proveda, David. "Language and Talk," in Shirley Steinberg, Priya Parmar, and Birgit Richard (eds.), *Contemporary Youth Culture, Volume I* (Greenwood Press, 2006), pp. 41-56.

Sawyer, Jessica. "Confessions of a Maine-iac: The Family, Academia, and Modernity." *Human Architecture, III*, 1&2, Fall 2004/ Spring 2005, 193-204.

Tanenbaum, Leora. *Slut! Growing Up Female with a Bad Reputation*, Perreneal. 2000.

Tatum, Beverly Daniel, "Why Are All the Black Kids Sitting Together in the Cafeteria?" Basic Books, 1997.

Willis, Susan. "Teens at Work: Negotiating the Jobless Future." *Generations of Youth*. Ed. Joe Austin and Michael Willard. New York University Press, 1998. 347-357.

Film:

Ghost World. Dir. Terry Zwigoff. Perf. Thora Birch, Scarlett Johansson, Steve Bucemi. MGM, 2001.

Sociological Reflections on My Work Experience

Brian Ahl

University of Massachusetts Boston

bri7654321@yahoo.com

Abstract: Work is a very complex concept. I always considered a paid job to be work, but in reality work can be a variety of activities. I have had a few paying jobs in my two decade lifetime but I never thought I did so much of what Braverman defines as work. Helping out around the house is just as much considered work as the job that my father does. The housework that gets done by my family members, the babysitting of little cousins, the process of helping neighbors, and the nurturing of children are all classified as work. Thinking about all the different jobs that I have had and different types of work that I have participated in, I have had time to reflect on what my work utopia would be. When I think of my work utopia, I think about waking up each morning and being happy going to work. I do not envision myself working in an office building crunching numbers. I want to work with children in some way. I desire to graduate from college in the next two years and work toward getting a job that helps children. In my perfect world I would be a guidance counselor and then coach a high school sport. I love children and love the idea of making the lives of other people better. I am a very social person and enjoy a good conversation so I feel like this would be a perfect fit for me.

Work is a very complex concept. I always just considered a paid job to be work, but in reality work can be a variety of activities. In his book *Labor and Monopoly Capital* (1974) Harry Braverman defines **work** as an activity with a goal that has an outcome or product. Work transforms nature by altering materials of nature that better support certain needs. I have had a few paying jobs in my two decade lifetime but I never thought I did so much of what Braverman defines as work. Helping out around the house is just as much considered work as the job that my father does. The housework that gets done by my family members, the babysitting of little cousins, the process of helping neighbors, and the nurturing of children are all classified as work.

Since the day I was born I was surrounded by this idea of work. I was born in 1987 during a period that is known as the postmodern era, a time of a new economy often referred to as **Post-Fordism.** Post-Fordism started around the early 1970s and is still considered to be contemporary. Post-Fordism is characterized by the rise of the service sector and the decline of industry, manufacturing, non-unionization, and low wages. More importantly, there has been a steady advancement in computer technologies. Finally, there is an emphasis on flexibility, i.e., the idea that a company should be flexible in the production of products and also in hours for the workers.

Brian Ahl is an undergraduate freshman at UMass Boston, whose major is undeclared. He wrote this paper while enrolled in the course Soc. 232: "Sociology of Work," instructed by Anna Beckwith (Lecturer of Sociology at UMass Boston) during the Spring 2008 semester.

The idea of mass production was the key component of the era before Post-Fordism which is known as **Fordism**. In his article titled "The Evolution of the New Industrial Technology," Stephen Meyer III describes Fordism as an economic philosophy. From this philosophy widespread prosperity and high profits can be achieved by high wages, thus allowing the workers to purchase products like the ones they produce (73). During the time when Fordism was prevalent, it was all about market demand. And once a company found out what and how much were needed then the company could increase or decrease the price accordingly. Fordism was known to be associated with the modern era and occurred during the 1920s up until the early 1970s. There was an emphasis on interchangeable parts and an assembly line. There was also a division of labor and standardized parts. A lot has changed over the past eighty years or so, but one thing that has not is the work. There is always plenty of work out there that needs to be done.

There are five people in my family. Although a **household** can be a family, the concept is not just limited to a family. A household can also consist of roommates, unmarried partners, or just friends who live together. In their book titled *Fast Forward* Dickinson and Schaeffer state that people in a household "join together different workers and separate kinds of work." Dickinson and Schaeffer go on to write that "the household is the organizational unit for working people the world over" (23). The people in a household, however, may work together to do many tasks and jobs necessary to survive. Most of the work that is done in the household goes unnoticed and is usually always unpaid. These unpaid jobs are very important, so work by members of a household is not limited to just **wage work**.

Wage work is work that a person does to receive some type of payment for. Payment can be made in, most commonly, a paycheck and can sometimes be accompanied by perks (Dickinson and Schaeffer 26). When I worked at a golf course as a caddy, one perk that I received was having the luxury of playing the course every Monday afternoon. The course both paid you and guaranteed that you would be able to play the course at least one time a week.

It is very rare that a household survives on strictly wage work. In my family my father works and my mother stays at home and does what Dickinson and Schaeffer define as **unpaid work.** Dickinson and Schaeffer say that "women do most of the world's work, and much of it is unpaid" (25). My mother is definitely responsible for the unpaid work in my family. She is a stay at home mother who does the cooking and cleaning. She runs the errands and takes care of the children. Although she is not rewarded with a paycheck every other week like my father is, the work she does is vital and necessary for the survival of the household. Without her the household would not succeed because it is a lot of work to care for the children and make sure that they are accommodated for.

As my brother, sister and I grew up we started helping out with the work around the house. This is what Dickinson and Schaeffer call **subsistence work.** They state that "to subsist as workers, people work together" (29). The act of working together on tasks and getting them completed is known as subsistence work. Around the house nearly all the chores and household tasks can fit into the subsistence work category. For example helping out with cooking dinner, setting the table, doing the laundry, taking out the garbage, caring for a younger sister, and even babysitting are all what Dickinson and Schaeffer consider subsistence work. The idea of people working together to complete a task is usually unpaid. Just like the unpaid work done by my mother and the wage work done by my father, subsistence work, in which nearly all the members of a household participate, is vital to the success and well being of a household.

My beliefs about work are closely connected with the reality of what happens in my family today. On such issues as whether

I would want my wife to work or be a stay at home mother, I would choose for her to stay at home and take care of the children and the work around the house. The idea of a **dual wage earner household** is not one that interests me. A dual wage earner household is one in which two people, usually adults, work. They both are engaged in wage work and bring home some type of paycheck to better support their household. I do not want to have to pay for my child to go to daycare and not be with one of his or her parents during the day. I believe that my views on this particular subject are due to the **ideologies** that have shaped my parents' and my views.

In her article titled "Household Structure and Family Ideologies; The Case of Vietnamese Refugee" Nazli Kabri writes about the role of ideologies. She states, "These family beliefs or ideologies define norms and expectations about household activities and relations that affect the household economy" (60). Because my household has always had my mother at home to watch after the kids and do the unpaid work around the house I hold the same beliefs about what is right. I also believe that a child's most important job is to do well in school. While they are in school they should concentrate on their studies and should not be worrying about having an important job and making money. I have this in my belief system because it is an ideology that is held by my mother and father, one that they have passed on to my siblings and me while we were growing up.

I think it is common that in a household there are certain children that do more work than other family members. I do believe that all family members, in some way or another, do a handful of different kinds of work. In my household, I have noticed that my brother and I do much more work than my younger sister. This may be because of her age or her gender but it is definitely true. My brother and I are always helping out around the house, doing yard work, doing laundry, and just giving an extra hand wherever it seems necessary.

In their article titled "Children's Share in Household Tasks" Frances Goldscheider and Linda Waite write about a household that is non-egalitarian. A **non-egalitarian household** is one in which family members are treated differently when it comes to specific types of work around the house. For example, my father would never expect my sister to help mow the lawn or help clean out the garage. I feel the explanation is because of her gender. She is a girl so it is more common for her to be doing work with my mother and not the outdoor work like my brother, father, and me. Furthermore, my mother would never expect nor want my brother, father, or myself cooking dinner or doing anything in the kitchen. Also, I feel there is more expected out of my older brother than what is expected out of me, and also my parents expect more help from me than they do of my younger sister. All of these ideas feed into Goldscheider and Waites's idea of a non-egalitarian household (813).

Some families depend a lot if not totally on the children to provide for the family. They do everything because the parents are not able. This was highlighted in the movie *What's Eating Gilbert Grape* when the main character Gilbert, played by Johnny Depp, and his two younger sisters have to take care of his obese mother and autistic younger brother. It is a lot of work to be the head of a household and these three kids do a good job at it. At times they struggle with it and want to have more of a social life, but they know that their household will not survive without their help. There are many families in today's world that are dependent on their children to help provide for their household. It is almost unheard of for a household that does not need the support and help from the children to be successful.

Although my work experience has not been that extensive, there is plenty to write and talk about. I have done a variety of work, including wage work, subsistence work, and unpaid work, all of which is considered human work. **Human work** is not instinctual, but rather, it is thoughtful (Braverman, 1974). A person who is doing human work is imagining what he or she wants to do and then does it. Human work has a purpose and meaning in a person's life. It is conceptual, creative, and conscious. Just the other day I was outside doing some yard work and noticed my old neighbor mowing her lawn. She is old and was struggling to complete the task. So I consciously thought about what the right thing to do was and decided that I would walk over and help her out. This is both an example of subsistence work and human work.

I started working when I was around the age of thirteen or so, delivering papers to some houses around my street. It taught me about responsibility and meeting deadlines and overall was a good working experience. As I grew a few years older I started getting bored of working as a paper delivery boy so I decided to seek a job that was compatible with my interests. Of course I decided to be a golf caddy at a local private golf course. The money was decent but also if I caddied I was allowed to play golf there on Monday afternoon. There was no definite wage that I would be given every time I caddied, but rather it would be up to the person I caddied for. In his article titled "The Capitalist Firm in the 21st Century" Walter Powell discusses the type of work that caddying would fall into. He defines it as **craft-based work.** Powell says in this article "Sociologists have long studied 'craft-based work' in which workers labored on short-term projects and were paid for specific performances (88)." I would caddy for a different member of the country club nearly every time I showed up at the golf course. After the round I was given a certain wage based upon how I did. The golfer decided whether or not the caddy did a good job and this would later be revealed to the caddy based on how much money he was given at the end of the round.

It was a common occurrence that even if you did caddy as well as you could, your tip was still dependent on how well the golfer played that day. So if the golfer played really poorly but you still gave him the appropriate clubs, told him the correct yardages, and took out and put back in the flag there was still a good chance that you would not be paid the appropriate wage because of his performance. Even though I would do my job as well as I possibly could, I would not be rewarded for it. There was a sense of **alienation in the labor process,** as Karl Marx would call it. In his writing of "Alienated Labour" Marx talked a lot about the labor process and how people would contribute to the product, in this case on the assembly lines, and then not see the final product (46). This is similar to my experiences as a caddy. I would be alienated from the group of players and only be asked to give him a club or take out the flagstick. There was this sense that I was contributing to the good of the golfer and to improving his playing ability, but at the same time I could only express my individuality through the shots that he made. In both cases, assembly lines and caddying, you helped produce the final product but you are alienated from affecting the true outcome because you are only one piece of the final product.

I stuck with the caddying job for a few years and it carried me into the summer of my junior year of high school. I never worked during the school year because I was concentrating on my studies. My only wage work during the school year was homework and getting good grades. As the summer of junior year approached I was going to work. But this was going to be different kind of work. In order to graduate from high school all the graduates must complete a one hundred and fifty hour community service project. Community service is one type of **share work.** For my project I worked at a summer camp for kids from the inner city of Boston. It was a stay away camp, where the kids would take a bus to a small town in Massachusetts and stay for about a week. They would participate in group activities, go the beach, make new friends, and hopefully learn a thing or two about the world from the people who are employed there. After I completed my one hundred and fifty hours of community service at the camp I got a job for the following summer and am still employed there today.

Having the opportunity to work at this camp has made me look at life from a completely different perspective. I never really sat down to think about how lucky I am or about the types of opportunities I have had. The children that go to this camp will most likely never have the same types of opportunities as I have had unfortunately. This issue has really had a strong influence on my life and no longer take anything for granted.

C. Wright Mills wrote about the **sociological imagination** (1959) and stated how important it is to learn about society by developing the ability to relate one's personal troubles to larger social forces and public issues. He noted that a trouble is a private matter and an issue is a public matter. He went on to say "This distinction is an essential tool of the sociological imagination and a feature of all classic work in social science" (350).

I can see that there are both personal trouble and public issues associated with the camp I mentioned above. In terms of personal troubles, these kids deserve better and deserve more opportunities. These kids deserve to go to a good school and get a good education, instead of going to a school that does not prepare them for the world. Because they come from poor neighborhoods they do not have the same resources that wealthier children have. In terms of public issues, I wonder how something can be done to fix the problem at the institutional level. These children should have the same opportunities as every other kid in the state, country, and world. It is troubling to think about the struggles that these kids might and most likely will encounter later on in their lives. And they do not deserve not having an equal chance. These children are starting behind other children because of the house they were born into. And a big part of the larger social forces that are causing troubles for these kids is the government and its policies. It is about time they figured out a way for it to be an equal playing field for all the children. Although I do not feel as though it will happen anytime soon, I think the world needs to start taking strides, at least, in the right direction.

At the beginning of my volunteer year at camp things were going all right. It seemed as though everyone working there had already found his or her niche, and I was the new guy. At times I felt a little bit awkward, but I tried to be outgoing and ultimately made a few good friends, friends that are still friends now and will be for the rest of my life. I learned the ropes of the camp pretty fast too. I asked many questions and learned the right things to do and the wrong things not to do. When you are dealing with kids they will take advantage of the new guy whenever they can. I didn't want to be "that" new guy.

In his article "Jobless Poverty: A New Form of Social Dislocation in the Inner-City Ghetto" William Wilson writes about these skills that I was acquiring. He calls them **hard skills.** Wilson describes hard skills as the specific skills that can be easily taught. I learned these skills easily. I would float around to the different groups and pick up tidbits from the various counselors who were on duty. Examples of hard skills at camp would be knowing the times of activities, what time bed time is, and things the kids are allowed to do and not allowed to do. I seemed to pick up these skills pretty fast and by the end of my first summer I was pretty confident in my hard skills.

Wilson also talks about soft skills in the aforementioned article. **Soft skills** are much more complex and harder to learn. They take time to develop and master. Soft skills are intangible. I have now been at the camp for four years and still do not know if I have completely mastered all the soft skills that are necessary to be perfect at this job. But then again, no one is perfect. Examples at the camp of soft skills would include leadership qualities in relation to your fellow employees but also to the idea of being a leader or role model to the children. Most of these kids do not have a good role model at home and it is always nice to know that these children look up to you. Other soft skills may include dealing with certain situations at camp, like how to discipline a child for poor behavior or reward a child for appropriate behavior. It is also important to gain the respect of the children. They sometimes think that because they are away at home they do not need to behave so it is important to make the kids understand that the counselors need to be treated with the utmost respect.

I soon realized that the camp is pretty much a routine that takes place every day, week, or summer; but there are some changes every time it is held. It was hard to notice at first but after four years at the place you begin to realize the routine and understand how to handle a certain situation. It is extremely similar to the routine that people working at MacDonald's deal with. This idea of **routinization** is discussed in Robin Leidner's article "Over the Counter: McDonald's," where she states that "McDonald's had routinized the work of its crews so thoroughly that decision making had practically been eliminated from the jobs" (346). The decisions were made by the higher-ups in the company to make it easier for the people working and dealing with the people who come to the restaurant. This is similar to the camp, such as the idea that my boss has set up a schedule to follow during the day and it is given to the counselors. The boss has taught the counselors how to discipline the children and how to treat certain situations and make certain decisions. The camp is run on the same routine every week of every summer I have been there. There are the same meals, activities, and it is very rare that a situation comes up that has never occurred in a previous summer. Once a counselor has developed his or her soft skills they will notice that camp is a routine that has a few changes every so often.

When thinking about Leidner's article about McDonald's I thought of the movie *Fast Food Nation*. This movie shows the working crew that is in charge of preparing the beef for the restaurants. The **conditions of work** are extremely unsanitary as shown and were pretty repulsive. I have worked in the kitchen at camp and made sure to wash my hands before and after and also always wore gloves and a hat while working there. The idea that a worker's hands must be washed and that they need to wear gloves is institutionalized. If the camp is audited and a person working in the kitchen is not wearing gloves or a hat, the camp could get into trouble. The working conditions at the camp are immaculate, pristine, and completely different from the working conditions that were shown during the movie *Fast Food Nation*.

The working hours of someone that works at the camp are very similar from day to day. A counselor will either work the morning shift, starting when the kids arise, and going until the mid afternoon or they will work the mid afternoon until the kids go to sleep. And then it is required that two counselors sleep in each bunkhouse. In her article titled "Toward a 24 Hour Economy" Harriet Presser writes about the idea of **rotating.** Rotating is a concept that deals with a person's schedule being different from day to day. At the camp there really is not any rotating. A counselor rotates with another counselor at the break in the mid afternoon and then they are on until night and perhaps the next morning or afternoon when the other counselor is scheduled to show up and relieve that person of their duties for the day (462).

Working at the camp is a lot of fun, but at the same time it is very stressful at times and also it is tremendously tiring. The days are generally very long and after three or four straight days of working there without a break it becomes tiring. The camp is generally understaffed. There should be more counselors so that the ones working there are not as stressed. The counselors that are working there are what Barbara Brandt would call **overworked.** The counselors are worked extremely hard for far too long for the entire summer. In her article titled "An Issue for Everybody" Brandt writes, "in order to maximize profits, companies are "downsizing"—laying off as many employees as possible and overworking those who remain" (Brandt 13). While this is a non-profit camp and profits aren't a priority, overworking takes place because there is a lack of people wanting to be counselors, or perhaps a combination of both. Because of these problems the camp is suffering and the counselors are overworked.

Because the camp is overworked, it is obvious that the camp is **understaffed**. The camp really struggles because of the under-employment. This is the sole reason for the overworking. There is only one benefit for the workers of being underemployed. In most cases when there is not enough workers there is usually always plenty of work that has to be done (Brandt 14). This leaves an ample amount of what Brandt calls **overtime.** Overtime is extra working time that you get paid for usually in the form of "time and a half." Or you are rewarded in a different way that is clearly stated in your working statement. Overtime can be either voluntary or mandatory. At the camp I had plenty of overtime hours which made my paychecks much larger.

All of these extra hours took a lot out of the counselors. I feel they might have taken away from the children's experiences at the camp. Working at the camp is extremely fun, though. One thing that all the people have to remember is to always be happy and jolly around the kids. Even if you are having a bad day, it is part of the job to make the best out of the day and make sure that the kids have the most fun. Going to camp is like a vacation for most of the children that go there so it would be unfair to take away from their get-a-way. Arlie Hochschild describes this type of work as **emotional labor.** In her article titled "The Managed Heart" she states that emotional labor is the "management of feeling to create a publicity observable facial and bodily display" (73). It is part of the "job." Working with these kids who are all full of energy and loving every minute of being away from the city are not, it is necessary to be emotionally affectionate. A big part of being a counselor is being happy and joyful and trying to make the lives of the children a lot better and richer because they are having fun.

Although it is nearly impossible at times during the job, a counselor must not get attached to the children. They are there for the week and you have to make it the best experience for them. When there is a problem that happens at camp each counselor or the boss has to look at the issue with an even eye. There can be no playing of favorites. Each child should get the same opportunities as every other child. Hochschild also talks about this. She defines the term **feeling rules**. According to her article feeling rules "are standards used in emotional conversation to determine what is rightly owed and owing in the currency of feeling" (74). There is a difference between how we actually may feel about something and how we are supposed to feel about it. Feeling rules are how we are supposed and expected to feel according to prevalent cultural norms. If there is a situation where a camper is involved in a fight, for example, a counselor can't just listen to one side of the story because that child has a better reputation than the other. Rather the counselor must sit down and leave all emotions out of it, and figure out what happened and what went wrong and figure out the appropriate way to solve the problem.

At camp often many problems arise. It is inevitable to have a group of about sixty kids or so in one camp and go through a week without an astronomical number of problems. This can be very stressful for the workers and at times very hard to handle. Simon Head, in his article "On the Digital Assembly Line," writes about dealing with this type of problem. He talks about **management of stress** and how a worker deal with some of the pressures and anxiety a job may have. I feel like the best way to cope with some of the stresses at the camp is to have a good weekend or have a good night when you have the evening off. Taking the job everyone understands that it is going to be a very stressful job, but at the end of the day you are a better person for what you are doing. So managing the stress while working at camp it is good to have a strong, close-knit staff that can talk about problems and discuss the day or week. The staff is a veteran staff and everyone knows each other very well so it is easy to go to the other counselors or the boss to get through a difficult time.

Thinking about all the different jobs that I have had and different types of work that I have participated in, I have had time to reflect on what my work utopia would be. When I think of my work utopia, I think about waking up each morning and being happy going to work. I do not envision myself working in an office building crunching numbers. I want to work with children in some way. I desire to graduate from college in the next two years and work toward getting a job that helps children. In my perfect world I would be a guidance counselor and then coach a high school sport. I love children and love the idea of making the lives of other people better. I am a very social person and enjoy a good conversation so I feel like this would be a perfect fit for me. As for work at home I will help with the housework and cleaning and dishes but I am not a good cook so hopefully I can find someone that can keep my hunger satiated.

WORKS CITED

Wharton, Amy S. *Working in America*. 3rd ed. New York, New York: McGraw-Hill Companies, 2006.

Brandt, Barbara. *An Issue for Everybody* in John de Graaf (ed.), *Take Back your Time; Fighting Overwork and Time Poverty in America.* Berret Koehler Publi. Inc., 2003 pp.6-36.

Braverman, Harry. *Labor and Monopoly Capital: The Degradation of Work in the Twentieth Century.* New York: Monthly Review Press, 1974.

Dickinson, Torry and Robert Schaeffer. *Fast Forward* Rowman and Littlefield Publ. Inc., 2001.

Goldscheider, Frances K. and Linda J. Waite. *Children's Share in Household Tasks* in Karen V. Hansen and Anita Garaey (eds.) *Families in the U.S. Kinship and Domestic Politics* Temple University Press, 1998. Pp 809-817

Head, Simon. 2006. "On the Digital Assembly Line." Pp. 125-136 in *Working in America: Continuity, Conflict, and Change* Third Edition, edited by Amy S. Wharton. Boston: McGraw Hill.

Hochschild, Arlie Russell. 2006. "The Managed Heart." Pp. 69-78 in *Working In America,* edited by Amy Wharton. New York, New York: McGraw Hill Companies, Inc.

Kibria, Nazli. *Household Structure and Family Ideologies: The Case of Vietnamese Refugees* in Karen V Hansen and Anita Garey (eds.) *Families in the U.S. Kinship and Domestic Politics* Temple University Press, 1998 pp 55-68

Leidner, Robin. 2006. "Over the Counter." Pp. 345-357 in *Working in America: Continuity, Conflict, and Change* Third Edition, edited by Amy S. Wharton. Boston: McGraw Hill.

Marx, Karl. 2006. "Alienated Labour." Pp. 44-51 in *Working in America: Continuity, Conflict, and Change* Third Edition, edited by Amy S. Wharton. Boston: McGraw Hill.

Meyer, Stephen III. 2006. "The Evolution of the New Industrial Technology." Pp. 31-44 In *Working In America,* edited by Amy Wharton. New York, New York: McGraw Hill Companies, Inc.

Mills, C. Wright. The Sociological Imagination. Oxford University Press,1959. Excerpted in *Social Theory: The Multicultural and Classic Readings* edited by Charles Lemert, Boulder: Westview Press

Powell, Walter W. 2006. "The Capitalist Firm in the Twenty-First Century: Emerging Patterns in Western Enterprise." Pp. 80-94 in *Working in America: Continuity, Conflict, and Change* Third Edition, edited by Amy S. Wharton. Boston: McGraw Hill.

Presser, Harriet B. 2006. "Toward a 24 Hour Economy: The U.S. Experience and Implications for the Family." Pp. 459-465 in *Working in America: Continuity, Conflict, and Change* Third Edition, edited by Amy S. Wharton. Boston: McGraw Hill.

Wilson, William Julius. 2006. "Jobless Poverty: A New Form of Social Dislocation in the Inner City Ghetto." Pp. 178-186 in *Working in America: Continuity, Conflict, and Change* Third Edition, edited by Amy S. Wharton. Boston: McGraw Hill.

Films:

Fast Food Nation. Houghton Mifflin. 2006.

What's Eating Gilbert Grape. J & M Entertainment. 1993.

Omar Khayyam Center for Integrative Research • in Utopia, Mysticism, and Science •

About OKCIR

The Omar Khayyam* Center for Integrative Research in Utopia, Mysticism, and Science (Utopystics) is an independent research, pedagogical, and publishing initiative dedicated to exploring, in a simultaneously world-historical and self-reflective framework, the human search for a just global society.

Since the world's utopian, mystical, and scientific movements have been the primary sources of inspiration, knowledge, and/or practice in this field, OKCIR aims to critically reexamine the limits and contributions of these world-historical traditions—seeking to clearly understand why they have failed to bring about the good society, and what each can integratively contribute toward realizing that end.

OKCIR posits that human failures at building a de-alienated global society stem from a spatiotemporal distanciation between the "self" and the "social" foci of efforts at world-historical change in favor of the good society. This distanciation, deepened through the disintegrating clashes of settled vs. nomadic life-styles during long eras of political, cultural, and economic imperialism ever since the fall of ancient civilizations, has itself been a result of the polarization of ideologies of change into religious and scientific modalities rooted in a common, philosophically perpetuated, idealist vs. materialist dualism. The combined perpetuation of this dualism in methods, theories, and praxes of change has become manifested in the world-historical fragmentation of the creative humanist paradigm into its predominantly Western utopian and Eastern mystical variants, both eschewed by a globally hegemonic scientific movement. OKCIR posits that a fundamental explanation for the above is to be sought in the problem of habituation, i.e., the human propensity to become subconsciously attached to sensations, ideas, feelings, things, relations, and processes. It is from this that the dualisms of mind/matter, self/society, and theory/practice—still fragmenting the methods, theories, and praxes of change—fundamentally emanate.

The diverse projections of human creative powers onto objective laws of motion of nature or history, supernatural agencies, or a wise few, represent the degree to which the very agencies of human liberation have themselves grown alienated from one another. The failed conscious and intentional shocks of the two major humanist renaissances of the 6th-4th centuries B.C. and of the 13th-15th centuries A.D. in bringing about a lasting synthesis of the three polarized and failing fragments of the humanist endeavor, have given rise in the modern period to the "antisystemic" mode of seeking social change which by its very nature of spatiotemporally distanciating the actual means from the promised ends of change has also proven to be an exercise in failure. OKCIR posits that the way out of this world-historical impasse is that of inventing a new humanist renaissance involving a far-reaching self-critical dialogue across the utopian, mystical, and scientific traditions.

The need for OKCIR as an alternative spacetime for our alchemical journey arises from the limits of hitherto invented structures of knowledge—East and West. OKCIR provides an integrative spacetime for knowledge production as an alternative to the fragmented and disciplined landscapes of philosophy, religion, and science. OKCIR is a utopystic university, a virtual research center existing simultaneously in personal and world-historical spacetimes—a makeshift "campus," personal and global. It is built on the premise that the search for alternative human realities cannot advance without alternative research and educational landscapes. The ends and the means of the search, in other words, must coincide. Utopystics cannot advance without utopystic universities.

OKCIR is utopystic in the sense that it continually seeks new personal and world-historical realities that did not previously exist—thus it must by nature be self-critical and self-transcending against its own habituating inertia. OKCIR is a movement in the arts—more specifically, a movement in architecture: the spatiotemporal art of becoming human. Temporally, it is not projected onto a distant future or past, but is undertaken in the everyday life of the present here-and-now. Spatially, it is situated not in retreat from but in the midst of the mainstream culture. It is concerned with the spatiotemporal dialectics of world-historical self-knowledge and change.

OKCIR's central purpose is to develop new conceptual (methodological, theoretical, historical), practical, curricular, inspirational, and disseminative structures of knowledge whereby the individual can radically understand and determine how world-history and her/his selves constitute one another.

---

* Persian poet, astronomer, and mathematician of the 11th/12th centuries A.D. Author of the Rubaiyat. "Khayyam" means "tent-maker" in Persian.

www.ingramcontent.com/pod-product-compliance
Lightning Source LLC
LaVergne TN
LVHW081937200926
848484LV00009B/2453